AF842281

Embryonic Stem Cells and the Law

Joshua Weiser

Embryonic Stem Cells and the Law

Crafting A Humane System of Regulation

Springer

Joshua Weiser
Boynton Beach, FL, USA

ISBN 978-3-031-59749-7 ISBN 978-3-031-59750-3 (eBook)
https://doi.org/10.1007/978-3-031-59750-3

© The Editor(s) (if applicable) and The Author(s), under exclusive license to Springer Nature Switzerland AG 2024
This work is subject to copyright. All rights are solely and exclusively licensed by the Publisher, whether the whole or part of the material is concerned, specifically the rights of translation, reprinting, reuse of illustrations, recitation, broadcasting, reproduction on microfilms or in any other physical way, and transmission or information storage and retrieval, electronic adaptation, computer software, or by similar or dissimilar methodology now known or hereafter developed.
The use of general descriptive names, registered names, trademarks, service marks, etc. in this publication does not imply, even in the absence of a specific statement, that such names are exempt from the relevant protective laws and regulations and therefore free for general use.
The publisher, the authors and the editors are safe to assume that the advice and information in this book are believed to be true and accurate at the date of publication. Neither the publisher nor the authors or the editors give a warranty, expressed or implied, with respect to the material contained herein or for any errors or omissions that may have been made. The publisher remains neutral with regard to jurisdictional claims in published maps and institutional affiliations.

This Springer imprint is published by the registered company Springer Nature Switzerland AG
The registered company address is: Gewerbestrasse 11, 6330 Cham, Switzerland

If disposing of this product, please recycle the paper.

This book is dedicated in loving memory of my Grandmother Ruth Roberts, my Grandfather Michael Weiser, and my Grandmother Sylvia Weiser.

Preface

Embryonic stem cells have been at the forefront of much controversy since their initial discovery and description. While the unknown still remains tantalizing, what is known is nonetheless incredible as well. As the source embryonic stem cell is in fact an embryo, there have been many who have called for a total ban on this type of research, yet there are many who champion expanding our collective knowledge of these powerful cells so that people can benefit from what can be discovered. Whenever there are opposing sides to an argument, there are many issues that follow; this is especially true when there are people on one side who believe they are fighting to save lives and on the other believe that further research will lead to cures and treatments for some of the most debilitating issues and diseases that plague humanity. In instances such as these, many look to their governments to create and enforce laws to either allow and support or ban such research.

This book uses the New Haven approach in order to fully understand the opposing views while also looking at all relevant laws on the national, local, and even international levels as well as their conditioning factors in order to create a prediction as to what the future holds and any recommendations in an attempt to ensure that the views of all are taken into account. The end goal of this approach is to create policy that at a minimum is based on the values that all humans share. While this is not an easy task for any topic, it is especially difficult for such a naturally polarizing one.

To obtain the desired results, one must fully understand the science behind embryonic stem cell research, including the known and potential benefits and risks, as well as all of the arguments in favor of and against this area of research. It is also crucial for one to know about the pertinent laws coming from the international community as a whole such as those created through treaties, laws in place in other countries, the laws of the USA, as well as state and local laws, and the reasons behind their creation. Only in viewing these laws in light of their conditioning factors can one appreciate them; for example, one can see that certain laws were created based on the potential revenue certain research can attract, and there are also laws that were created based strictly on religious scripture. Once all of this is understood, one can create policy based on the shared interests of all.

Boynton Beach, FL, USA Joshua Weiser

Acknowledgments

This book is an edited and updated version of my J.S.D. thesis in Intercultural Human Rights at St. Thomas University in Miami, Florida.

This book and degree have been a true labor of love. There have been so many people that I would like to include, but first I must thank my mother Laurie Weiser, father Alan Weiser, sister Michele Weiser, and grandfather Stanley Roberts. All of you have been instrumental in helping my dream become a reality and there are not enough words that I can use to show my gratitude. Your help with ideas, edits, and inspiration pushed me past the finish line. It was your belief in me that gave me the courage to tackle this in the first place. Without your patience and support, I would have been completely lost. I love and appreciate you all so much. When I think about my family and the love and support you all give, I truly understand the meaning behind Grandpa explaining how some people are just "born lucky!"

To my beautiful, smart, funny, amazing, and helpful girlfriend Alexis Intriago, I can't thank you enough for all of your support and help. You have been there offering your help and support since we met and I can't express to you how grateful I am for you to be in my life.

To the best family and friends imaginable: Aunt Ellen, Uncle Mitch, Blair, Jordan, Giff, and Ali; Aunt Karen, Marc, Scott, Jackie, Tracy, and Derek; Uncle Neil and Aunt Laura; Aunt Carol, Richard, Corey, and Danielle. You have always been there in my corner for as long as I can remember. My best friends who are the best a person could ever ask for, Rick and Andreina Ottman and Andrew and Samantha Agnini. You guys have been there since I started this and your continued support means the world to me. I also want to recognize some of my closest friends and my family's closest friends, Harvey, Debra, Samantha and Erica Daniels, Jill, Carly and Jesse Todd, Shawn Shore, Charlie Gandia, Bruce and Julie Baron, Larry and Nira Traster, Leo and Marla Schwartz who have been at my side for as long as I have known you.

Professor Doctor Siegfried Wiessner, I will forever be grateful for your guidance, patience, inspiration, and mentorship. It was such an honor to look to you and your publications to help in making my own. After attending your much deserved Festschrift I will always remember and repeat to myself that we must do everything with "surgical precision" and that our actions can bring real change into this world. I will take both of these lessons with me in everything I do. This whole experience has been eye-opening in ways that I never imagined. I truly thank you for being the best mentor and guide on this journey.

A very special thank you to another inspiration who helped guide and focus me for this project my very own surgeon and good friend Dr. Chad Perlyn. Not only have you given me the ability to feel great, you also helped inspire me to write this book. Thanks to your innovative mind and spirit, I was able to experience what people can go through when using stem cells and their amazing potential benefits.

Professor Doctor Roza Pati, I can't thank you enough for all of your help and recommendations you provided me with throughout this whole process.

I want to express my gratitude to Haydee Gonzalez and the whole St. Thomas School of Law, especially those involved with Intercultural Human Rights program. You have all been an amazing support system.

I am also extremely grateful to those at Springer who have helped me get to this point and guided me along the way, Dr. Brigitte Reschke and Peter Pushpanathan.

I cannot skip over all of the amazing scientists, doctors, patients, writers, and researchers who have helped me in writing this book just in doing what they do.

Contents

Chapter 1
Introduction

This book will seek to strike a balance in order to give the greatest amount of freedom in stem cell research, while not being disrespectful of cultural and religious traditions and values. To do so, it is best to use a method known as policy-oriented jurisprudence, or as the "New Haven School," or "New Haven" approach.[1]

> The New Haven School defines law as a process of decision that is both authoritative and controlling; it places past such decisions in the illuminating light of their conditioning factors, both environmental and predispositional, and appraises decision trends for their compatibility with clarified goals; it forecasts, to the extent possible, alternative future decisions and their consequences; and it provides conceptual tools for those using it to invent and appraise alternative decisions, constitutive arrangements, and courses of action using the guiding light of a preferred future world public order of human dignity... To facilitate actual decision making, it proposes a praxis of five intellectual tasks: goal formulation, trend description, factor analysis, projection of future decisions, and the invention of alternatives. A public order of human dignity is defined as one which approximates the optimum access by all human beings to all things they cherish: power, wealth, enlightenment, skill, well-being, affection, respect, and rectitude.[2]

This is all crucial for an complete picture as to the issue at hand. One cannot be expected to propose a solution to an issue without first looking to the arguments and stances held by all sides of the argument. Equally important is to understand and explain the laws and decisions that have come in the past that can be viewed as impacting or giving guidance on the topic at hand.

Since the discovery of embryonic stem cells, controversy has surrounded these cells described as "powerful."[3] These cells are considered controversial because they have historically required the destruction of an existing, discarded, surplus embryo,

---

[1] Reisman et al. (2007), pp. 575–582; Wiessner (2010), pp. 47–48 (explaining that this form of jurisprudence uses a method of analysis that can look to various societal problems and offer thought out alternatives and solutions that can apply to all).

[2] *Id.*

[3] Patil (2014), pp. 188–194; Jensen et al. (2009), pp. 513–519.

© The Author(s), under exclusive license to Springer Nature Switzerland AG 2024

J. Weiser, *Embryonic Stem Cells and the Law*,
https://doi.org/10.1007/978-3-031-59750-3_1

yet due to their nature as a "blank" cell that can be changed into any other makes their potential tantalizing.[4] As a result, many countries and individual states have put limits, restrictions and have outright outlawed embryonic stem cell research as a whole or even its funding.[5] There are so many potential benefits that researchers and scientists have already hypothesized would revolutionize medicine with treatments and cures for Parkinson's, Alzheimer's disease, multiple sclerosis, diabetes, liver failure, heart disease and countless others.[6] Simply outlawing certain stem cell-based therapies will lead to more and more people suffering needlessly when there are options to help them. To go one step further, promoting, and even funding all forms of stem cell research by the federal government can lead to great new strides in medical science. Not only does this topic deal with the scientific aspects, but it also has a main focus on the human rights issues that arise from the withholding of lifesaving medical care, and research which should be afforded to all. Using countries' laws as a starting point, as well as looking to the various unique state laws, we can dissect them and determine the various motivations behind their crafting. This is a very important issue, because even though it would seem easy to alter our laws to better recognize and support the technology, we must be very careful. It is also imperative to understand that this issue is in fact one in which directly implicates human life and human rights on both sides of the argument. Many seeking beneficial laws that enable protections for embryonic stem cell research argue that not doing so violates human rights in that it does not afford people the opportunities to better their lives, and on the other hand, many of those who are against implementing laws allowing for this form of research want to protect the human rights of the embryo where the stem cells would have to be derived from. Generally, laws are drafted to reflect the culture and beliefs of nations. For example, in China, there are very few restrictions on any form of stem cell research.[7] China actually seeks to expand its research capabilities to the farthest extent, yet in doing so, China has actually forced women to give up embryos for science. This is an extreme practice, as studies regarding this issue have shown that the motivation for China's decisions has been the amount of money that their research will yield. This seems due to the fact that culturally, women are not given the necessary and important freedoms to choose what to do with their embryos.

There are countries with no restrictions at all, such as Mexico. In Mexico, there are no laws that deal with the regulation of stem cell research. This has caused many problems. There are "stem cell clinics" where people can go to receive various treatments which are usually conducted in unsterile environments. Also, there are no proper studies to document the success, or lack thereof, of the treatments. This has led to many injuries and other problems for people who visit these "clinics."

---

[4] *See generally* Mitalipova et al. (2003), pp. 521–526.

[5] Walters (2004), pp. 3–5.

[6] Hoang et al. (2022), p. 272.

[7] *See generally* Liao et al. (2007), pp. 1107–1112.

There are even issues that have to be dealt with which are completely new, and until only very recently, seemingly impossible to resolve. This relates to the very concept of embryonic stem cells. Embryonic stem cells are effectively "cells which give rise to all somatic cell types in the embryo... the indefinite self-renewal ability and plasticity of ESC allows for in vitro generation of an unlimited number of distinct cell types, and has opened new avenues for regenerative medicine."[8] What this all means is that these specialized cells have the ability to be altered into any other type of cell, and can be used to repair damaged or missing tissue.[9] It is now widely believed that the main alternative to embryonic stem cells, known as "adult stem cells" cells can be transformed into embryonic-like stem cells through scientific manipulation.[10] Adult stem cells are considered an alternative to embryonic stem cells because they naturally occur throughout the adult human body and are continually created and are easily harvested for study.[11] This is important because the abilities and uses of embryonic stem cells far outweigh the benefits of adult stem cells. Even though this has not yet happened, it is a goal of many research scientists, and it is believed that this new advancement can occur at any time. The only embryonic stem cell laws that exist are primarily put in place so that needless rampant destruction of embryos does not occur. When scientists are able to successfully manipulate adult stem cells in this way, the newly created stem cells may possibly avoid the ethical issues raised by the destruction of life. There are also cells known as induced pluripotent cells which are another form of adult cell that is converted to something more akin to an embryonic stem cell. Both of these cell types will be discussed.

Another method which too is heavily restricted, if not more restricted, is therapeutic cloning which removes the nucleus from a cell and replaced with a different somatic cell nucleus, creates a "non-human embryo".[12] It is considered non-human because absent placing said cells in a uterine environment they can literally never become a human to begin with, so they are essentially pre-human cells.[13] To better understand the science behind this method, scientists produce embryonic stem cells by taking an unfertilized egg, and transplanting a patient's own cellular nucleus into the unfertilized egg with the intent to produce the desired cells so that they could be transplanted back into the patient creating compatible replacement tissues to help with degenerative or inherited diseases.[14] However, this area of science is typically banned due to the fact that in order to obtain the required cells, the created "embryo" would first have to be destroyed.[15] As mentioned above, scientists and doctors

---

[8] Vazin and Freed (2010), pp. 589–603.

[9] *See generally Id.*

[10] Ye et al. (2013), p. 63.

[11] *See generally* Zakrzewski et al. (2019), pp. 1–22.

[12] Hansen (2022), pp. 86–88.

[13] *Id.*

[14] *Id.*

[15] Foury (2007), pp. 112–120.

working on such procedures believe that its success can likely treat Huntington's chorea, Parkinson's disease, multiple sclerosis or cystic fibrosis, among others.[16] Many of the same laws dealing with human embryonic stem cell harvesting and testing apply to this method as well, as well as many of the same potential medical issues.[17]

Stem cells are the only known cells that have the potential to develop into completely different cells, to remain stem cells.[18] Stem cells also have the ability to repair damaged tissue, and they can divide to create more stem cells, or other cells for as long as a person lives.[19] For this reason, stem cells are truly unique and have been termed "pluripotent" which is the actual characteristic of stem cells to be changed to any other form of cells in the body.[20] There are two main types of stem cells which come from two completely different sources.[21] There are adult stem cells which many people have not heard of, and there are the extremely controversial, yet most beneficial form which are embryonic stem cells.[22] There are also further subdivisions regarding the two forms of stem cells, which is discussed in depth below.[23] While in the human body, stem cells act as a repair system to either fix or replace damaged cells through a process that the body undertakes on its own.[24] Since discovering the existence of stem cells, scientists and doctors have found that in order to artificially insert stem cells into areas of the body which need repair or help, these stem cells can be removed from either the host who requires the repairs, or from an embryo.[25] Stem cells are thus important to study and to further understand because they contain two properties which no other known cells contain at the same time.[26] First, when stem cells are in their most basic form, they are considered "unspecialized" which means that they are essentially a blank slate which has the ability to self-replicate for as long as the host is alive.[27] The second and most important specialized function comes from scientists altering the stem cells to become a different type of cell with a special function.[28] The function tends to depend on what the scientists are looking to create.[29] The full potential of what stem

[16] Hansen (2022), pp. 86–88.

[17] *See generally* Dhar and Hsi-en Ho (2009), pp. 113–115.

[18] National Institute of Health (2016).

[19] *Id.*

[20] Constitutional Rights Foundation (2006).

[21] Poliwoda (2022), pp. 1–9.

[22] *Id.* pp. 4–5.

[23] *See* Zhao and Moore (2018), pp. 1–10 (discussing Neural stem cells, which are specialized stem cells which can only differentiate into other neural cells).

[24] Malam and Cohn (2014), pp. 7–8.

[25] *See generally* Jin et al. (2023), pp. 1–31.

[26] National Institute of Health (2016).

[27] *Id.*

[28] *Id.*

[29] *See id.* (explaining the distinct abilities, and possibilities that are unique to stem cells).

cells can do has not yet been fully realized, and many scientists believe that the studies and research in existence only merely scratch the surface.[30] The reason that the United States should change their laws to comport with most of the international communities' laws is that stem cells can heal and cure so many diseases and problems that many people in the world suffer from.[31] There is also a subdivision of adult stem cells which are induced to alter their state to revert them to something akin to embryonic stem cells.[32] These cells are called "induced pluripotent stem cells, and their creation comes about through a process where embryonic genes are placed with a somatic cell which causes the mix to create something that resembles a stem cell.[33] A somatic cell is any cell in the body that is not a sperm or egg cell, and contain chromosomes that are directly inherited from the parents.[34] Further, a chromosome are the genetic structure that carries the genetic information that is inherited from the parents.[35] The problem with this form though is that there is a much decreased success rate, and there is very little research being done to create the proverbial "o-negative" stem cell which could be used much like an embryonic stem cell and can help repair or replace the problem site.[36] Many view this form of converting somatic cells into stem cells as a more ethical variation which can still provide many of the same benefits as the more powerful embryonic stem cells, but fail to account for the new discoveries that are made which use only embryonic stem cells obtained in an ethical manner where no embryo has to be destroyed or harmed.[37] Yet the introduction of these cells into a human body poses some very big health risks as they are inserted "anywhere" into a human's genes through the use of a retrovirus which can cause any number of infinite cancer causing genes to be expressed.[38] The National Human Genome Research Institute describes a retrovirus as one in which uses ribonucleic acid, or RNA and once infected, the infected cell converts the RNA into DNA (deoxyribonucleic acid) which will then be produced by the infected cell to do what its set goal was.[39] Studies have found that recipients of stem cell transplants can face some serious health risks such as various forms of cancers, and random tumors.[40] There are many unknowns in this area as to why these

---

[30] Spivack LaRosa (2021).

[31] The New Atlantis (2012), pp. 129–133.

[32] National Institute of Health (2016).

[33] *See generally* Kim et al. (2009), pp. 472–476.

[34] *See*: National Human Genome Research Institute (2023a); Yu et al. (2007), pp. 1917–1920.

[35] *Id.*

[36] *Id.*

[37] Briceno (2016) (discussing Dr. Alan Moy's research into an ethical alternative to using embryonic stem cells in light of working at a Catholic research institute (John Paul II Medical Research Institute) which has banned embryonic stem cell research).

[38] *See generally* Mitsui et al. (2017), pp. 51–58.

[39] *See*: National Human Genome Research Institute (2023b).

[40] Zhang et al. (2017), pp. 75756–75766.

situations arise, yet one major reason is that stem cells self-proliferate and self-renew, allowing for them to reproduce endlessly.[41]

Through experiments that have occurred all over the world, a lot of potential has been shown for diseases, and medical ailments likely cured or to be alleviated through the use of stem cells. Diabetes, different cancers, heart disease, spinal cord injuries including paralysis, Parkinson's disease, Alzheimer's disease, Lou Gehrig's disease, degenerative joint disease, blindness, deafness, sickle cell anemia, mitochondrial diseases, lung disease, and even auto-immune diseases can all likely be helped through the study and eventual use of stem cells.[42] There are many other methods that also show promise, but most of which will have a more difficult time and take much longer to work out, but would avoid such pitfalls, if only adult stem cells are allowed to be used. Another problem persists if federal funding is so restricted because in order to get federal funding, there needs to be a showing of the expected benefits, and this is practically impossible to do if the lab cannot get funding on their own.

1.1 Observational Standpoint

As you will read within this book, I have a personal connection to stem cells. After suffering with a recurring tumor for over three decades, my doctor offered me a potential solution in the form of stem cells. I am in the rare position to have been able to see firsthand what benefits stem cells can confer, and this further has caused me to question what else they could be used for. This led to my curiosity in exploring the topic of the laws embryonic stem cells under the laws of the United States, and the world as a whole. In fact, I have even been fortunate enough to see first-hand how quickly the treatments can work. I was unsure about the cost of the treatment and worried if my insurance company would be willing to pay for this, or if I would have to pay myself and take the risk. To my surprise, once I was offered experimental adult stem cell treatments, the insurance company was unquestioning, and paid the hospital immediately. I later learned that due to the substantial promise offered by this treatment, and the likely permanence of their nature in treatment, it would save them a lot more money for years to come because they would not have to pay for the more extensive and short working operations. As explained above, I have had over 60 operations and procedures on my leg. There is a lot of information that I can write on this issue, but for the sake of time, I will explain the overall issue and how and why stem cell were used. I was born with a congenital issue which was misdiagnosed which caused my body to produce an excess of blood and lymph vessels in my leg which would grow into a bundle and cause severe pain and limit my activities and movement. The surgeries prior to the stem cell treatments were utilized at first to

[41] *Id.* p. 75757.

[42] Lyon (2013).

remove the tumors, but after numerous years and operations, doctors realized that in removing the tumors, they were not treating the issue and only temporarily stopping the situation. After only a short period, the tumors would return and again have to be removed. Once the correct diagnosis was found, doctors than decided to "kill" the tumor by injecting each vessel with different forms of chemicals. This was successful but unfortunately left me with a huge void in my leg that was filled in with scar tissue. Scar tissue is formed after certain forms of trauma occur in the body which creates collagen fibers to replace the injured tissue.[43] In my case, the scar tissue inhibited at least as much as the tumor itself and caused a great deal of pain, so my current Doctor removed all the scar tissue which left a void that would likely be replaced by new scar tissue if no further action was taken. My Doctor gave me the choice to be a "guinea pig" of sorts and try to use my own body's stem cells to form new tissue which would be there absent all of the operations. As a test subject, I had to give informed consent, and before doing so, I asked as many questions as possible and read up on the processes thoroughly. I was told that these treatments usually do not succeed after the first attempt because the human body automatically re-absorbs around 40% of the stem cells, and due to the trauma only so much can be taken and reinjected into a person at once. Immediately after the first procedure I actually felt wonderful for the first time in around 25 years. The feeling was short lived though, and I had to have two more stem cell operations to finish filling up the void with the proper tissue, yet once everything was done, the process worked.

Due to the fact that I was still suffering from substantial pain, my same doctor offered me another trial which would be similar to a common procedure known as Platelet-rich plasma or PRP. PRP is often used for athletes who have some sort of injury, and the process is similar to the process my doctor uses for the stem cell injections. First, blood is drawn from the patient, which is then placed in a centrifuge where the platelets are separated from the rest of the blood. The doctor then is able to inject the platelets back into the patient and typically the healing process occurs dramatically faster than other methods of treatment. For my new stem cell treatment, the same methods are applied but the stem cells are isolated and mixed with PRP and finally reinjected into the treatment site. My doctor explained that a typical patient must undergo three stem cell injection procedures before obtaining the desired result, but there is a chance that the first or second could work. Because this has not been done before in Nicklaus Children's Hospital (where I am a patient), my doctor gave me the option to pay for the entire kit that he would have to use, and in turn, I would be afforded the three injections as normal, plus a fourth because the first was a clinical trial for the hospital, my doctor, and the company who supplied the kit. I accepted the chance to try this procedure, and without a doubt, this was an insanely easy and practically painless experience. After the unmixed stem cells are taken, they are placed in the centrifuge for four minutes, and then four needles are filled with the now mixed stem cells and plasma. The only painful part of the procedure itself is the reinjection of the stem cells, which I was told was due to the solution in the mix, and

[43] Whaling (2020).

not from the platelets or stem cells themselves. As it turns out, cosmetic surgeons and those who use stem cell injections for cosmetic reasons give patients the option to pay extra to have a painless solution, or the doctors choose to use the painless one and charge the patients more money. The following day there is discomfort at the sites of the four injections, but there is relief from the pain that is being treated fairly quickly. After my first set, I was told to wait four weeks before trying the second one, which was slightly more painful due to the sensitive area that was being worked on. I was to wait on parts three and four, and I was told I must wait five to seven weeks for the third treatment, yet the first two were successful and the rest were not needed. Throughout my life I have had to rely on pain medicine to get me through the days which led me to live a very sedentary life, yet now I ride a bike daily, lost 85 pounds and have not had any issues at all in the area. The most exciting aspect of being the test subject for this treatment, and benefitting myself, I was very happy to learn that since my successful first procedure, other doctors at the hospital have been using the same methods to help children who are suffering from various forms of pain. When I asked my doctor what would have happened had we used embryonic stem cells, he was very honest and explained that he does not truly know. He said that there still needs to be a lot more research, but he did believe that I would have had the same success, yet said he was happier we did not go down that route.

References

Briceno C (2016) Ethically advancing scientific research and medicine. Available via the Deacon. https://www.the-deacon.com/2021/08/15/ethically-advancing-scientific-research-and-medicine. Accessed 26 Mar 2024

Constitutional Rights Foundation (2006) Stem cell research: the promise and pitfalls. Available via Teach Democracy. https://teachdemocracy.org/online-lessons/bill-of-rights-in-action/bria-22-4-a-stem-cell-research-the-promise-and-the-pitfalls-4. Accessed 26 Mar 2024

Dhar D, Hsi-en Ho J (2009) Stem cell research policies around the world. Yale J Biol Med 3:113–115

Foury CK (2007) Therapeutic cloning: promises and issues. McGill J Med 2:112–120

Hansen JES (2022) Embryonic stem cell production through therapeutic cloning has fewer ethical problems than stem cell harvest from surplus IVF embryos. J Med Ethics 28:86–88

Hoang DM, Pham PT, Bach TQ, Ngo ATL, Nguyen QT, Phan TTK, Nguyen GH, Le PTT, Hoang VT, Forsyth NR, Heke M, Nguyen LT (2022) Stem cell-based therapy for human diseases. Signal Transfict Target Ther 7:272

Jensen J, Hyllner J, Björquist P (2009) Human embryonic stem cell technologies and drug discovery. J Cell Phys 219:513–519

Jin Y, Li S, Yu Q, Chen T, Liu D (2023) Application of stem cells in regeneration medicine. MedComm 4:1–31

Kim D, Kim C-H, Moon JI, Chung Y-G, Chang M-Y, Han B-S, Ko S, Yang E, Cha KY, Lanza R, Kim KS (2009) Generation of human induced pluripotent stem cells by direct delivery of reprogramming proteins. Cell Stem Cell 4:472–476

Liao L, Li L, Chunhua Zhao R (2007) Stem cell research in China. Philos Trans R Soc Lond B Biol Sci 362:1107–1112

Lyon L (2013) Stem cells: 10 diseases they may–or–may not–cure. Available via U.S. News. U.-S. News & World Report. http://health.usnews.com/health-news/family-health/diabetes/

articles/2009/03/13/stem-cells-10-diseases-they-may%2D%2Dor-may-not%2D%2Dcure. Accessed 26 Mar 2024
Malam Z, Cohn RD (2014) Stem cells on alert: priming quiescent stem cells after remote injury. Cell Stem Cell 15:7–8
Mitalipova M, Calhoun J, Shin S, Wininger D, Schulz T, Noggle S, Venable A, Lyons I, Robins A, Stice S (2003) Human embryonic stem cell lines derived from discarded embryos. Stem Cells 21:521–526
Mitsui K, Ide K, Takahashi T, Kosai K-I (2017) Viral vector-based innovative approaches to directly abolishing tumorigenic pluripotent stem cells for safer regenerative medicine. Mol Ther Methods Clin Dev 5:51–58
National Institute of Health (2016) Stem cell information home page. Stem cell information. Available via National Institute of Health [hereinafter NIH]. https://stemcells.nih.gov/clinical-trial-info. Accessed 18 Jan 2024
National Human Genome Research Institute [hereinafter NHGRI] (2023a) Somatic cells. Definition. Available via National Human Genome Research Institute [hereinafter NHGRI]. https://www.genome.gov/genetics-glossary/Somatic-Cells. Accessed 26 Mar 2024
NHGRI (2023b) Retrovirus. Definition. Available via NHGRI. https://www.genome.gov/genetics-glossary/Retrovirus. Accessed 26 Mar 2024
Patil AM (2014) Embryonic stem cell research ethical and legal controversies. Indian Acad Forensic Med 36:513–519
Poliwoda S, Noor N, Downs E, Schaaf A, Cantwell A, Ganti L, Kaye AD, Mosel LI, Carroll CB, Viswanath O, Urits I (2022) Stem cells a comprehensive review of origins and emerging clinical roles in medical practice. Orthop Rev (Pavia) 14:1–9. Available via NLM. https://www.ncbi.nlm.nih.gov/pmc/articles/PMC9404248. Accessed 26 Mar 2024
Reisman MW, Wiessner S, Willard AR (2007) The new haven school: a brief introduction. Yale Int Law 32:575–582
Spivack LaRosa S (2021) The untapped potential of stem cells. Available via Cedars-Sinai. https://www.cedars-sinai.org/blog/untapped-potential-stem-cells.html. Accessed 19 Jan 2024
The New Atlantis (2012) Appendix E: an overview of international human embryonic stem cell laws. 34:129–133
Vazin T, Freed WJ (2010) Human embryonic stem cells: derivation, culture, and differentiation: a review. Restor Neurol Neurosci 28:589–603
Walters L (2004) Human embryonic stem cell research: an intercultural perspective. Kennedy Inst Ethics J 14:3–38
Whaling M (2020) Understanding scar tissue: what it is + how to care for it. Available via Mind Body Green. https://www.mindbodygreen.com/0-18152/understanding-scar-tissue-what-it-is-how-to-care-for-it.html. Accessed 26 Mar 2024
Wiessner S (2010) The new haven school of jurisprudence: a universal toolkit for understanding and shaping the law. Asia Pac Law Rev 18:45–48
Ye L, Swingen C, Zhang J (2013) Induced pluripotent stem cells and their potential for basic and clinical sciences. Curr Cardiol Rev 9:63–72
Yu J et al (2007) Induced pluripotent stem cell lines derived from human somatic cells. Science 318: 1917–1920
Zakrzewski W, Dobrzyński M, Szymonowicz M, Rybak Z (2019) Stem cells: past, present, and future. Stem Cell Res Ther 10:68. Available via National Library of Medicine. https://www.ncbi.nlm.nih.gov/pmc/articles/PMC6390367/. Accessed 26 Mar 2024
Zhang C-L, Huang T, Wu B-L, He W-X, Liu D (2017) Stem cells in cancer therapy: opportunities and challenges. Oncotarget 8:75756–75766
Zhao X, Moore D (2018) Neural stem cells: developmental mechanisms and disease modeling. Cell Tissue Res 371:1

Chapter 2
Delimitation of the Problem

The medical promise of stem cell-based therapies is both vast and alluring. Everyday new studies are being conducted, and the collective knowledge of the benefits that stem cells can and will provide grows. However, there are many issues surrounding this exciting field of study which not only tend to impair the further study of stem cells, but also seeks to ban it all together. To better understand the unique problems that the study of stem cells poses, it is first necessary to understand the basic science behind stem cells.

The purpose of this book is to analyze the problem in depth, explore first the scientific and technological parameters of stem cell-based therapies and then present the various claimants and their claims regarding the desirability, or undesirability of furthering the collective understanding of stem cells. Next, it will proceed to describe the past trends and decisions in the law across the United States, and throughout the rest of the world in light of their conditioning factors. Based on the changing and changed conditioning factors, future trends in decision will be presented, and, ultimately, recommendations will be developed to effectuate a closet approach as possible to a legal regime regarding this issue that forms part of a world order of human dignity.

The current laws of the United States which regulate the study, use, and funding of stem cells are grossly inadequate. This is best shown through a lens which looks to different international laws which are in place to regulate the same studies. As a "developed" country, and a ratifying state party to the International Covenant on Civil and Political Rights ("ICCPR"), the United States should be viewed as having the obligation to explore all methods that can be helpful to a large number of its citizens in improving their qualities of life, expanding one's lifespan, and keeping someone alive.[1] Even though the ICCPR does not contain the "right to health" as the International Covenant on Economic, Social, Cultural Rights (the "ICESCR") does, it still contains the right to life which has been explored in the area of bioethics, and

[1] Clados (2012), p. 70.

© The Author(s), under exclusive license to Springer Nature Switzerland AG 2024
J. Weiser, *Embryonic Stem Cells and the Law*,
https://doi.org/10.1007/978-3-031-59750-3_2

viewed as encompassing life protecting, and life expanding.[2] While this right, which is found under Article 6 is deemed a negative right, meaning that the obligation on the ratifying country lies in what they cannot do, the Human Rights Committee's General Comment No. 36 states: "The duty to protect life also implies that States parties should take appropriate measures to address the general conditions in society that may give rise to direct threats to life or prevent individuals from enjoying their right to life with dignity."[3] This sentence seemingly turned what was once viewed only as a negative law into an obligation the State to take action, at least arguable so.[4]

Further, certain states have taken stricter approaches when dealing with embryonic stem cell research, and equally as problematic, some states are allowing their own laws to remain in perpetual limbo to avoid having to deal with this very pressing and important issue.[5] Much can also be said as to the fact that there is a major discrepancy when it comes to federal law and state law which puts the two in direct conflict with each other.[6] Embryonic stem cells are controversial as they were initially only obtained through the destruction of an embryo, yet scientists have hypothesized and found success in repairing what was once deemed unrepairable tissue using embryonic stem cells.[7] In fact, many believe that some of the most deadly and debilitating diseases and ailments can be treated using this approach.[8] While experiments using embryonic stem cells are banned in some states, they are promoted in others, this is mainly due to the nature of the embryonic stem cells.[9] The arguments against embryonic stem cell research typically are based in the fact that an embryo has to be used, whether destroyed or not, to obtain this specialized cell, and many states and governing bodies, as well as individuals feel that an embryo is a life in being, and as a result, there is an inherent duty to protect it from harm and to maintain its rights as a living being.[10]

2.1 Stem Cell Primer

Before delving into specifics, and because this topic is at the forefront of science, a stem cell primer should be provided. In order to better understand the everything, an overview of the relevant basics is required. First, one should understand what stem

---

[2] *Id.* p. 71.

[3] *See* Human Rights Committee, ICCPR, Gen. Comm. 36, 124th Session, 26, (Sept. 3, 2019) GE.19-15012(E).

[4] *Id.*

[5] Lee (2018).

[6] Acosta and Golub (2016), pp. 435–436.

[7] Lo and Parham (2009), pp. 204–213.

[8] *Id.*

[9] Acosta and Golub (2016), pp. 433–434.

[10] *Id.*

cells are and what they do in the human body. It is important to explain that the human body is made up of mostly specialized cells that serve specific functions (of which there are hundreds). Examples include skin cells, heart cells, blood cells, muscle cells, etc. although many scientists group cells into four categories by type. One must also understand that all cells begin their "life" as a stem cell, which is then automatically converted by the body to become whatever specialized cell the body requires.[11] There are two main sources of stem cells, as well as embryonic and adult.[12] Stem cells can be viewed as specialized cells, that start out as a blank, "undifferentiated" cell, meaning that they are created without the specialized function in which they will develop into possessing.[13] As to embryonic stem cells, one must understand its "life-cycle". First, in nature, when a sperm fertilizes an egg, something known as a "zygote" is formed which is a single cell which then divides, and form a multitude of cells on its own, a so-called blastocyst, a forerunner of an embryo it is made up of around 150–200 cells.[14] The blastocyst is essentially a bundle of cells and forms approximately 3–5 days after fertilization.[15] Once at the blastocyst stage, embryonic stem cells form in what is known as the "inner cell mass".[16] These cells can be safely removed by skilled scientists and researchers who then have the job of coaxing these newly removed stem cells into the desired cell type in a laboratory setting.[17] If the cells in the inner mass were to be left alone, to grow as they were originally intended to, they would eventually, naturally differentiate into a specialized cell.[18] Once properly changed or "differentiated" into the desired cell type, the new cell or cells can be placed in the location that is in need of repair or area at issue, and these cells should be able to accomplish what they were intended to do. As discussed above, adult stem cells originate and can be found in all parts of a developed human body.[19] They too have the ability to self-renew and create countless more stem cells.[20] Scientists have found numerous methods to obtain these cells, yet actually finding them in the body can be a difficult process as they can essentially remain hidden until the body requires some form of repair.[21] Adult stem cells have the role of replacing old and/or damaged tissue in the body.[22] For example, when someone gets a cut on their skin, epithelial stem cells are utilized

---

[11] *See generally* Chagastelles and Nardi (2011), pp. 63–67.

[12] *See generally* Zakrzewski et al. (2019).

[13] *Id.*

[14] *Id.*

[15] Kalra and Tomar (2014), pp. 923–926.

[16] *Id.* at 927.

[17] *Id.*

[18] *Id.*at 919.

[19] *See:* Körbling and Estrov (2003), pp. 570–582.

[20] *Id.*

[21] Brazier (2018).

[22] *Id.*

to replace the damaged tissue with fresh, new skin cells.[23] There are also newer forms of stem cells that will be discussed such as the induced pluripotent stem cell, and the mesenchymal stem cell, both of which are technical terms for adult stem cells that has been altered in a laboratory setting to take on more characteristics of an embryonic stem cell, so as to make them more effective and have less drawbacks of regular adult stem cells.[24] To accomplish this task, scientists must first locate and remove adult stem cells from a person's tissue, and then alter them while out of the body.[25] Unfortunately, there is still a lot of research this is needed before declaring this form of stem cell as superior, or equal to the abilities of embryonic stem cells.[26]

There are fears that the development of stem cell research will lead to major changes in the world in a way that is negative, such as cloning, or cross-hybridization. Scientists have stated that most forms would likely not be even possible within the near future. The other issue that many people have with embryonic stem cell research is that they believe that in order to retrieve stem cells from embryos, the embryos would have to be destroyed. There have been many newspaper articles and clips on the news on television that discuss the supposed near future of stem cell usage in regards to creating human/animal hybrids, some of which have been successful in a laboratory setting.[27] Most articles or clips fail to mention the actual reason for doing so, the potential benefits, and the fact that there is never any human/animal crosses actually allowed to grow past the embryonic stage.[28] Further, similar news stories are published regularly which discuss questionable "stem cell clinics" which can be found both in the United States, and abroad.[29] These stories tend to highlight unproven and completely unapproved procedures which have little to no scientific basis to justify their use without any qualifying, or discretionary notice to potential patients regarding the fact that these "clinics" operate outside of science and the law.[30]

A very similar issue arises from vaccination fears.[31] A researcher named Dr. Wakefield, who found and published the supposed links between autism and vaccinations was later found to have received a fortune of money to fudge numbers in order to "discover" a link between the two.[32] His findings were extremely

---

[23] *Id.*

[24] *Id.*

[25] *Id.*

[26] *Id.*

[27] Blakemore (2017).

[28] *Id.*

[29] *See* de Jesus Medina Arellano (2012), pp. 63–78, July–December 2012; Roget (2017).

[30] *Id.*

[31] Wente (2010).

[32] *Id.* (discussing autism research scientist Dr. Wakefield of Britain was secretly paid £400,000 by a rival vaccination company in order to steer peoples' belief that the current vaccinations were responsible for the rise in autism diagnoses so that when the rival company's new vaccination was released, people would choose the rival's version as opposed to the current one out of fear of exposing their children to the higher potential of autism).

damning and published in Britain's Lancet.[33] In fact, groups emerged championing this research and further propagating the fear in parents with respect to vaccinating their children.[34] This has actually led to much worse health outcomes for the unvaccinated children thus literally achieving the opposite benefits of avoiding the vaccine in the first place.[35] When the news first broke that Dr. Wakefield altered his "study's" results to achieve a benefit for his secret employers, Paul Offit, chief of infectious diseases at the Children's Hospital of Philadelphia made a very poignant statement which was: "It's very easy to scare people; it's very hard to unscare them…"[36] These scare tactics tend to further perpetuate misinformation and generally use ear or eye catching language to attract viewers. Many studies have been undertaken to prove the link between fear-based media stories, and attracting viewers' attention.[37] Unfortunately laws do not tend to prohibit or criminalize this fear mongering strategy, so what we are left with, is a portion of the population living in fear of science that can actually help save them.

Alarmingly, while speaking with many contemporaries that are around my age, their only basis of knowledge regarding stem cells comes from popular culture.[38] Many television shows have portrayed embryonic stem cells in a humorous fashion such as South Park, and Family Guy.[39] Each went to an extreme in showing their "take" on stem cells. South Park has actually aired numerous episodes that both discuss and try to joke about the process and the abilities of stem cells.[40] In the episode titled "Krazy Kripples," which originally aired on March 26, 2003, Christopher Reeve arrives to the fictional town and in order to cure his paralysis, he trucks in tons of human embryos which he then breaks open and sucks the life out of to regain his ability to walk, and eventually gain super powers.[41] In another episode titled "Kenny Dies," originally airing on December 5, 2001, one of the main characters is dying due to complications with muscular dystrophy, and another of the main characters is trying to have embryonic stem cell research legalized so that his friend could be saved. In the end, they are unsuccessful and the child dies, thus illustrating the need for the legalization of embryonic stem cell studies to prevent needless deaths.[42] The TV show Family Guy also tackled the subject in an episode titled "McStroke," which originally appeared on air on January 13, 2008, when the main character Peter suffers a stroke, he suffers relentlessly until he enters a stem cell clinic where he then gets treatment and immediately his symptoms of the stroke are

---

[33] *Id.*

[34] *Id.*

[35] *Id.*

[36] *Id.* (citing the retraction made by The Lancet in regards to Dr. Wakefield's supposed findings).

[37] Deborah Serani (2001).

[38] Donne (2017).

[39] *See* Myers (2016).

[40] *Id.*

[41] South Park Archive (2019a).

[42] South Park Archive (2019b).

gone.[43] The character makes a comment about why the country is not actively funding this area of science in reference to the problematic laws and policies preventing embryonic stem cell research.[44] With these as many of their only guides, many are unsure what to believe, and as a result, tend to take an improper stance, or have improper beliefs as to the overall subject.[45] Even though there seems to be positive messages behind the episodes in regards to stem cells, they all make it seem that stem cells are purely fanciful and joke-worthy.

The next largest challenge to allowing and creating workable stem cell laws is our own government and its citizens. As explained in detail below, there are groups of citizens who strongly oppose embryonic stem cell research and use. These groups are typically political conservatives who tend to fall into one of the religious groups that oppose embryonic stem cell studies and research. Understandably, there is an argument as to when life technically exists. Many of these religious organizations typically believe that life begins at conception, while there are varying beliefs among others.[46] Because these religious organizations have a great number of members, political conservatives tend to accept this religious view in an attempt to gain more voters.[47] This is not atypical of all political groups, but in this case, it is the conservatives who fight against embryonic stem cell use and studies because their voters essentially demand so. Aside from the usual separation of church and state laws and issues, these views as to when life begins are given legitimacy by the conservatives in voting against stem cell laws or crafting unworkable laws that tend to achieve nothing.

Besides the fact that there are legitimate arguments that can be made opposing this idea as to when life begins, there are also scientific methods which are ignored that would maintain the dignity of the embryos and not risk their lives to begin with. Science is an evolving discipline which finds new methods and new techniques very often.[48] There are many laws that establish hurdles and limitations for embryonic stem cell research, leaving virtually no room for evaluating potential advancements. Many citizens are oblivious to the overall facts and both actual benefits and potential benefits that stem cells provide as well as their numerous risks. If people knew that utilizing embryonic stem cells in studies can lead to life extending findings, and discoveries of cures to some of the many horrific medical issues gripping much of the population, it is highly likely that their view would either change or be created and expressed in supporting this important research.[49] Scientists believe that embryonic stem cells will eventually be used to treat or alleviate diseases like Parkinson's,

[43] Family Guy Fandom (2019).

[44] *Id.*

[45] *Id.*; International Movie Database (2010).

[46] Bukuras (2022); *see generally* Provost (2016).

[47] *See generally* Lienesch (1982), pp. 403–425.

[48] Malik (2020).

[49] Mayo Clinic Staff (2023); Izrael et al. (2018), pp. 152–153, 168–169.

diabetes, Alzheimer's, various autoimmune disorders and much more.[50] Even if the people are not suffering themselves, typically they know others such as loved ones, or friends that can benefit from such studies and treatment options.[51]

To better understand the arguments as discussed in detail below, it is helpful to delve into the specific forms of stem cells, where they come from, and their unique abilities. The most "powerful" form of stem cells is called embryonic stem cells. This is because they can be taken directly from embryos that are in the earliest stages of development and are found to be the most pluripotent out of all forms of cells.[52] They are so powerful because they have a greater ability, and likelihood to be successfully differentiated into the desired cell.[53] This form of stem cells can also be found in the blood and bone marrow of fetuses which are also in a very early form of development[54] The next type are known as "adult stem cells" and as the name suggests, comes from already formed beings from birth, to old age.[55] They can also be found in umbilical cords, and in the placenta.[56] The best way to understand how and why adult stem cells exist is to look at it from an economic standpoint such as supply and demand, where there are is a plethora of adult stem cells in each area of the body where they can be found, waiting for the body to exhaust, or to be in need of repair for a certain area, and then have a call for adult stem cells to be used to replenish or fix the problem.[57] A perfect analogy was used by Doctor David Borenstein to help understand how and what adult stem cells are and compared the human body to a gold mine and stated "There's tons of gold in there," he says, pointing to my body. "The problem is, it's sitting in the ground, doing nothing. You gotta go in there and get it."[58] The doctor cited in this article is Dr. David Borenstein who is a rehabilitative specialist located in New York who advocates the use of adult stem cells to help rehabilitate damage done to one's own body by using the same person's own stem cells.[59]

Currently, the United States limits the number of human embryotic stem cell lines ("hESC") through its stem cell laws.[60] These hESC lines are the only ones that are currently allowed to be used in any form of research and are the original stem cell

---

[50] Mayo Clinic Staff (2023).

[51] *Id.* (discussing the potential benefits of embryonic stem cell treatments for ALS, and a particular drug's effectiveness in both prolonging the life in mice and allowing them to be much more comfortable with lab induced ALS).

[52] Bailey (2019).

[53] *Id.*

[54] *Id.*

[55] National Institute of Health (2022).

[56] *Id.*

[57] *See* Boston Children's Hospital (2009).

[58] Joseph Hooper (2017), "Does Stem Cell Therapy Actually Work?", available at: https://www.mensjournal.com/health-fitness/does-stem-cell-therapy-actually-work-w437788/ (Sept. 6, 2014).

[59] *Id.*

[60] MIT Committee on Assessment of Biohazard & Embryonic Stem Cell Research Oversight (2008).

lines that come from the National Stem Cell Bank and had to be in existence in 2001.[61] Currently, there are only a very small amount of hESC lines, and the problem with the low number comes from the fact that each of the hESC lines contain their own limits and abilities which require further exploration and the creation of new lines to determine their limits and abilities.[62] An example of this would be a stem cell line which was found to have a greater ability to transform into another type of cell.[63] There is a common myth, or misunderstanding which also gets in the way of changing laws which is that embryos have to be destroyed in order to retrieve stem cells.[64] While this was true for a number of years, scientists have now created a method where stem cells can be collected from very early stage embryos without destroying them at all.[65] To accomplish this feat, the researchers essentially removes cells from an embryo, without harming the embryo, and then perform a biopsy on the cell to remove individual stem cells from the embryo, which at this point is made up of approximately 10 cells, and now known as a "blastomere."[66] This method is actually almost identical to the techniques used in preimplantation genetic studies which are used to determine the status of the health of an embryo after *IVF* fertilization, yet allows the embryo the full ability to develop further.[67] Studies such as these are done very often as they are used to screen embryos for potential diseases.[68]

Adult stem cells contain certain limitations that embryonic stem cells do not. They are not as easily differentiated into the desired cell, and they are actually harder to find in tissue.[69] The only possible advantage that adult stem cells have over embryonic stem cells is that some believe that because they tend to come from the host who requires repair, they are less likely to be rejected.[70] In other words, adult stem cells seem to lack the inherent abilities of embryonic stem cells such as the ability to be altered into becoming any cell type where adult stem cells are limited to fewer cell types.[71] This book will primarily focus on embryonic stem cells, but it will also cover various issues where laws are in existence that pertain to both forms.

It is also important to understand the difference between therapeutic cloning, and reproductive cloning. Reproductive cloning is the most controversial and stands for the process of creating an embryo using stem cells from another embryo, leaving the

---

[61] Murugan (2009), pp. 101–103.

[62] Carpenter et al. (2004), pp. 253–254.

[63] Univ. of Wash. - Health Sciences/UW News (2014).

[64] Chung et al. (2008), pp. 113–117; Dittrich et al. (2015), pp. 1239–1242.

[65] *Id.*

[66] Dittrich et al. (2015), p. 1241.

[67] *See* Van der Aa et al. (2013), p. 71 (2013).

[68] *Id.*

[69] Bailey (2019).

[70] *Id.*

[71] *See generally* DeWitt and Knight (2002).

newly created embryo to be genetically identical to the original.[72] After creation, the newly cloned embryo is inserted into the "uterine environment" for development purposes.[73] A common example of this type of cloning technique can be seen in "Dolly" the cloned sheep, who was the first animal to be a result of this cloning process.[74] Therapeutic cloning involves using stem cells from an embryo and creating a genetically identical embryo but the embryo is never allowed to develop, and remains in a laboratory setting.[75]

New techniques are created very often to resolve the embryonic stem cell "problem" which effectively alter an adult, or at times a specialized cell, such as a red blood cell, to transform it into something very close to an embryonic stem cell.[76] In doing so, there are certain benefits and limitations.[77] For instance, scientists have discovered that there are certain times and conditions where the host body automatically converts blood vessels into two distinct forms of stem cells, one for new blood vessels and the other for the blood system.[78] This was no easy task but the study definitively showed that this process naturally occurs in the host body.[79] Even more recently, scientists have used new technology that allows them to alter skin cells and turn them back into stem cells.[80] This too is a very difficult task but can eventually offer a new means of obtaining stem cells in a less controversial manner.[81]

There is so much hope placed within this field by those who are in need of treatments and cures, and those who seek to further our collective understanding on the subject to help those in need. As a result, much of the major pushes to further this area of science comes from doctors and scientists. Due to the medical implications that would come to reality if the American laws were changed as to allow for federal funding, and expanded laws that enable scientists and doctors to conduct experiments, procedures, and operations using stem cells, which would not only benefit science and medicine as a whole, but also greatly improve the lives of the patients who are currently suffering, and while learning what stem cells could be used for, help people who will require medical intervention in the future.[82]

Many scientists view the current laws as very restrictive in terms of allowing for experimentation, and receiving federal funds. One of their main objections to the laws pertain to the rules that only allow for existing hESC lines to be used. By only having a small number of hESC lines to work with, scientists cannot further explore

[72] New York State Stem Cell Science Dept. of Health (2015).

[73] *Id.*

[74] *Id.*

[75] *Id.*

[76] Bergiers et al. (2018), pp. 1, 18.

[77] *Id.*

[78] *Id.* pp. 15–18.

[79] *Id.*

[80] Gladstone Institutes (2018).

[81] *Id.*

[82] de Wert and Mummery (2003), p. 680.

the potentials of stem cells. This law also causes a problem due to the fact that certain of the lines have already been found to be problematic and unfit for human use due to the initial way that they were derived.[83] Finally, scientists argue that adding new stem cell lines would enable them to discover new abilities, and better lines that tend to allow for cellular change than those that are currently used.[84] Much of the arguments put forth by this group are based on the already known possibilities and abilities of embryonic stem cells, and the need for further funding.

Both scientists and doctors work for much of their lives in a field that tends to work towards helping others in need. In fact, many doctors in the United States and throughout many other countries require new doctors to take a form of the Hippocratic Oath which effectively imparts on the new doctors that they have certain duties owed to their patients.[85] This oath was created in ancient Greece by the "father of medicine" Hippocrates to create a set of ethical principles on doctors that all doctors should follow, such as, doctors should: treat the sick to the best of one's ability, preserve patient privacy, teach the secrets of medicine to the next generation...[86] The above principles were created in antiquity, and to this day the oath is regularly updated to further impart the principles on each generation of doctors.[87] When dealing with the original oath, the above has for the most part remained the same, as for this book, it is obvious why doctors and scientists want to have access to the most cutting edge techniques to help those in need.[88] Particularly the principle of treating the sick/those in need of medical help to the best of one's ability is literally hampered by outdated laws which were for the most part created without full knowledge of the issues that stem cells present.[89] Doctors who advocate for the expanding of the federal embryonic stem cell guidelines are effectively told that even though there is a known cure, or a potential cure to any number of diseases and ailments, they cannot help their patients to the best of their abilities and/or knowledge.[90]

Knowing that embryonic stem cells can effectively heal/cure patients or at the very least relieve patients of chronic and horrific pain, but being constrained by laws created by those who only have some information on stem cells is very frustrating.[91] A great example of this issue comes from my own personal doctor who introduced me to a patient whom he could no longer help. A young child had a muscle disease that required many surgeries before he was 8 years old. The scars severely limited

[83] *Id.*

[84] Van der Aa et al. (2013), p. 71 (2013).

[85] Tyson (2001).

[86] *Id.*

[87] *Id.*

[88] Parsa-Parsi (2017), pp. 1971–1972.

[89] *Id.* (discussing the newly revised version of the Hippocratic Oath which was adopted by the World Medical Association (WMA) General Assembly on October 14, 2017, in Chicago).

[90] Murugan (2009), pp. 101–103.

[91] Cyranoski (2013).

the child's movement and unfortunately, there was nothing more my Doctor could do to help him. The child's parents were luckily very aware of the use of embryonic stem cells in other countries, and were fortunate enough to have the means to take action based on their knowledge. First, they asked my Doctor what his thoughts were on using embryonic stem cells, and next what his thoughts were on going out of the country to China for the procedure. My Doctor was skeptical due to China's overly loose laws (discussed further in this book) which he explained to the child's parents, but stated that he knew that embryonic stem cells have in fact been able to help people with all different forms of ailments. With this knowledge, the parents brought their child to China where they spent around two months which included a small number of procedures to treat the child with the stem cells. When he finally returned to Miami, Florida, they visited the Doctor and the Doctor was shocked to see just how well the treatments worked. The young boy now has a much wider range of motion where he could barely move before, his scars actually looked better, and he can now live life as a "normal" child. Had this child's parents not been aware and not had the means to bring him to China for these treatments, the child would still be suffering due to his pain, limited range of motion and mental trauma from his original scars.

There has also been a lot of news over the past two years pertaining to the potential in using embryonic stem cells has created, such as restoring eye sight in patient with various eye degenerative issue; restoring movement in paralyzed patients, and curing various other diseases.[92] In fact, based on these discoveries, investors have sought to build up laboratories to further research the benefits of embryonic stem cells in many different states, however, due to the difficult to adhere to laws, very little embryonic stem cell experiments have been conducted.[93] Due to the laws in place, these laboratories have had to use other species to test their theories which have had astounding success which they tend to believe would work similarly in humans, yet cannot test these treatments on those in need.[94] To make matters worse, investors in private laboratories have called for immediate proof of success or they will stop funding these experiments, which of course has led to the untimely shutting down of these labs.[95]

As a scientist or doctor who is already practicing and studying in the field of embryonic stem cells, it must be very frustrating to know that there are means to help people which are vastly overlooked due to certain peoples' moral and ethical beliefs. To many of them it would seem that the costs of the experimentation, at any level can be vastly outweighed by the potential benefits. One has to understand that the chief complaint and reason for the essential ban of embryonic stem cells is due to the risk of destroying said embryos. This argument is discussed at length below, but to contrast this argument, pro-embryonic stem cell believers point to the fact that a

---

[92] Smith (2017).

[93] *Id.*

[94] *Id.*

[95] *Id.*

frozen embryo only has a certain amount of time before they become completely unusable.[96] It is extremely important to discuss this issue even though science seems to have grown beyond the need because many of the strict laws were created to protect the embryo. Before new experimental procedures were discovered that protect the embryo while removing an embryonic stem cell, the main argument for their When a person seeks out a doctor for in vitro fertilization (*IVF*), there are typically three or four unused embryos after the successful *IVF*.[97] The typical options for the remaining embryos is to keep them frozen for potential future *IVF* procedures, discard the extra embryos, or to allow the embryos to be donated for scientific purposes or to families in need.[98] When freezing embryos, there is only a short amount of time, approximately eight to ten years, and there are storage fees associated with embryo storage.[99] In fact, many state laws only allow embryos to be stored for a maximum of ten years due to the risks of using older embryos in *IVF* procedures.[100] With these facts at hand, scientists and doctors have argued that the benefits that can be conferred by unused embryos far outweigh the choice of simply discarding them.

Embryonic stem cells have also shown promise in the arena learning how blood cells are created, and attempting to understand the nature of converting embryonic stem cells into new blood cells.[101] Blood transfusions are extremely common, especially in situations where there has been a traumatic injury, or one in which there were issues during an operation.[102] To give one an idea for as to just how common these situations are, one must look to some of the most staggering statistics provided by The Community Blood Center: 4.5 million Americans require a blood transfusion yearly; a person is in need of a transfusion every two seconds; blood shortages occur in an almost cyclical nature; blood centers often run short of the rarer blood types; and currently, there is no substitute for human blood.[103] It is easy to overlook blood transfusions in times like today because they are so routine, and rarely have issues, unless there is a large scale event such as the case of Covid-19.[104] In these times there are shortages in the blood supply because fewer people are donating, yet transfusions remain just as commonplace as they were prior to the pandemic.[105] There are also other times when blood is needed when there is a devastating event such as a catastrophic earthquake or hurricane that causes injuries

[96] Beil (2009).

[97] Gurevich (2020).

[98] Bowen (2008).

[99] Baldwin (2018).

[100] *Id.*

[101] Slukvin (2010), pp. 1621–1623.

[102] *Id.*

[103] *See* Community Blood Center (2007).

[104] Marks (2020).

[105] *See id.*

to a larger than normal amount of the population.[106] With all of this knowledge, it is easy to see that creating a blood cells from stem cells, or finding a blood cell substitute is imperative to help people on a global scale.

Scientists have already had some success in converting non-embryonic, adult stem cells into blood cells with some ease, thanks to a groundbreaking study by Luc Douay, and his team, but issues quickly were noted due to the fact that non-embryonic stem cells do not self-proliferate, or the ability to self-reproduce, in a virtually unlimited manner, which makes them difficult to use for actual industrial production.[107] This is where embryonic stem cells have a total advantage, not only do they require less coaxing, can self-proliferate, and by their very nature, already transform into blood cells.[108] As they are essentially blank cells, the blood cells converted from embryonic stem cells can be further ensured to be free from disease, and ensure that there is less of a risk of an inflammatory response that causes a rejection by the receiving host body.[109]

To take this one step further, in explaining what types of therapies could be possible in the near future using embryonic stem cells, scientists and doctors believe that there will be blood therapies that use embryonic stem cells, making them completely compatible with the patient.[110] Some examples include the ability to create new red blood cells for those in need of blood transfusions; as well as bone marrow to be used in transplants.[111] All of these potential blood therapies will also have the added bonus of being virus free, and the patients in need will also not need to find a match for a transplant.[112] While viral contamination in blood used for transfusions is a relatively low occurrence at this point in time, it still occurs and poses a great risk to patients who are in need.[113]

There are also those who would like to see the potential of stem cells reached to benefit their own lives. It is believed that the study of embryonic stem cells will lead to discoveries that can extend the life of average human beings. As explained above, stem cell treatments, when conducted properly can lead to much better lives for those who utilize them.[114] If some of the desired, hypothesized outcomes come to fruition, people would live without the same pains that come with aging, without having to suffer from serious conditions, diseases, or other age related issues.[115] One major issue regarding treatments using embryonic stem cells is that there are so few when there are essentially unlimited diseases and ailments, so until more testing is allowed

---

[106] The American Red Cross (2019).

[107] Marks (2020).

[108] *Id.*

[109] See *Id.*

[110] Kaufman (2004).

[111] *Id.*

[112] *Id.*

[113] American Red Cross (2024); Buerger and Jain (2023).

[114] Nedelman (2018).

[115] Chang et al. (2022), pp. 901–910.

and done, the scientific community is left guessing or testing on animals.[116] However, even with the limited amount of information that is currently known, scientists and doctors believe that embryonic stem cells can lead to longer life spans, and can help people age in a more comfortable manner.[117] Everyday our population is constantly bombarded with scary information regarding so called "super bugs" and races for various cures for other life threatening diseases, yet the answer to helping those suffering may very well lay with embryonic stem cells.[118] For example, researchers have found that embryonic stem cells have a natural resistance to viral infections.[119] Even adult stem cells have shown similar promise when used in relation to "West Nile virus, cytomegalovirus and myxoma virus," which "are also unable to infect adult stem cells efficiently."[120] Many who desire further research into this area can cite various trials conducted using embryonic stem cells that have shown a great deal of progress.[121] Scientists have even used human embryonic stem cells to help repair heart tissue in monkeys after suffering from heart attacks.[122] Scientists are always seeking better treatments for heart disease and other blood flow issues when dealing with the human heart, and in this study scientists found that blood flow increased by 10% in many of the test monkeys.[123] This number may seem insignificant, but due to the fact that heart disease in all its forms are the leading cause of death in humans, makes this study all that more important.[124] Scientists have also used adult stem cells to help treat those suffering from heart failure, and see similar improvements, but these studies are conducted much more regularly than using embryonic stem cells.[125]

To take this one step further, people around the world have been amazed by the fact that there are animals that can survive cellular damage on a profound scale and actually come out without any limitations or issues.[126] Scientists studied planarian flatworms which are an interesting species because unlike humans, they maintain a larger number of stem cells as they reach adulthood than humans.[127] Humans typically have their stem cells that exist while in the embryo slowly turn to particular specialized cells thus losing their pluripotency.[128] Scientists chose the planarian

---

[116]Cyranoski (2018), pp. 428–430.

[117]*Id.*

[118]*Id.*

[119]*See generally* Wu et al. (2019), p. 50.

[120]*Id.* pp. 50–51.

[121]*See* Fuchs and Segre (2012), pp. 143–155.

[122]Nedelman (2018).

[123]*Id.*

[124]*Id.*

[125]Sheldrick (2018).

[126]Zeng et al. (2018), pp. 2–3.

[127]*Id.*

[128]*See generally* Romito and Cobellis (2016).

flatworm because they are essentially unkillable.[129] For decades scientists have known that these flatworms can regenerate essentially any part of their body, including their head, and survive the most catastrophic of injuries, for the first time, scientists have identified the reason and it all revolves around stem cells.[130] As explained above, the planarian flatworm maintains a stockpile of stem cells and through testing, scientists have actually tested their abilities by blasting a group of worms with intense amount of dangerous radiation which caused extreme damage, yet once the scientists injected stem cells into the flatworms, their bodies were rebuilt and replaced the damaged tissue with new, healthy tissue.[131] While many may wonder why a flatworm was used to better understand embryonic stem cells, it should be noted that it is "[b]ecause planarian stem cells share characteristics with those of humans, the work will aid scientists striving to understand how stem cells can be used to completely repair damaged tissues and organs."[132] Due to this flatworm species regenerative abilities, researchers believe that there is much we can learn to aid in our own tissue repair.[133]

In an another study, scientists have discovered a jellyfish, or *Hydra* jellyfish, that is essentially immortal thanks in large part to their own stem cells.[134] In a highly publicized study, scientists announced that they have discovered how the *Hydra* can live forever through the use of stem cells.[135] Similarly, scientists have discovered another form of immortal jellyfish called *Turritopsis dohrnii, which also uses stem cells, but instead of preventing aging like the Hydra does, this form of jellyfish transforms its own organs and body to an earlier stage of their lifecycle.*[136] *To better understand what this means, this form of jellyfish like many others starts out as a larva, than grows into a polyp, than to a budding polyp, than an ephrya and finally a medusa.*[137] *When there are times of struggle, such as lack of food, raising or lowering water temperatures, or just plain aging, the Turritopsis dohrnii through using its own stem cells is able to create an entirely new body plan and reverts back to a polyp stage.*[138] *Scientists choose to study this species and others like the Hydra to learn "how some embryonic developmental programs can be repurposed," to the benefit of all.*[139] *Excitingly, scientists already know what would have to be done to humans for a similar result, which would require a "reprogramming" of sorts to change our epigenomes which control parts of our DNA code and tells when to turn*

[129] Puiu (2023).
[130] Shiroor (2023), pp. 7–8.
[131] *Id.*
[132] Newmark (2006).
[133] *See generally* Bohr et al. (2021).
[134] Sito (2017).
[135] *Id.*
[136] *Id.*
[137] Piraino et al. (1996), pp. 302–303 (1996).
[138] *Id.* p. 311.
[139] Kalomiris (2022).

different genomes on and off.[140] *To do so involves what are called* "Yamanaka factors" which are "[H]ighly expressed in embryonic stem (ES) cells, and their over-expression can induce pluripotency in both mouse and human somatic cells, indicating that these factors regulate the developmental signaling network necessary for ES cell pluripotency."[141] There is a major hurdle however which comes from the fact that the Yamanaka factors have to be turned on an off at the correct times or we risk the cells reverting to into a mass of undifferentiated embryonic stem cells which could turn into the wrong form of cells.[142] Logically, if instead of altering the Yamanaka factors, and directly using embryonic stem cells, the risk would be extremely less, with amazing potential. Others see potential in the area of fighting or controlling diseases and viruses which have been left to proliferate and at times torture those who suffer from them without having any ability to find help. As a result, much research and speculation has surrounded what types of benefits using embryonic stem cells can bring. While there seem to be nearly endless possibilities as to what people believe these stem cells can do, a lot of focus has been on curing paralysis, diabetes, Parkinson's disease, stroke recovery, curing blindness, regrowing lost limbs, the healing of difficult wounds, heart disease, regrowing teeth, among many others.

2.2 What Can Embryonic Stem Cells Treat?

2.2.1 Paralysis Treatments

As touched on above, there are many people who believe that paralysis can be minimized, and potentially eradicated through the study and use of embryonic stem cells. A major champion of embryonic stem cell use was actor Christopher Reeve who himself suffered from paralysis.[143] Christopher Reeve was injured and paralyzed in a horse riding accident in 1995, and as a result, became a "champion" for embryonic stem cell research in the area of paraplegics.[144] In fact, Mr. Reeve was invited to speak at "Patients Press Conference for the United Nations" in October of 2004.[145] The idea behind stem cell treatments to end paralysis has been discussed for many years, and there have been great successes in this field.[146] There have also

---

[140] *See* National Human Genome Research Institute (2020).

[141] *Id.*

[142] Sito (2017).

[143] Williams (2004), pp. R907–R908.

[144] *Id.* p. R908.

[145] *See* Christopher Reeve's Speech to the United Nations, (October 2004).

[146] *See* Connor (2010) (discussing the benefits of using embryonic stem cells to help a paralyzed man walk again).

been major successes in the treatment of paralyzed rats.[147] In 1999, scientists were able to transplant embryonic stem cells into a rat with a damaged spinal cord, and have them differentiate successfully to promote recovery of the rat's spine.[148] Unfortunately Mr. Reeve passed away due to a heart attack that arose from the use of antibiotics to treat an infection, yet his legacy as a champion and advocate of stem cell usage is still just as relevant and important today.[149] Even before many of these studies published their findings, there was wide speculation as to the potential of embryonic stem cells and this condition, and in 2003, The Christopher Reeve Paralysis Act was introduced in the US House of Representatives and Senate which was an attempt to pass a law which would increase funding from the US National Institutes of Health to develop treatments for paralysis, including embryonic stem cells.[150] Even though this Act did not pass, it did not prevent further attempts to increase funding to this important research.[151]

There have been many ailments and diseases that studies have shown can be alleviated or even cured thanks to embryonic stem cell-based therapies.[152] The following information comes from many groundbreaking studies that have shown promising results for diseases that can affect all people, and have had no known cures, and can benefit all.

2.2.2 *Diabetes*

Scientists and researchers have had some major successes using embryonic stem cell transplantation to aid those who suffer from various forms of diabetes.[153] In a promising study, scientists used two different methods for coaxing embryonic stem cells into pancreatic cells that could produce insulin.[154] With this knowledge, scientists and researchers will soon be able to create targeted, specialized therapies that aid in regenerating damaged or problematic cells.[155] Studies have shown positive results for all types of diabetes, including Type-1 which is most prevalent in children and young people, and is very hard to manage.[156] To give an idea of how widespread type 1 diabetes is, one just needs to look at the figures to see how many people are currently suffering; approximately 13,000 new cases of Type-1 diabetes

---

[147] *See generally* Deshpande et al. (2006), pp. 32–44.

[148] McDonald et al. (1999), pp. 1410–1412.

[149] Williams (2004), pp. R907–R908.

[150] *See* H. Rept. 110-378 - CHRISTOPHER AND DANA REEVE PARALYSIS ACT.

[151] *Id.*

[152] *See generally* Park et al. (2008), pp. 877–885.

[153] Wei et al. (2013), p. 7.

[154] *Id.* p. 1.

[155] Miyazaki et al. (2004), pp. 1030–1037.

[156] Freeborn (2017), pp. 188–194; American Diabetes Association (2023).

are diagnosed yearly in the United States; approximately 352,000 people under the age of 20 currently suffer from this disease; and it is extremely important to note that there is not necessarily one single gene that causes this form of diabetes.[157] It is also important to understand the process by which Type-1 diabetes causes such health issues, and the simplest explanation is that the body wrongfully attacks its own pancreatic insulin-producing β cells.[158] Without these β cells, the body cannot produce insulin appropriately.[159] These genes have presented a unique issue when dealing with this disease because there is not necessarily an "on" or "off" switch that are the root cause, thus making it much harder to create a cure or permanent treatment.[160] Other studies using embryonic stem cells have also shown promise in creating benefits and treatments of patients suffering from Type-2 diabetes.[161]

Embryonic stem cells have also been tied to creating new therapies that will treat degenerative diseases in the elderly, something which typically has to happen to every human being as we age.[162] Studies have found that using embryonic stem cells can help those in need of donor tissue, which there is a shortage of, and can virtually remove the issue of implant rejection because the embryonic stem cells are effectively blank and can be used without much worry.[163] Due to the constraints in the current laws of the United States, these tests had to be conducted on mice with induced diseases or those that were already older, yet the results were extremely promising, and point to a high likelihood of success if used in humans.[164] To better understand what diseases the referenced study worked to alleviate include Parkinson's disease, spinal cord injuries, myocardial infarction and heart failure, and diabetes which essentially covers such a broad spectrum that most if not all people will eventually have to deal with in one way or another.[165]

2.2.3 *Parkinson's Disease*

The National Institute of Health's National Institute of Aging describes Parkinson's Disease as a "...brain disorder that leads to shaking, stiffness, and difficulty with walking, balance, and coordination."[166] The disease typically affects the elderly, usually occurring in people over 60 years old, but can occur in younger people, and

---

[157] *See generally* Steck and Rewers (2016), pp. 176–185.

[158] Melton (2011), p. 2309.

[159] *Id.* p. 2307.

[160] *Id.* pp. 2309–2310.

[161] Shroff (2016), pp. 223–230.

[162] He et al. (2003).

[163] *Id.* p. 279.

[164] *Id.*

[165] *Id.* pp. 274–283.

[166] National Institute Of Aging, Parkinson's Disease Information (2023).

most surprisingly seems to occur randomly in people, with only a small percentage being due to a genetic mutation or familial history.[167] Parkinson's Disease has been found to occur due to the "...death of dopaminergic neurones in the substantia nigra of the midbrain."[168] Effectively, for an unknown reason, the dopamine producing neurons, found in a person's midbrain, die and can no longer produce dopamine, and this loss of dopamine is directly tied to the loss of physical movement and the motor functions that become impaired due to Parkinson's Disease.[169] Thankfully, there has been promise in curing this disease thanks to embryonic stem cells in various models and experiments.[170] Recent experiments have had very promising results using embryonic stem cells to generate neurons which work to repair the lesions of the brain.[171] Astonishingly, experiments have shown in mice, that once the lesions have been repaired using these cells, lost behaviors were able to return.[172] Many scientists see promise in using embryonic stem cells, yet agree that more studying must be conducted for there to be anything that could ever be used on human beings as cures.[173]

After years of research and experimentation which showed that using embryonic stem cells in patients with Parkinson's disease can "fully reverse motor impairments in experimental models," a company created a new line of embryonic stem cells that are inserted into the patient once the company's scientists have pre-differentiated them into the desired cell type, in this case a mid-brain dopamine neuron.[174] As a result, 8 patients in Sweden suffering from Parkinson's disease were chosen to undergo an experimental procedure using these lab altered embryonic stem cells, and in February of 2023 the first patient in Sweden received these cells to treat his Parkinson's disease. To minimize any risks of the injected stem cells from moving to places other than what the research required, the scientists placed them directly into the brain in areas that can be as small as four millimeters.[175] Unfortunately, as this is an ongoing study, most results will not be made public until "a sufficient amount of material from the clinical trial has been collected and analyzed, in adherence with health care confidentiality regulations."[176]

In another study conducted by BlueRock Therapeutics, which is a subsidiary of Bayer, is also using embryonic stem cells in what has been described as one of the costliest experiments using embryonic stem cells to implant lab altered versions into

---

167 *Id.*

168 Chinta and Andersen (2005), p. 311.

169 Schapira (1999), pp. 311–312.

170 Correia et al. (2005), pp. 487–498.

171 Kirkeby et al. (2012).

172 Steinbeck et al. (2015), pp. 204–209.

173 *See* Ben-Hur et al. (2004), pp. 1254–1255.

174 Kirkeby et al. (2023), p. 1301; Norris and Korc (2010), pp. 2783–2798 (explaining that cell differentiation is a process where cells become other types of more specialized cells).

175 Lund University (2023).

176 *Id.*

twelve Parkinson's disease patients.[177] There are a lot of similarities as the Swedish experiment described above, as this study too has inserted pre-differentiated embryonic stem cells into Parkinson's patients, but their main goal here is to prove that these procedures are safe, and that they can lead to the creation of cells needed to mitigate or treat Parkinson's disease.[178] This study was discussed on August 28, 2023 at the International Congress for Parkinson's Disease and Movement Disorder in Copenhagen by the neurologist who is spearheading this study, Dr. Claire Henchcliffe who stated that not only did these stem cells survive implantation, they have also been successful in reducing the patient's symptoms thanks to an increase in dopamine producing cells.[179] Even though these results sound promising, so much more is still needed to definitively prove that the success came from the implanted stem cells, and that no health risks or issues arise at any time due to the implantation.[180]

2.2.4 *Strokes*

Strokes are typically caused when something causes a block that prevents the normal flow of blood to the brain (usually a blood clot), or when there is a sudden onset of bleeding in the brain.[181] Without the normal flow of blood, the brain cells in the brain are deprived of oxygen and necessary nutrients which leads to the cells dying off, and when there is a bleed in a brain, the leaked-out blood causes pressure on the cells of the brain which kills them off.[182] Currently, treatments typically use various medications to ensure that there are no further blood clots as well as some surgical options for brain bleeds that attempt repair the damaged blood vessel and a lot of hard work at rehabilitation.[183] Those who survive a stroke typically can experience another for some time after, and tend to suffer some paralysis, weakness, cognitive impairments, pain, depression, and other difficult issues.[184] Strokes are considered to be the "second e leading cause of death and a major contributor to disability worldwide," and for those undergoing treatment, the goal is to restore as much blood flow as possible to the brain and treating the neurological problems that the stroke caused.[185] Due to the prevalence strokes, especially in developing countries

---

[177] Regalado (2023).

[178] *Id.*

[179] *Id.*

[180] *Id.*

[181] National Institute of Health (2023).

[182] *Id.*; Kuriakose and Xiao (2020), pp. 2–6.

[183] *Id.* at 12.

[184] National Institute of Health (2023).

[185] *Id.*

has led to a lot of pre-clinical and clinical trials using stem cells.[186] Scientists have found success in using embryonic stem cells in not only repairing the damaged blood vessels but also have shown "improved sensorimotor function."[187]

2.2.5 *Blindness*

Another crucial area where embryonic stem cells have shown promise is in curing different forms of blindness.[188] The reason that multiple causes of blindness have been cured or helped is because many forms have a similar effect which is the death of the eye's retinal neurons, which can also be combined with another leading cause, which is the death of retinal epithelial cells.[189] In any event, the same form of treatment would be needed which would have to restore the dead or damaged cells.[190] In many of the various studies using mice and/or rats, the use of embryonic stem cells proved that the damaged or dead cells could be restored, along with eye sight in certain cases.[191] These studies finally led to various human trials, many of which were successful in restoring at least some sight by converting embryonic stem cells.[192] In 2018 scientists funded by California's Stem Cell Agency; which operates under California Institute for Regenerative Medicine ("CIRM"); published an exciting paper which made international headlines.[193] In this experiment the doctors created what they described as a "patch," made up of embryonic stem cells on a synthetic membrane, and carefully placed it into what is known as the "subretinal" space of one eye for each of the patients, both of whom suffered from severe age-related macular degeneration.[194] The study had 10 participants, and two of the patients (A woman 60s, and a man in his 80s.) had a significant portion of their vision successfully restored to where they could read using glasses, from being unable to read at all.[195] Unfortunately, even with such a huge success, there were three patients that had to be treated for serious side effects from the treatment, but

---

[186]Kuriakose and Xiao (2020), pp. 14–15.

[187]Oyamada et al. (2008), p. 7; Liu et al. (2021), pp. 1, 7 (explaining that although promise has been shown in using embryonic stem cells to treat stroke patients, ethical concerns in using this form of stem cells have weighed heavily on many and prevented a lot from further studies).

[188]Translational Vision Science & Technology Panel (2014), pp. 37–38.

[189]*Id.* pp. 35–37.

[190]*Id.* pp. 34, 38.

[191]*See* Carr et al. (2009), pp. *1–12* (discussing the use of embryonic stem cells in a rat that had damaged cells; Lund et al. (2006), pp. 189–199 (reporting that their experiment proved successful in restoring eye sight in rats).

[192]*See generally* Tucker et al. (2014), pp. 9–16.

[193]*See* Gallagher (2018); Welch (2018).

[194]*See generally* da Cruz et al. (2018), pp. 8–12.

[195]*Id.* pp. 4, 34.

were treated.[196] It is of note that age-related macular degeneration is one of the leading causes of blindness in people, and that it affects over 6 million people globally.[197] CIRM has funded other studies aimed at restoring eye sight, one in particular, aimed at restoring eye sight in patients suffering from retinitis pigmentosa.[198]

2.2.6 Limb Regeneration

Stem cell treatments are not just potentially groundbreaking for their possible curative abilities, but may also work to regrow limbs. One topic that has been a "stand in" for many science fiction movies and books comes in the form of limb regeneration, which literally means restoring a missing limb which could have occurred through genetic mutations, a surgical procedure or even through an accident.[199] We know how this process works thanks in part to studies conducted on various vertebrates, whom, when in their embryo stages where their "limb cells and tissues are still undifferentiated" can be regenerated.[200] This astounding finding has been shown to exist in animals as diverse as tadpoles, to mice, and chickens.[201] However, for the time being, scientists have only successfully regenerated limbs when the embryo is still in a stage of development, and only limbs that have still yet to undergo differentiation and are still in their stem cell forms.[202] It is worth noting that scientists and researchers are still actively researching invertebrates as well, who even as adults can maintain the ability to regenerate their missing limbs, but thus far the only vertebrates that exhibit this ability is newts and salamanders, both of which are amphibians.[203] Looking at these special vertebrates and the invertebrates that maintain this regeneration ability into adulthood, scientists still see promise in learning more about their stem cells, and how they might, if ever benefit mammalian stem cells.[204]

---

[196] Ring (2018).

[197] *Id.*

[198] McCormack (2020).

[199] *See generally* Bryant et al. (2002), pp. 887–896.

[200] *Id.* at 888.

[201] *Id.*

[202] *Id.*

[203] *Id.* p. 895.

[204] *Id.*

2.2.7 Wound Healing

For decades scientists and researchers investigating the unique ability of early-stage fetuses to heal from a wound without leaving a scar.[205] It has long been known that a combination of embryonic stem cells, as well as other natural responses from the body such as "…distinct growth factor profile, an attenuated inflammatory response,…" the presence of type III collagen and other natural processes, yet obtaining similar results in a laboratory setting have been mostly elusive.[206] One of the most identifiable issues as to why there has been so little success is due to the fact that many researchers have been constrained in the types of stem cells that are available for study.[207] "Despite these promising findings, the use of embryonic stem cells has remained controversial. The cells could be the most suitable ones over adult stem cells for skin tissue regeneration owing to their capacity of self-renewal… for treating cutaneous injuries."[208] In a study that looked to compare the potential benefits of the various types of stem cells in the healing process, the researchers found that embryonic stem cells were the most potent in being able to heal, in fact, six patients with "non-healing" wounds were part of the study and all six benefitted from their wounds healing after receiving the embryonic stem cells, yet the authors cautioned that more still needs to be known as to possible negative effects of this form of therapy.[209] This is not to say that there was no success with other forms of stem cells, researchers found that using induced pluripotent stem cells does produce some healing benefits, but pose further extreme health risks, many of which are unknown, and their creation is very costly due to their inefficiency in reprogramming.[210] In using adult stem cells, researchers found success in wound healing, and noted the benefit of not having any ethical considerations as they are found throughout the body, but noted that in some cases, there while there was healing even in "non-healing" patients, it was not a total success with all patients having their wounds healed.[211]

2.2.8 Cardiovascular Repair

In 2018 scientists were able to perform a clinical trial using embryonic hESC in an attempt to repair a malfunction causing severe ischemia which is an issue where

[205] *See* Liu et al. (1997), pp. 1852–1856.

[206] See *generally* Leung et al. (2012), pp. 371–378; Liu et al. (1997), pp. 1852–1856 (explaining that type III collagen is found in early embryos).

[207] Dehkordi et al. (2019), pp. 6–11.

[208] *Id.* at 7.

[209] *Id.*

[210] *Id.*

[211] *Id.* at 8; Ravari et al. (2011), pp. 705–711.

there is an insufficient blood supply.[212] In total, six patients were part of this study, one died from an unrelated health issue, while the other five recovered without any issues.[213] The study followed the patients for a period of 18 months, and the results were very promising as they showed that the embryonic stem cells were able to successfully differentiate into two specific types of heart muscle cells, cardiomyocytes (the cells responsible for contracting the heart) and nodal pacemaker cells (control electrical pulses in the heart).[214]

2.2.9 Dentistry

Surprisingly, embryonic stem cells seem to also be able to regenerate teeth.[215] In multiple. similar studies that look to what the future holds from embryonic stem cells and teeth, many scientists have agreed that in the near future, they will be able to grow all of the parts that make up a single tooth, or perhaps an entire tooth itself.[216] These novel studies have already begun in earnest and scientists currently believe that they must first "mimic the natural tooth development process either in vitro or in vivo using stem cells," and go on from there.[217]

2.2.10 Other Forms of Regeneration/Slowing Down the Aging Process

Limb regeneration and healing a serious wound is not the only form of regeneration scientists have been studying, as this is essentially an end goal, combing all other forms of regeneration, because these other forms include regenerating nerves and blood vessels, skeletal tissues, muscles and connective tissues.[218] The current difficulties come from the fact that many of these areas contain multiple types of cells, which make up more complex forms, such as the formation of a muscle itself, but this is not a problem for scientists who have taken up the cause to work out this mystery.[219] Many scientists and researchers believe that using embryonic stem cells, or potentially other forms of stem cells will work to slow down and potentially

[212] Menasché et al. (2018), pp. 429–432.

[213] *Id.*

[214] *Id.* p. 434.

[215] *See generally* Otsu et al. (2014), pp. 1–10.

[216] Thesleff and Tummers (2003), pp. 43–50.

[217] Otsu et al. (2014), pp. 1–10.

[218] Bryant et al. (2002), pp. 889–891.

[219] *Id.* at 891.

reverse the natural aging process.[220] Dr. von Schwarz who is a well-known doctor, who is considered a leader in modern technologies for serious diseases with extensive knowledge on stem cells made headlines recently when he was interviewed for an article discussing the use of stem cells to slow down the aging process and stated that he believed that living to 120 will be possible by the end of the decade, and to 150 by the end of the century using stem cells.[221] It is important to note however, that Dr. von Schwarz stated that stem cells alone will not be enough to achieve such a lofty goal, but that people will need to treat their bodies well by exercising and eating healthy.[222]

2.2.11 General Curative Abilities

Embryonic stem cells are considered amazingly powerful cells that if explored thoroughly enough, will yield impressive results that can benefit all people. In fact, there are so many theories as to what embryonic stem cells will be able to achieve in the near future, which include curing forms of blindness, uncurable by today's standards; engineering tissue; replacing tissue; as well as curing many, currently uncurable diseases and viruses.[223]

Some potential and interesting areas have emerged which can be helped through embryonic stem cell treatments which do not involve horrific diseases but mundane, run of the mill life changes. As a result another interesting benefit that has been getting extra attention lately is that stem cells seem to allow for hair regrowth.[224] There is an estimated 50 million men and 30 million women who are currently suffering from partial to complete hair loss in the United States alone, making this potential treatment, all the more desirable.[225] Currently there are various methods that have been utilized to regrow hair, such as stem cell injections.[226] These injections work by harvesting stem cells by separating them from a person's own fat tissue, or bone marrow, they are then mixed with the same person's platelets, spun in a centrifuge, and finally are reinjected into the area being worked on.[227] While there has been much media attention for this treatment, there are two substantial hurdles to obtaining this benefit: first, insurance companies tend to not pay for stem cell injections of this kind, nor will they reimburse the patient, and second,

---

[220] Yu et al. (2023), p. 87.

[221] Powers Maurice (2023); Lanum (2023).

[222] *Id.*

[223] *See generally* Salgado et al. (2013); Vergara and Tsonis (2013).

[224] Talavera-Adame et al. (2017), pp. 11–19.

[225] Sipherd (2017).

[226] *See* Gentile et al. (2015), pp. 1317–1323.

[227] *See generally* Peters (2017).

there have still been mixed results.[228] As of now, only adult stem cell treatments of this kind have been published and practiced due to the substantial hurdles placed on embryonic stem cell research.[229] It is also important to note, that in speaking with various doctors who offer this procedure, there are tremendous issues with people with significant hair loss, or that have lost their hair more than a few years prior to seeking treatment. I specifically asked if using embryonic stem cells would yield better results, and unanimously, all stated that they believed that there would be a strong possibility, but that they could not be sure until conducting their own research.[230] My doctor also joked with me about my own surgical site. He stated that because scars cannot allow for hair growth, that I should look to the stem cell injection sites to see if any new hair growth occurs on my scar. It appears that there are many potential breakthroughs on the horizon for such treatments, but the current treatments are mainly aimed at increasing hair density rather than complete hair regrowth.[231]

2.3 Methods of Extraction

With all of this knowledge or even potential for knowledge at issue, there are still many who are against the study of embryonic stem cells, or anything that directly impacts the embryo. But with so much emphasis placed on the potential of utilizing embryonic stem cells, there have been drastic advancements in technology which must also be taken into account because some may directly impact the arguments of those in opposition in relation to the destruction of embryos, or at least help make finding a solution a little easier. Scientists have devised ways to obtain stem cells from embryos without harming them.[232] There are also methods of changing other cells such as adult stem cells into stem cells, albeit with their inherent limitations.[233] This fact may bring about the first true change in the anti-embryonic stem cell religious minds because there is no longer the risk of damaging or even worse, destroying the embryo.[234] The method of extraction is actually the most interesting concept because scientists, in theory would utilize a safe method which is already used on embryos in utero when determining if there are any damaged genes, or if any health conditions exist, and once a single human embryonic stem cell was successfully removed, scientists have a method of cultivating the single cell by making them

---

[228] *See* Knoepfler (2018).

[229] *See generally* Dickerson (2014), pp. 48–49.

[230] *See generally* Gentile et al. (2017), pp. 1–10.

[231] *See id.* pp. 1, 10, (concluding: "[w]e have shown that the isolated cells are capable to improve the hair density in patients affected by androgenetic alopecia (AGA).").

[232] Kaufman (2004).

[233] Pittenge et al. (1999), pp. 143–147.

[234] *Id.*

self clone which could likely create as much as needed at the time.[235] This is a great revelation because prior to scientists finding methods of extracting embryonic stem cells without destroying an embryo, the common method employs a process known as somatic cell nuclear transfer, where a nucleus is removed from a somatic cell, or a cell that is neither a sperm nor an egg, and then implants the somatic cell into a donor egg which essentially tricks the donor egg into believing that the egg has been fertilized and the process of dividing, as if there was a fertilized egg begins until an embryo is created.[236] Once there is an embryo, the relevant cells are removed from the inner cell mass and treated in a way to provide embryonic stem cells, but this process requires the destruction of the donor embryo.[237] Once complete, the embryonic stem cell is placed into a solution where they are allowed to self-proliferate and create countless more embryonic stem cells and form entire new embryonic stem cell lines.[238]

Even though this breakthrough seems like a potential problem solver, Rev. Tadeusz Pacholczyk, who is an ethicist at the National Catholic Bioethics Center in Philadelphia, stated that it was his belief that harvesting embryonic stem cells in this manner could end up being "a starting source for harvestable raw materials, in a gesture that reduces young humans to commodities."[239] While this argument is not without merit, it does assume that this method is not the end point of the embryonic retrieval process, and treats it as though this method will be used to harvest other potential tissues from embryos. This method was discovered during President George W. Bush's presidency with its extra restrictive laws, stem cell expert, Russell Korobkin stated that it was his belief that this form of extraction "should be permissible…" and even states that the National Institute of Health would fund research involving embryonic stem cells, so long as they were obtained using this method.[240]

2.4 Potential Harmful Issues

While so much promise has been discussed in relation to what embryonic stem cells are capable of in the healing and curing process, the dangers should also be known. Collectively, we know little in regards to embryonic stem cells, and many of the risks have been shown in animal models, with the belief that they can occur in humans as well.[241] Scientists have seen what is known as *tumorigenesis*, which creates

[235] *Id.* pp. 145–146.
[236] Murnaghan (2020).
[237] *Id.*
[238] *See generally id.*
[239] Kaplan (2012).
[240] *Id.*
[241] *See generally* Draper and Andrews (2002), pp. 309–315.

teratomas, irregular growth leading to incorrect cellular changes, including changing to the wrong cell type, as well as the stem cell transplant just failing to function as it was intended to.[242]

2.4.1 *Teratomas*

A very big issue arises from the fact that if embryonic stem cells were placed into an area without first pre-differentiating, or coaxing them into the desired type of cell, the receiving patient will more than likely develop tumors known as teratomas, which would likely lead to the death of the patient.[243] However, it is also important to note that many who are against embryonic stem cell treatments due to this reason, neglect to recognize that induced pluripotent stem cells have a faster and more efficient teratoma response than what would occur had embryonic stem cells been used instead.[244] Another important factor that most studies involving embryonic stem cells and the creation of teratomas is that most researchers agree that further studies of embryonic stem cells are required to learn how to further prevent this issue, and many have found that pre-implantation coaxing of the cells into the desired cell type can prevent further teratoma formation.[245] To further illustrate this fact, scientists have found that keeping the embryonic cells *in vitro* for a prolonged time, while in their pre-differentiated phase leads to a healthier subject, free of teratomas.[246] For example, in a study on Parkinson's syndrome in rats, scientists had three sets of pre-differentiated cells which were kept *in vitro*, the first for a period of 16 days, and the second set for a period of 20 days and the final for 23 days.[247] The rats that had the 16 day pre-differentiated cells grafted into them died of horrible teratomas, while the rats from both the second and third group remained perfectly healthy until the end of the experiment.[248] So, while this fear is well grounded, and can be seen in many research journals, it is also crucial to learn how to prevent the formation of teratomas, so that embryonic stem cells could be used to benefit those in need, without risking their health, lives or safety.

---

[242] Bailey (2019).

[243] Freed (2002), pp. 1755–1757.

[244] Gutierrez-Aranda et al. (2010), pp. 1568–1570.

[245] Hentzea et al. (2009), pp. 198–208.

[246] Brederlau et al. (2006), pp. 1433–1440.

[247] *Id.* p. 1434.

[248] *Id.*

2.4.2 Cellular Issues/Off-Target Migration

Those working with embryonic stem cells also fear potential harms that can come with changes to the transplanted cell. Some of these come in the form of an inappropriate cellular change, and others come from the migration of the transplanted cells to parts of the body that were unintended to be impacted by such treatment.[249] While researchers have witnessed spontaneous changes in implanted embryonic stem cells, which have led the transplanted cells to seemingly change to a different cell type as opposed to the type intended, there is still very little known as to how or why this process happens.[250] Cellular migration occurs naturally in all living beings, but when a transplanted stem cell ends up in the wrong area of the body from what was actually intended it can lead to a huge host of issues depending on where the cells end up, and cell type they end up changing in to.[251]

One must also realize that using an alternative to embryonic stem cells, such as adult stem cells also pose their own dangers.[252] Many of the risks are similar to embryonic stem cells, but there is also a risk that the adult stem cells have been corrupted by the aging process which has allowed for "environmental hazards, such as toxins" to damage the existing cells [253] Adult stem cells are also just as much at risk of cellular migration and change.[254]

2.4.3 The Unknown

Due to the legal, moral and ethical constraints, there is much that is still unknown to scientists and researchers. It is believed that one potential issue should not occur, but it has been seen in certain research settings with non-human test subjects. Embryonic stem cells are by their definition essentially blank, yet there is still a chance that the recipient of the donor embryonic stem cells can have them rejected.[255] While much is still left in ambiguity as many of these studies that have shown the potential for rejection occurred in studies involving embryonic stem cells being transplanted into mice or other laboratory testing animals.[256]

While embryonic stem cells are essentially blank, there is still the potential for a chance that the recipient of the donor embryonic stem cells can have them

---

[249] Khan et al. (2018), pp. 1–8.

[250] Gould Halme and Kessler (2010), p. 1734.

[251] *Id.; see generally* de Lucas et al. (2017), pp. 746–754.

[252] Bailey (2019).

[253] *Id.*

[254] de Lucas et al. (2017), p. 748.

[255] Rong et al. (2014), pp. 121–130.

[256] *Id.* pp. 122–124.

rejected.[257] While much is still left in ambiguity as many of these studies that have shown the potential for rejection occurred in studies involving embryonic stem cells being transplanted into mice or other laboratory testing animals.[258]

2.5 Potential Alternatives

There are also arguments which base their beliefs on potential alternates that they believe are at the least extremely similar to embryonic stem cells in their potential for treatments, and cures. Due to the subject of embryonic stem cells being such a divisive topic, it is no wonder that scientists looked to other methods of obtaining stem cells without having to even look to an embryo. This led to Dr. Shinya Yamanaka of Kyoto University, as well as a number of his colleagues to create a procedure that eventually allowed them to turn an adult skin cell into a stem cell.[259] Since this time, there have been many advances in harvesting adult stem cells, such as taking them from blood, bone marrow, most organs, the brain and many other parts that make up the human body.[260]

It is extremely important to note however, that many leading scientists believe that adult stem cells should not be considered a replacement for embryonic stem cells, as they are two different studies all together.[261] Dr. Angelo Vescovi, who is a leading researcher and pioneer in stem cells in Italy explained that the study of embryonic stem cells looks to "how pluripotent cells gradually differentiate to become tissue-specific stem cells;" and that adult stem cells work somewhat in reverse, and look to how an already existing cell can revert back to an earlier developmental stage and further develop into another form of tissue.[262] Taking this belief further, Yale University's Dr. Diane Krause explained how in certain experiments, injecting adult bone marrow stem cells into rats that have had their bone marrow destroyed through radiation, the adult stem cells transformed into many other forms of tissue, and not just the expected bone marrow cells.[263] Both Dr. Vescovi and Dr. Krause are of the opinion that the different forms of stem cells need to both be studied as complementary to each other instead of treating one as a substitute for the other.[264]

---

[257] National Research Council (US) and Institute of Medicine (US) Committee on the Biological and Biomedical Applications of Stem Cell Research (2002), pp. 4, 10–12, 24.

[258] *Id.*

[259] Tucker (2007).

[260] *See* Passier and Mummery (2003), pp. 329–332.

[261] Gavaghan (2001), pp. 800–801.

[262] *Id.* p. 801.

[263] *Id.*

[264] *Id.*

Of course the above stated information is not going to end this argument, and scientists have been hard at work to discover other forms of adult stem cells that can more closely resemble the power embryonic stem cells.[265] A study in 2006 found that the bulge of hair follicles contain adult stem cells that have been shown to closer resemble embryonic stem cells when cultured in a similar fashion to how embryonic stem cells are cultured.[266] Scientists were able to coax the adult stem cells taken into one of three more primitive cells, yet only include neurons, smooth muscle cells, and melanocytes (specialized skin cells that create protective melanin).[267] Even with a promising study such as the 2006 hair follicle study, the limits of adult stem cells are clearly noticeable, and lack the virtually unending abilities of embryonic stem cells to differentiate into almost any form of specialized cell.[268] Thusly, it would seem that studies have shown that embryonic stem cells differ from adult stem cells in that they are less susceptible to an autoimmune rejecting them from the new host.[269]

There are also those who have a similar belief in another form of stem cell which is said to better mimic embryonic stem cells known as induced pluripotent stem cells.[270] These cells are most commonly created when a somatic cell from an adult is introduced to embryonic genes to essentially re-write the somatic cell and bring it back to an embryonic like state of blankness, so that it could be altered to become another type of cell.[271] A great aspect of this form of cell is that there are fewer ethical issues implicated, in that the cells do not have to be taken from an embryo.[272] Induced pluripotent stem cells shares one of the most important abilities with embryonic stem cells which is that of virtually unlimited self-renewal.[273] The similarities between the two types of cells also involves some of the worst side effects and objectionable risks seen in using embryonic stem cells, prior to the 2006 discovery of induced pluripotent stem cells, such as the creation of teratomas, and as far as the risks, using these cells could lead to the creation of embryos, meaning cloning human beings, and informed consent will still be required.[274] These issues alone show that induced pluripotent stem cells are not a perfect answer or substitute for actual embryonic stem cells in regards to the ethical issues.[275] Some studies have also found that "reprogramming" these somatic cells do not result in a stem cell that is identical to an embryonic stem cell, and that a lot more research must be conducted

---

[265] *See generally* Yu et al. (2006), pp. 1879–1888.

[266] *Id.* p. 1879.

[267] *Id.* pp. 1885–1887.

[268] Drukker et al. (2006), pp. 221–229.

[269] *Id.* pp. 227–228.

[270] *See* Cyranoski (2008), pp. 406–408.

[271] Goldthwaite (2018).

[272] *Id.*

[273] Medvedev et al. (2010), pp. 18–19.

[274] *See generally* Zheng (2016), pp. 1277–1284.

[275] *Id.*

to truly learn the differences.[276] Due to this area of exploration being so new, most trials using induced pluripotent stem cells have been conducted using mice.[277] Other issues arose during these preclinical studies, such as the finding that these types of cell are more susceptible to genomic damage.[278] These issues were very hard to notice at first because of the need for extremely sensitive instruments to make certain findings, such as discovering the accidental creation of transgenes when using induced pluripotent stem cells.[279]

Scientists have actively been looking for something similar to embryonic stem cells that lack the moral and ethical issues that are inherent in embryonic stem cells, as well as due to the volatility of the subject and the complete fluidity of the applicable laws.[280] It stands to reason that pharmaceutical companies and researchers would choose to invest in options that have a high potential of being approved by the FDA or not prevented by our government to remain profitable.

Due to the fact that scientists are actively researching adult stem cells, and converting them into embryonic-like stem cells at a pace that far exceeds that of embryonic stem cell research, which is mainly due to the punishing laws and rules pertaining to embryonic stem cell research, it is highly likely that massive advancements will occur.[281] With the large amount of focus given to adult stem cells, there have already been tremendous strides in utilizing them.[282] This has been happening for a long time and the trend seems to continue on, in 2009 research involving stem cells derived from sources other than embryos received $58 million dollars in funding whereas embryonic stem cell research only received $23 million dollars.[283]

Scientists have known for many years that there is a potential for essentially curing virtually all moral and ethical considerations facing those who support embryonic stem cell research, and yet are still just as powerful as embryonic stem cells.[284] Known as "parthenote" stem cells, these cells are created during what is called "parthenogenesis" which is a common form of reproduction among lower forms of life, and is a process where live birth can occur even in the absence of sperm.[285] This means that no viable embryo is ever created to obtain embryonic level stem cells.[286] Even though this sounds like the perfect answer to the overall question, as it stands, medical science is not yet advanced enough to take advantage of these

---

[276] Narsinh et al. (2011), pp. 635–638.

[277] Okano et al. (2013), pp. 523–525.

[278] *See* Zhang et al. (2017), pp. 320–326.

[279] *See generally Id.*

[280] *See generally* Mertes et al. (2006), pp. 2749–2755.

[281] *See generally* PBS News Hour (2008).

[282] Tarne (2018).

[283] *Id.*

[284] Brevini and Gandolfi (2008), pp. 23–37.

[285] *See generally* Cibelli et al. (2006), pp. 117–135.

[286] *Id.*

powerful cells.[287] Parthenogenesis occurs in humans in a very different way than the other life forms where live birth occurs in a two very important ways; first, in humans it is a spontaneous event which is not able to be controlled, and second when it occurs in humans tumors are the result and come in the form of ovarian teratomas and hydatidiform moles.[288] Some controversial scientists currently believe there are humans who successfully undergone parthenogenesis naturally and have "...gone unnoticed to science."[289] This is not to say that it is not possible in a laboratory setting, as scientists have successfully brought about parthenogenesis on an unfertilized egg, causing cellular division by using electrical stimulation or a chemical to bring about the desired reaction.[290]

However, there are also others who believe that the near future will bring about a sort of medical revolution that will make the need for using embryonic stem cells unnecessary and site various examples for their beliefs. Some of the most relevant arguments look to:

2.5.1 Genome Editing

Many people have heard of various forms of Genome, or gene editing, which literally works to alter the DNA of a live organism, and comes about through a scientific process[291] In order to edit a gene, scientists must first create a "template" that will include the lab created, correct DNA sequence, which must match the DNA of the target site, and must then be inserted into cells to reconstruct the DNA without the mutated DNA.[292] To better understand the use and functions of real gene editing, one only needs to look to a study that seems more at home in a science fiction movie, as opposed to a real research or clinical setting.[293] In this real life example, parents were planning on having children, but the would-be father suffered from a dangerous heart disease, that was due to the coding of his DNA which caused a mutation that caused the disease, and his disease would have a 50% chance of being inherited by his would-be biological children, so in an attempt to prevent the children from suffering from the same disease, scientists used the man's sperm, to be used to fertilize oocytes *in-vitro*, and while fertilizing the oocytes, a new technique was used to correct the man's DNA that would be used in the fertilization.[294] This new technology is known as CRISPR-Cas9 ("CRISPR"), which was used to work in

---

[287] de Carli and Campos Pereira (2017), pp. 58–59.

[288] *Id.*

[289] de Carli and Campos Pereira (2017), pp. 58–59; Daughtry and Mitalipov (2014), pp. 290–298.

[290] *See generally* Revazova et al. (2007), pp. 432–449.

[291] *See generally* Musunuru (2018), pp. 2808–2821.

[292] *Id.*

[293] *See* Ma et al. (2017), pp. 413–419.

[294] *Id.* at 414.

tandem with a corrected, synthetic version of the problematic DNA, and then injected into the oocytes.[295] The "Cas9" in CRISPR-Cas9 is a special protein which is used to unwind DNA, so that fixed DNA strand can be inserted.[296] Once completed *in-vitro*, numerous embryos are created, without any change to the already good DNA, but also with the now "fixed" DNA, the embryos can be transferred to the mother's womb, and can be carried to term until natural pregnancy.[297] While this is just one example of genome editing, it is one of the most important, and talked about situations which requires further discussion, owing in part to the future promise of CRISPR.

An interesting example was published recently where mice with oral herpes (herpes simplex 1), a traditionally incurable disease that is reported to effect two out of three people under the age of 50, or 3.7 billion people globally, was treated using gene editing techniques.[298] The researchers in the above cited case were successfully able to destroy the herpes virus DNA using not one, but two different gene editing techniques, which worked by essentially cutting away the virus DNA, and the results were so positive, that scientists are going to now attempt to recreate their success on those suffering from herpes simplex 2.[299]

2.5.2 *CRISPR-Cas9*

As discussed above, CRISPR is a gene editing tool, that has been developed very recently, and shows great promise. In fact,

> [t]he introduction of CRISPR-Cas9 systems as genome-editing tools in early 2013 represented a watershed in the field of genome editing—and, in retrospect, in biomedical research—for three reasons: (1) their ease of use, (2) their efficacy compared to other tools, and (3) their adaptability to applications beyond genome editing.[300]

There is so much promise inherently tied to CRISPR, that research has been booming on a global scale.[301] One of the most promising areas that CRISPR is being researched in relation to is cancer.[302] China has been a major leader in this field, and has reported that 86 patients with various forms of cancer have been treated thanks to CRISPR technology, by removing a gene known as "PD-1" which enables tumors to bind to proteins and instructs them not to attack the mutated cancer cells, and in removing this gene, the body is able to recognize and attack the problematic

---

[295] *Id.*

[296] *Id.*

[297] *Id.* at 415.

[298] *See* Feltman (2015); Aubert et al. (2020), pp. 1–15.

[299] *Id.* pp. 12–13.

[300] *See* World Health Organization (2023); Musunuru (2018), pp. 2808–2821.

[301] Rodriguez Fernandez (2017).

[302] Stein (2018).

cells.[303] These studies have proven so successful, that similar studies are now being conducted in the United States with hopes for similar results.[304]

CRISPR has also shown promise for those suffering from sickle cell anemia, as well as another blood disorder, both of which an issue where oxygen cannot be successfully transported through one's blood.[305] In order for CRISPR to be used in these cases, a patient's stem cells are first harvested from their bone marrow and synthetically stimulated to produce a form of red blood cell that have extremely high levels of fetal hemoglobin.[306] Fetal hemoglobin is naturally produced in newborns, and is much more effective at binding oxygen to the red blood cells than adult hemoglobin.[307] Once this newly coded cells are ready, they are reinserted into the patient.[308]

Similar to embryonic stem cells repairing or curing blindness, CRISPR technology has shown great advances in correcting inherited forms of blindness by altering the mutated, problematic DNA that caused the blindness in the first place.[309] CRISPR has also shown benefits for those suffering from AIDs which is a disease which there is no known cure for, and works by removing the AIDs DNA, and inserting a protein that certain people are born with that gives them a natural ability to be resistant to AIDs in the first place.[310] Scientists in China made headlines by announcing that they were able to successfully edit human embryos using CRISPR to make the embryos naturally resistant to AIDs.[311] Many similar issues that have been brought up involving embryonic stem cell research were complained about by many people because human embryos were used, but unlike embryonic stem cells, CRISPR's side effects or the side effect of removing or adding strands of DNA to a human embryo can have unforeseen consequences.[312] Due to all of the unknown possibilities that can occur with such a new form of treatment, the doctor who conducted this study has promised to monitor the children till they reach the age of 18, but China has prosecuted the doctor for illegal medical practice and has sentenced him to three years in jail.[313]

Another astonishing feat accomplished using CRISPR has been shown on patients that are suffering from cystic fibrosis, which is a disease, caused by a gene mutation, which can implicate a number of different genes, and causes

---

[303] *See generally* Su et al. (2016), pp. 1–13.

[304] Stein (2019).

[305] Stein (2018).

[306] *Id.*

[307] *Id.*

[308] *See generally* Zivot et al. (2018), pp. 1–15.

[309] Rodriguez Fernandez, Stein (2018).

[310] *See* Azvolinsky (2017).

[311] Cyranoski (2019).

[312] *See id.*

[313] Wen and Dockser Marcus (2019).

problems in breathing, which gets worse over time.[314] Typically, patients suffering from this debilitating lung disease have an average lifespan of only 40 years, and is considered a fairly common disease compared to others.[315] The goal in using CRISPR in cases where patients have cystic fibrosis is to edit out the mutated genes that cause the disease in the first place.[316]

One of the most important and heavily impacting studies comes from scientists using CRISPR to correct mutations in people suffering from Duchenne muscular dystrophy, which is caused due to a mutated form of the protein known as DMD which normally would aid in muscle contraction, which causes progressive muscle deterioration and to which there is no known cure.[317] The end goal is similar to the above studies which seek to replace the mutated, problematic DNA with corrected versions, but the DMD gene seems to have multiple mutations which would all have to be corrected in order for CRISPR to be completely successful.[318] In this study, scientists were able to correct a majority of the 3000 mutations caused by the mutated DMD gene by using CRISPR at 12 "mutation hotspots."[319]

Yet, with all of its promise, CRISPR still has some major issues, one of which is known as "off-target frequency" which causes mutations in areas that scientists were not aiming for, and can cause major problems for a patient.[320] Scientists are fearful that these mutations could have a horrible effect for those suffering from "neurodegenerative disorders" such as Huntington's disease.[321] In response to these potential problems, scientists have had to further develop CRISPR and have made a version known as "KamiCas9" which can "self-inactivate", or stop itself once the process is complete, so as not to alter un-mutated, healthy genes.[322]

Genomic editing is absolutely here, and amazing, life changing discoveries are virtually sure to follow.[323] Yet, while one can see the enormous potential that CRISPR has, there are still severe limitations, and a lot more research must be conducted. The chance of wrongfully mutating healthy genes can cause various forms of cancer to develop, or that the newly implanted healthy cells become cancerous.[324] Other scientists have noted that in correcting one problematic gene, other healthy genes may be impacted and cause unforeseen problems.[325] However, many leading scientists and researchers think that with more study and time, these

---

[314] *Id.*

[315] *Id.*

[316] *See generally* Roper and Yilmaz (2019), pp. 841–842.

[317] Rodriguez Fernandez (2019).

[318] *See generally* Long et al. (2018), pp. 822–826.

[319] *Id.* pp. 823–825.

[320] *See generally* Fu et al. (2013), pp. 822–826.

[321] Merienne (2017), pp. 2983–2986.

[322] *Id.* pp. 2986–2987.

[323] *See generally* Rouanet et al. (2017), p. 1237.

[324] Begley (2018).

[325] Haapaniemi et al. (2018), pp. 927–930.

problems can be eradicated or at the least minimized.[326] It is also important to note that many health issues will still require a form of repair that CRISPR cannot grant, such as repairing tissue damage, or diseases not caused by a mutated gene.[327] In the above cited study, the authors make sure to state that: "[S]tem cells are indispensable for repair and maintenance of homeostasis. In terms of tissue repair and regeneration capacity, stem cell-mediated cell therapy and gene therapy are regarded as core components of human medicine."[328]

There have already been great successes when combining these two unique therapeutic approaches.[329] This does not mean that they should both be required to stand on their own, as scientists have actually found that using CRISPR on human embryos can bring about positive mutations that will help all forms of embryonic research.[330] For instance, using CRISPR, scientists believe they can successfully alter an embryos genetic code to either protect against some familial, inheritable disease, or prevent future bouts of sickness due to a particular virus, like H.I.V.[331] Researchers have also been able to create models of specific diseases using embryonic stem cells, and working on curing them using CRISPR in various animals, and without having the ability to use stem cells of this kind, scientists would have a much harder time testing their gene editing techniques, further slowing down progress.[332]

CRISPR is also not without its own ethical issues.[333] In 2017, the United States National Academies of Sciences, Engineering, and Medicine (the "NASEM") Committee on Human Gene Editing published their report which sought to review the "scientific, ethical (moral), and legal concerns" that are inherently brought about through gene editing experiments.[334] The NASEM found that the current technology should ban the altering the heritable germline.[335] The word "heritable germline" actually refers to altering a genetic sequence in a reproductive cell that will eventually become a part of the DNA in every cell in the offspring's body.[336] In the end, the NASEM's report stated that in order for there to be any exceptions to this rule, there must be a true "regulatory framework" in place that would include the following:

> (1) No reasonable alternatives to the trial exist; (2) The goal of the research is to prevent a serious disease or condition; (3) The research focuses only on editing genes that seem to predispose or cause the disease or condition; (4) Gene conversion is limited to only those versions associated with ordinary health and that are unlikely to cause adverse effects;

---

[326] *Id.*

[327] *See generally* Zhang et al. (2017), pp. 230–241.

[328] *Id.* p. 230.

[329] *See generally* Ovando-Roche et al. (2017), pp. 112–123.

[330] *See* Tangermann (2018).

[331] *See* Ben Jehuda et al. (2018), pp. 329–331.

[332] Geerts (2023).

[333] *See generally* Baumann (2016), pp. 139–159; Greene and Master (2018), pp. 327–335.

[334] Brokowski (2018), pp. 115–116.

[335] *Id.* pp. 122–123.

[336] *See* National Cancer Institute (2011).

> (5) Credible preclinical and clinical health risk data are available; (6) Institutions establish both ongoing monitoring of the health and safety of clinical research participants; (7) Long-term follow-up plans are defined and implemented; (8) Transparency and privacy protections are in place; (9) Societal risks are controlled; and (10) Mechanisms inhibit extension to uses other than preventing a serious disease or condition.[337]

One major issue that comes up when discussing CRISPR is its potential to be unsafe.[338] Off-target effects tend to occur when using any form of genomic editor, but essentially causes mutations in areas that were not meant to be effected by the use of CRISPR.[339] However, scientists are already hard at work to reduce the occurrence of this issue and have found multiple means to reducing the changes, yet currently, there is no guarantee as to the various efficacies of these means.[340] Second, an issue that arises in embryonic stem cells, is similarly effected here, which is that of informed consent.[341] This argument is actually quite problematic as it is prefaced on a situation which has yet to occur.[342] Essentially, the argument states that it would be impossible for a person who has yet to be created to give informed consent because they are not yet born at the time that CRISPR was used to alter their genes.[343] This is by no means a perfect argument though and should not be given such weight as parents make decisions, especially medical ones, for their unborn, and young children all the time without being required to obtain informed consent from them.[344]

As one can clearly see, claiming that embryonic stem cell-based research and treatments can be replaced by gene editing tools like CRISPR is not only incorrect, but fails to realize that the two independent areas of study share many ethical issues. The fact that the two separate areas can be joined together to further medical science should also be understood so that people do not make the mistake of thinking that one is more beneficial then the other.

CRISPR recently gained FDA approval in December of 2023 for the first time for people suffering from Sickle Cell disease.[345] In the press release announcing their approval, the FDA explained that this disease is a blood disorder, which sees a mutation in hemoglobin which is responsible for delivering oxygen throughout the body and cause the cells to change into a crescent shape leading to restrictions of blood flow causing oxygen not to get to where it needs to go, and excruciating pain for those that suffer.[346] Approximately 100,000 Americans, usually African

[337] Brokowski (2018), p. 121.

[338] National Human Genome Research Institute (2017).

[339] *See generally* Hajiahmadi et al. (2019), pp. 1–19.

[340] *See generally* National Human Genome Research Institute (2017).

[341] Smolenski (2015), pp. 35–37.

[342] Fu et al. (2013), pp. 822–826.

[343] *Id.*

[344] *See generally* Harrison (2004), pp. 99–102.

[345] *See* U.S. Food & Drug Administration (2023).

[346] *Id.*

Americans and Hispanic Americans, suffer from this potentially life-threatening disease.[347] This particular form of CRISPR has been titled "Casgevy" and works by editing a person's DNA in their stem cells, effectively knocking out a gene that allows for the gene to cause the blood cells to revert back, causing the cells to produce a fetal form of hemoglobin as it is the adult form that is the problem.[348] The FDA's approval allows for patients who are aged 12 or older to receive this form of treatment, and the data that was used by the FDA leading to approval saw 44 patients treated, with 93.5% success rate in meeting the goal, and zero issues of graft rejection.[349]

2.5.3 *Cybernetics*

Cybernetics is an area of science that studies human control functions and of electronic and mechanical systems that could be developed to repair, improve, or replace said human control functions through the use of mechanical communication.[350] Just recently, scientists announced that within five years, a bionic eye would be created that would not only restore vision to those who are blind, but would actually be able to do far more than the human eye.[351]

Researchers have shown great successes in aiding those who have some form of persistent paralysis using advanced cybernetics.[352] To accomplish this task, scientists would eventually have to use a neural implant that allows messages to travel both to and from the brain and the effected nerve fibers.[353] In a 2002 study, scientists created an implant that was connected directly to a computer to act as the brain, and to a test subject's left arm.[354] The test subject was right hand dominant, and to test their theory, two incisions were made into the subject's arm, both near the wrist, where 100 "needle electrodes" were placed, and then an electrode passer was placed in between where the two incisions were made.[355] In this experiment, scientists created a neural interface, by allowing the computer to both send and receive neural signals which would eventually allow the brain to control neural devices so that the patient could move his once paralyzed limb.[356]

---

[347] *Id.*

[348] Fliesler (2023).

[349] U.S. Food & Drug Administration (2023).

[350] *See generally* The Editors of Encyclopaedia (2023).

[351] Wooller (2020).

[352] Warwick et al. (2003), pp. 1372–1373.

[353] *Id.* p. 1369.

[354] *Id.* p. 1372.

[355] *Id.* p. 1369.

[356] *Id.* pp. 1370–1371.

The same can also be said for those who have had to have a limb amputated though a similar process.[357] Scientists have already had some success restoring the sense of touch in a person who no longer could due to paralysis.[358] To accomplish such a difficult task, scientists created an electric implant that is connected to a prosthetic arm that has touch sensors, and is connected directly to a patient's brain.[359] Interestingly, this study was paid for by the Defense Advanced Research Projects Agency, or DARPA which is an agency that is part of the United States Department of Defense.[360] In combining these studies, a cybernetic implant could likely allow for movement plus sensation in a person who is paralyzed and who would otherwise no longer be able to manipulate his limb, or feel any form of sensation.[361]

Even though there are many applications that have not yet been realized, medical science has already been using cybernetics in a multitude of ways to help better the lives of those who might otherwise be suffering.[362] One of the best examples is the pacemaker which is a battery powered, implanted device used to create or stimulate electrical impulses to the heart in order for the heart to beat, and therefore pump blood.[363] Another very common form of cybernetics, which is often combined with pacemakers) is the Implantable Cardioverter Defibrillator ("ICD"). An ICD is implanted into a person and is able to detect potentially dangerous abnormalities in the heart's natural rhythm and can send an electrical shock that will often recalibrate the heart's rhythm to its normal safe rate.[364]

A novel groundbreaking procedure was recently created that helped those who are paralyzed move, and feel again, and has been getting some attention through various forms of media.[365] In a recent story, a terror victim that was paralyzed at the age of 19, had a successful crowdfunding campaign to raise nearly one million dollars to cover the costs of the operation and the post-surgical rehabilitation.[366] While there is great hope for this young man, and for many others who will undergo this type of surgery, it is important to note that the initial operation costs around "$276,000," and the rehabilitation can cost almost twice that number.[367] Other studies using similar technologies have successful, such as in one study where

---

[357] *See generally* Davis et al. (2016), p. 15.

[358] Geddes (2016).

[359] *Id.*

[360] *Id.*

[361] Warwick et al. (2003), pp. 1372–1373.

[362] *See generally* Mehrali et al. (2018), pp. 1–39.

[363] *See generally* American Heart Association (2019).

[364] National Health Service (2018).

[365] Ben-Kimon (2023).

[366] *Id.*

[367] *Id.*

three patients with severe spinal cord damage regained "trunk and leg motor function."[368]

The current thought is that cybernetics will be the next form of human evolution, only instead of natural selection controlling the outcome, humans will be able to theoretically, decide for themselves.[369] However, because cybernetics look to replace human tissue with that of something mechanical or robotic, the newly created cybernetics would be vulnerable to hacking and other forms of outside interference.[370] Potentially even more problematic, this form of "enhancement" is so cutting edge that there are no truly applicable laws.[371]

2.5.4 *Bioprinting*

Bioprinting has been gaining a lot more traction as a means to create different forms of tissue to be used to repair damaged tissue, and possibly even repair or replace damaged organs.[372] Bioprinting is a process which uses cells and various growth factors to create structures similar to those naturally proceed by the body.[373] To accomplish this task, a printer using "bioink," which is made up of human cells, and usually a binding agent of some sort, to print the desired artificial tissue in layers, in a manner similar to how a 3D printer typically works.[374] Bioprinting has so many possible uses, but looking at examples will explain their uses best. If a person is missing an ear for any reason, a lab can now successfully bioprint an artificial ear, that will be made of human tissue, and thanks to 3D modeling, look almost exactly like it would have naturally.[375] Humanity is still not close enough to perfecting bioprinting, and more investigations and research will be required to tackle some of the toughest issues.[376] Currently, the printing process is so painfully slow, that it would not be commercially feasible to have a widespread operation printing tissue.[377] Even worse, our technology has yet to develop to a point where we can actually fully produce any organs, and scientists are working to find ways to solve the problems that would eventually lead to being able to do so.[378]

---

[368] Abbasi (2022), p. 909.

[369] Chan (2008), pp. 70–74.

[370] Groeneveld et al. (2019), p. 163.

[371] Wittes and Chong (2014).

[372] Mashambanhaka (2018).

[373] *Id.*

[374] Pires (2018).

[375] *See generally* Kang et al. (2016), pp. 312–319.

[376] Malkoc (2018), pp. 64–65.

[377] *Id.* at 64.

[378] *Id.*

Bioprinting has also shown benefits in testing drugs and or chemicals by building artificial tissue and/or organ models, which can then be used to determine the possible efficacy and dangers of using specific drugs, products and treatments on the artificially created tissue.[379] It is also important to note that instead of working in a form of competition with embryonic stem cells, they can be complimentary to each other.[380] In the above sited study, scientists were able to for the first time, bioprint using embryonic stem cells, and not only keep them alive once in their 3D structure, but also continued to self-proliferate and maintain their special attributes that makes them a pluripotent cell.[381] In the near future, studies that combine these two areas will likely be able to fabricate tissue and be used in experiments designed to test effects of drugs and/or compounds.[382] Scientists have also found that using stem cells of any sort, would be very useful in building the components needed to assist in bioprinting as they can turn into multiple types of cells.[383] The researchers in the cited article also specifically stated that embryonic stem cells, and induced pluripotent stem cells would be more beneficial as they "can turn into any cell type."[384]

2.5.5 Nanobots and Nanotechnology

Another topic that seems to come from "science fiction," is the use of micro-sized robots known as "nanobots" that will be able to be injected into a patient's blood stream and perform specific tasks, as well as other forms of micro-sized technology known as "nanotechnology."[385] Currently, scientists and researchers believe the nanotechnology can be used through various mediums, including nanobots, to diagnose, prevent and treat many different diseases and issues.[386] Diagnosis will come from micro implants or nanobots that are placed into a human body which will theoretically be used to identify problems prior to their full expression.[387] Where nanotechnology overlaps with stem cells, is the area of regenerative medicine.[388] Currently, something known as "nanoparticles" is used which can target specific parts of a body.[389] Nanobots, will be in the near future, and will likely be able to approach specific cells and stimulate them to produce or stop producing various

379 *See generally* Pati et al. (2016), pp. 4650–4665.

380 *See* Ouyang et al. (2016); Ong et al. (2018), pp. 223–231.

381 *Id.* p. 225.

382 *Id.*

383 Malkoc (2018), p. 64.

384 *Id.*

385 Bhat (2014), pp. 44–49.

386 Patil et al. (2008), pp. 34–40.

387 *Id.* pp. 34–36.

388 *Id.*

389 *Id.* pp. 36–37.

responses, depending on what is called for.[390] The future for nanotechnology is practically endless, and there are already studies and hypotheses for nanodentistry; drug dispensing nanobots; surgical nanobots; nanogenerators; and even surgical robots.[391] The future definitely seems bright for nanotechnologies, and scientists are extremely hopeful that nanobots will eventually be able to target deep tissue tumors, which would otherwise be impossible, or impracticable for doctors to do.[392] It is also believed that because they are able to work on such a small scale, they will be able to aid in cellular repair for damaged tissues, speeding up recovery, or making recovery possible.[393]

2.5.6 *Synthetic Blood*

As discussed above, replacing blood cells would likely be achievable using embryonic stem cells, and possibly adult stem cells, but scientists have been working on a new form of blood, created in laboratories, that can be used to replace human blood.[394] There is such a need for a replacement to blood that it is estimated that once synthetic blood is available for human use, its sales are likely to top "$7.6 billion in the United States alone."[395] Currently, there are multiple different types of artificial blood, but so far, there is an extreme limitation in all forms of their use due to the fact that they only last around 20 to 30 hours once inside the human body, and all lack the ability to fight diseases that red blood cells are usually inherently able to so.[396] With these serious flaws in the current synthetic blood creations, some of attempted to differentiate red blood cells from various types of stem cells which so far, has had great successes, but with the current technology, the costs are too high for this to be truly viable, at least for the time being.[397]

2.5.7 *In Vitro Gametogenesis*

There is something that is being actively worked on, yet has not yet fully come to fruition is known as in vitro gametogenesis ("IVG") which allows scientists to essentially grow their own embryos by reprogramming almost any other type of

---

[390] *Id.* pp. 37–38.

[391] *Id.;* Emerich and Thanos (2005), pp. 655–663.

[392] Andhari et al. (2020), pp. 12–15.

[393] Mitra (2017), pp. 218–222.

[394] Sarkar (2008), pp. 140–144.

[395] *Id.* p. 140.

[396] *Id.* p. 142.

[397] Moradi et al. (2016), p. 39.

cell in a body into a sperm and egg cells.[398] For example, a person can have a skin cell or hair cell taken and then changed into a sperm or egg cell which can be combined in a lab to create an embryo.[399] While our collective scientific knowledge has not really fully allowed for this new area of science to be used on humans, and we are still some time away from it, it is absolutely being worked on, and comes with some major implications.[400] The reason for this area of science's development lies in the fact that it will be one of the ultimate answers for couples who are infertile.[401] It will also likely allow for same-sex couples and even individuals to "procreate without the genetic contribution of another individual."[402] Science has even found that such a technique could help prevent animal extinction or even bring some back.[403] Although, true perils come in when one realizes that IVG will also allow for a potentially endless supply of embryos being created for scientific research, without many of the previous constraints because there are currently very few regulations in place at this time.[404] This technique has already been used to create mice egg cells from what was once male mice skin cells, which then led to the creation of some healthy mice which were able to procreate and even have "grand-pups."[405] It is important to note that IVG is different from *IVF* in that *IVF* requires already existing sperm and egg cells where IVG can use any other, non-sperm or non-egg cell to create an embryo.[406]

2.5.8 Synthetic Human Embryos

In June of 2023 embryonic stem cells were back in the headlines as scientists had successfully reprogrammed them into cells that would eventually form the "placenta, yolk sac and the embryo itself."[407] While these cells are called "synthetic embryos," they actually are more reminiscent of an extremely early embryo, and currently their use is of interest for many reasons.[408] Researchers will theoretically be able to see further into the embryonic development process then what is currently possible.[409] With our current level of understanding, and scientific capabilities, much of our

[398] *See generally* Cohen et al. (2017), p. 1.

[399] *Id.*

[400] Weintraub (2023a).

[401] *Id.*

[402] Suter (2015), pp. 87–119, 88.

[403] *See generally* Bolton et al. (2022), pp. 121–146.

[404] Suter (2015), p. 116.

[405] Weintraub (2023a).

[406] Maune (2022).

[407] See Devlin (2023); Wilson (2023).

[408] *Id.*

[409] *Id.*

knowledge in embryonic development stops at around day 14 due to an early ethical rule that limited researchers abilities to keep an embryo "sustainable," or viable before it could ever reach any level of sentience, and because the synthetic embryo would never be considered "natural" the same limiting rules would not apply.[410] Currently, the period after day 14, and before day 28 in embryonic development is known as a "black box" as scientists had to stop at day 14 and day 28 is when tissue from aborted fetuses is allowed to be studied. Having information from the "black box" period could help explain problems in pregnancies, birth defects, and other developmental issues. There is also the potential for synthetic embryos to eventually be able to grow healthy cells which will be able to be removed and transplanted into a person who is in need of such help.[411]

2.5.9 *Xenotransplantation*

A rather unique focus of research known as "xenotransplantation" has been studied for quite some time and is which is defined as: "any procedure that involves the transplantation, implantation, or infusion into a human recipient of either (a) live cells, tissues, or organs from a nonhuman animal source, or (b) human body fluids, cells, tissues, or organs that have had ex vivo contact with live nonhuman animal cells, tissues, or organs."[412] How the above definition relates to stem cells is due to part 'b' of the above definition due to the manner in which the study is undergone.[413] The process for which xenotransplantation with stem cells it requires that the scientist co-culture human embryos with non-human animal cells outside of a human body which is known as "ex vivo" and once done successfully, the co-cultured cells are placed into a female recipient.[414] While sounding like the plot to a bad science fiction movie, the actual potential of xenotransplantation are indeed impressive.[415] Currently, the National Kidney Foundation states that once on the waiting list, there is a three to five year wait for most patients in need of a kidney transplant, yet this time could be reduced to barely any wait at all if doctors were able to source a workable kidney from an animal.[416] Even though this too sounds far-fetched, in February of 2023 scientists were able to transplant genetically

---

[410] *Id.*; Comerford (2020).

[411] Appleby and Bredenoord (2018), p. 9437; Connor (2016).

[412] U.S. Food & Drug Administration (2009).

[413] Cooper et al. (2015), pp. 205–210.

[414] U.S. Food & Drug Administration (2009).

[415] Cooper et al. (2015), pp. 205–210.

[416] National Kidney Foundation (2024); Locke et al. (2023), pp. 1106–1108 (explaining that as of August of 2023, approximately 40% of patients waiting for a kidney transplant will die within five years of first getting on the list, and that using modified pig kidneys can alleviate the wait and save many people).

modified pig kidneys into a man who was clinically brain dead, and within four minutes, the pig kidneys were functioning to the point of making urine.[417] The study was only set to last for a one week period to determine if this idea was indeed viable, and once the study was complete, the researchers determined that the kidneys operated perfectly, and were still functioning as they would had it been the person's own natural kidneys.[418]

There was also a report in 2022 of a successful heart transplant, where the donor heart was sourced from a pig.[419] The recipient was a man in his 50's who was at risk of dying due to having life threatening heart disease, and even with a full understanding of the risks, the patient decided to have the genetically modified heart transplanted into him.[420] He managed to live 61 days with his new heart, and ended up dying of heart failure as opposed to what everyone else suspected he would have which was organ rejection.[421]

A second patient, another man in his 50's as well received a genetically modified pigs' heart the following year. While the heart functioned successfully for six weeks without any rejection, yet the patient did end up dying from "a sudden onset of heart failure."[422] The doctors specifically found that his cause of death was likely due to his body's condition prior to having the donor heart transplanted into him, but also found that there was a pig virus detected in the heart, but scientists could not determine if the virus caused the death.[423]

While these two examples seem to point a negative future for xenotransplantation, doctors have been very excited about the potential for the future, as once more research is conducted, they believe that this will be a viable life-saving solution to so many people in dire need of an organ donation.[424] The current method of deciding whether a person can obtain a donor heart require a litany of determinations that must be made first, that look to the patient's current health status; levels of antibodies; blood type; body size; and once all of this is considered, the person's geographic area becomes another determinant, with xenotransplantation, there would be so much less needed, and there would be more than enough hearts to go around.[425]

It should also be noted that doctors and researchers are working with and have worked with xenotransplantations of virtually all other organs into people in need

[417] *Id.* p. 1108.

[418] Greer (2023).

[419] Rabin (2022).

[420] Madhusoodanan (2022), pp. 199–2001.

[421] Fieldstadt (2022).

[422] Weintraub (2023b).

[423] *Id.*

[424] Ekser et al. (2017), pp. 513–521.

[425] Penn Medicine (2024).

such as livers, pancreases, lungs and even eyes.[426] Even when patients end up passing after such transplantations those working in this area often find promise that the future of xenotransplantation is solid and should be further researched.[427] In a conclusion to a recent study the author summarized what the future of this area of science holds, stating:

> With limited human results and unlimited potential, xenotransplantation shines a beacon of hope for a brighter future. However, we must navigate through the complexities of balancing scientific progress and patient welfare, avoiding being blinded by xenotransplantation's unquestionable potential.[428]

However, one will quickly notice that most, if not all of these "alternatives" are not directly interchangeable with stem cells at all, especially not embryonic stem cells, and even further that many of these alternatives have their own ethical issues. It would seem more prudent to combine all of these technologies together with stem cell research, or even combine each "alternative" with stem cells separately to obtain the best, most efficient results. Even though these alternatives do not necessarily implicate the same ethical issues as embryonic stem cells, they all share the same concerns. All are emerging technologies, with very little data, so the main concerns come from safety issues, yet at the same time the issue of equality necessarily comes up because an unfortunate side-effect of new technologies is that they are very costly, and with such steep prices, could lead to a division in who can afford to be helped and who cannot.[429]

In seeking to find the crucial balance in crafting new stem cell laws, it is necessary to take into account the various beliefs of the claimants and how other laws have been crafted abroad. The legal systems of many countries have already addressed the concerns raised by the value systems of their religious, cultural and ethical traditions. These concerns focus particularly, though not exclusively, on embryonic stem cells. In categorizing the various laws of the international community, we can identify three main groups: those with less restrictive laws, those with more restrictive laws, and those with no laws whatsoever. It would seem easy to just read and attempt to understand the laws of other countries, but to do so in a beneficial manner, it is important that the cultures, beliefs, and ideals of the people and the government of each of the countries be understood. In other words, the conditioning factors—predispositional and environmental—of these laws, administrative decisions and court decisions will have to be identified as well. In the end, this book seeks to craft a new set of guidelines that will regulate stem cell laws. In its creation, all opinions and beliefs of the various groups who are both for and against stem cell law expansion will be respected, and utilized so that the end result will be a very well-balanced recommendation not infringing on anyone's beliefs. Due to the fact that

---

[426] *See generally:* Cross-Najafi et al. (2023), pp. 1–10; Weaver (2020), pp. 1–15; Sahara et al. (2019), pp. 541–548; Yoon et al. (2020).

[427] Nunes dos Santos (2023), pp. 287, 296–297.

[428] *Id.* at 297.

[429] *See generally* Malkoc (2018), pp. 597–579.

there are so many races, religions, and cultures represented in the United States and on both sides of the issue, these new guidelines can possibly lead to the formulation of a universal solution.

2.5.10 *Biobots/Anthrobots*

In a recent study published in November of 2023, scientists created tiny Anthrobots, which are also known as "biobots," that are described as "tiny, self-assembling robots made from human cells" and are made from "human tracheal cells without any genetic modification."[430] These cells are known as "progenitor seed cells" which are cells that are similar to stem cells in that they are able to differentiate into a specific cell type only as opposed to stem cells which, depending on the type can differentiate into any number of cells.[431] Once developed and implanted the cells are able to move on their own to the target site, and self-assemble into what the authors of the study call a "superbot."[432] Once these biobots form a superbot, they have the ability to promote nerve cell growth in areas that had been damaged.[433] There is still much to research in this area as the researchers who conducted the experiments have only done so in a petri dish, and the researchers cannot explain what causes these biobots to promote the type of healing explaining in their research that they are still unaware as to which of the many potential aspects of the Anthrobot are required.[434] The future seems hopeful for this particular alternative, as it is believed that other forms of cells can likely create these biobots, enabling for other forms of repair and that further research will determine what other healing capabilities they may have.[435]

References

Abbasi J (2022) Patients with complete paralysis walk with spinal cord implant. JAMA 327:909

Acosta ND, Golub SH (2016) The new federalism: state policies regarding embryonic stem cell research. J Law Med Ethics 44:419–436

Afifi-Sabet K (2023) These astonishing biobots can help neurons regrow — but researchers have no idea how. Available via Live Science https://www.livescience.com/technology/robotics/astonishing-biobots-called-anthrobots-can-help-neurons-regrow. Accessed 26 Mar 2024

American Diabetes Association (2023) Statistics about diabetes. Available via Diabetes.org. https://diabetes.org/about-diabetes/statistics/about-diabetes. Accessed 26 Mar 2024

---

[430] Afifi-Sabet (2023).

[431] *See generally* Microscope Master (2021).

[432] Gumuskaya et al. (2023), p. 12.

[433] *Id.*

[434] *Id.* at 14.

[435] *Id.* at 15.; Afifi-Sabet (2023).

American Heart Association (2019) Pacemaker. Available via the American Heart Association. https://www.heart.org/en/health-topics/arrhythmia/prevention%2D%2Dtreatment-of-arrhythmia/pacemaker. Accessed 26 Mar 2024

American Red Cross (2024) What happens to donated blood? Blood transfusions. Risks and complications. Available via the American Red Cross. https://www.redcrossblood.org/donate-blood/blood-donation-process/what-happens-to-donated-blood/blood-transfusions/risks-complications.html. Accessed 13 Feb 2024

Andhari SS, Wavhale RD, Dhobale KD, Tawade BV, Chate GP, Patil YN, Khandare JJ, Banerjee SS (2020) Self-propelling targeted magneto-nanobots for deep tumor penetration and pH-responsive intracellular drug delivery. Sci Rep 10:1–16. Available via NCBI https://pubmed.ncbi.nlm.nih.gov/32170128/

Appleby JB, Bredenoord AL (2018) Should the 14-day rule for embryo research become the 28-day rule? EMBO Mol Med 10:9437. Available via NCBI https://www.ncbi.nlm.nih.gov/pmc/articles/PMC6127884

Aubert M, Strongin DE, Roychoudhury P et al (2020) Gene editing and elimination of latent herpes simplex virus in vivo. Nat Commun 11:1–15. Available via Nature. https://doi.org/10.1038/s41467-020-17936-5. Accessed 26 Mar 2024

Azvolinsky A (2017) Resistance to HIV engineered via CRISPR. Available via the Scientist. https://www.the-scientist.com/daily-news/resistance-to-hiv-engineered-via-crispr-31137. Accessed 26 Mar 2024

Bailey R (2019) What are stem cells? Available via ThoughtCo. http://biology.about.com/od/biotechnologycloning/ss/stem-cells.htm. Accessed 26 Mar 2024

Baldwin K (2018) Freezing eggs: a straightforward, expert guide best results. Available via the Special Broadcasting Service. https://www.sbs.com.au/news/the-feed/article/freezing-eggs-a-straightforward-expert-guide-to-best-results/h0yhzflqr. Accessed 26 Mar 2024

Baumann M (2016) CRISPR/Cas9 genome editing – new and old ethical issues arising from a revolutionary technology. Nanoethics 10:139–159

Begley S (2018) CRISPR-edited cells linked to cancer risk in 2 studies. Available via Scientific American. https://www.scientificamerican.com/article/crispr-edited-cells-linked-to-cancer-risk-in-2-studies. Accessed 26 Mar 2024

Beil L (2009) What happens to extra embryos after IVF? Available via Cable News Network [hereinafter CNN]. https://www.cnn.com/2009/HEALTH/09/01/extra.ivf.embryos/index.html. Accessed 26 Mar 2024

Ben-Hur T, Idelson M, Khaner H, Pera M, Reinhartz E, Itzik A, Reubinoff BE (2004) Transplantation of human embryonic stem cell–derived neural progenitors improves behavioral deficit in Parkinsonian rats. Stem Cells 22:1246

Ben-Kimon E (2023) Israel terror victim may walk again after groundbreaking surgery. Available via Ynet News. https://www.ynetnews.com/magazine/article/b1fpkcnro. Accessed 26 Mar 2024

Bergiers I, Andrews T, Vargel Bölükbaşı O, Buness A, Janosz E, Lopez-Anguita N, Ganter K, Kosim K, Celen C, Itır Perçin G, Collier P, Baying B, Benes V, Hemberg M, Lancrin C (2018) Single-Cell transcriptomics reveals a new dynamical function of transcription factors during embryonic hematopoiesis. Elife 7:1. Available via NCBI https://www.ncbi.nlm.nih.gov/pmc/articles/PMC5860872/

Bhat AS (2014) Nanobots: the future of medicine. Int J Eng Manage Sci 5:44–49. Available via Research Gate https://www.researchgate.net/publication/277475210_Nanobots_The_future_of_medicine

Blakemore E (2017) Human-Pig hybrid created in the lab—here are the facts. Available via National Geographic. https://www.nationalgeographic.com/news/2017/01/human-pig-hybrid-embryo-chimera-organs-health-science. Accessed 26 Mar 2024

Bohr TE, Shiroor DA, Adler CE (2021) Planarian stem cells sense the identity of the missing pharynx to launch its targeted regeneration. eLife 10:e68830. Available via eLife Sciences. https://elifesciences.org/articles/68830

Bolton RL, Mooney A, Pettit MT, Bolton AE, Morgan L, Drake GJ, Appeltant R, Walker SL, Gillis JD, Hvilsom C (2022) Resurrecting biodiversity: advanced assisted reproductive technologies and biobanking. Reprod Fertil 3:121–146

Boston Children's Hospital (2009) Where do we get adult stem cells? Available via Boston Children's Hospital. http://stemcell.childrenshospital.org/about-stem-cells/adult-somatic-stem-cells-101/where-do-we-get-adult-stem-cells. Accessed 26 Mar 2024

Bowen A (2008) Extra sperm, eggs, embryos mean parents must make decisions. Available via Chicago Tribune. https://www.chicagotribune.com/2018/06/14/extra-sperm-eggs-embryos-mean-parents-must-make-decisions. Accessed 26 Mar 2024

Brazier Y (2018) What are stem cells, and what do they do? Available via Medical News Today. https://www.medicalnewstoday.com/articles/323343. Accessed 26 Mar 2024

Brederlau A, Correia AS, Anisimov SV, Elmi M, Paul G, Roybon L, Morizane A, BergquistIlse Riebe F, Nannmark U, Carta M, Hanse E, Takahashi J, Sasai Y, Funa K, Brundin P, Eriksson PS, Li J-Y (2006) Transplantation of human embryonic stem cell-derived cells to a rat model of Parkinson's disease: effect of in vitro differentiation on graft survival and teratoma formation. Stem Cells 24:1433–1440

Brevini TAL, Gandolfi F (2008) Parthenotes as a source of embryonic stem cells. Cell Prolif 41:20–30, 23–37

Brokowski C (2018) Do CRISPR germline ethics statements cut it? CRISPR J 1:115–125

Bryant SV, Endo T, Gardiner DM (2002) Vertebrate limb regeneration and the origin of limb stem cells. Int J Dev Biol 46:887

Buerger CS, Jain H (2023) Infectious complications of blood transfusion. In: StatPearls [Internet]. StatPearls Publishing, Treasure Island (FL); 2024 Jan-. Available via the National Library of Medicine https://www.ncbi.nlm.nih.gov/books/NBK585035/. Accessed 2 Jan 2024

Bukuras J (2022) What the U.S. bishops have said about Dobbs. Available via Catholic World Report. https://www.catholicworldreport.com/2022/06/27/what-the-u-s-bishops-have-said-about-dobbs/. Accessed 26 Mar 2024

Carpenter MK, Rosler ES, Fisk GJ, Bradenberger R, Ares X, Miura T, Lucero M, Rao MS (2004) Properties of four embryonic stem cell lines maintained in a feeder-free culture system. Dev Dyn 229:243, 253–254

Carr A-J, Vugler AA, Hikita ST, Lawrence JM, Gias C, Chen L, Coffey PJ et al (2009) Protective effects of human iPS-derived retinal pigment epithelium cell transplantation in the retinal dystrophic rat. PLoS One 4:1–12

Chagastelles PC, Nardi NB (2011) Biology of stem cells: an overview. Kidney Int Suppl 1:63–67

Chan S (2008) Humanity 2.0? EMBO Rep 9:70–74

Chang L, Fan W, Pan X, Zhu X (2022) Stem cells to reverse aging. Chin Med J 8:901–910

Chinta SJ, Andersen JK (2005) Dopaminergic neurons. Int J Biochem Cell Biol 37:942

Christopher Reeve (2004) Speech to the United Nations. Available via YouTube. https://www.youtube.com/watch?v=pVUKP2-T1rs. Accessed 26 Mar 2024

Chung Y, Klimanskaya I, Becker S, Li T, Maserati M, Lu S-J, Zdravkovic T, Ilic D, Genbacev O, Fisher S, Krtolica A, Lanza R (2008) Human embryonic stem cell lines generated without embryo destruction. Cell Stem Cell 2:113–117

Cibelli JB, Cunniff K, Vrana KE (2006) Embryonic stem cells from parthenotes. Methods Enzymol 418:117–135. Available via NCBI. https://pubmed.ncbi.nlm.nih.gov/17141033/

Clados MS (2012) Bioethics in international law: an analysis of the intertwining of bioethical and legal discourses. Dissertation, LMU München: Faculty of Philosophy, Philosophy of Science and the Study of Religion at 70. Available via Deutsche Nationalbibliothe. https://d-nb.info/1031380159/34. Accessed 26 Mar 2024

Cohen IG, Daley GQ, Adashi EY (2017) The promise and peril of emerging reproductive technologies. Sci Trans Med 9:1. Available via Harvard Law. https://petrieflom.law.harvard.edu/resources/article/the-promise-and-peril-of-emerging-reproductive-technologies. Accessed 26 Mar 2024

Comerford F (2020) Stem cells open developmental 'blackbox'. Available via Medium. https://fynncomerford.medium.com/stem-cells-open-developmental-blackbox-78503cda3a51. Accessed 26 Mar 2024

Community Blood Center (2007) Blood facts. Available via GivingBlood.org. https://www.givingblood.org/about-blood/blood-facts.aspx. Accessed 26 Mar 2024

Connor S (2010) Paralysed man is first to be treated with stem cells from embryo. Available via the Independent. https://www.independent.co.uk/news/science/paralysed-man-is-first-to-be-treated-with-stem-cells-from-embryo-2103986.html. Accessed 26 Mar 2024

Connor S (2016) Inside the 'black box' of human development. Available via The Guardian. https://www.theguardian.com/science/2016/jun/05/human-development-ivf-embryos-14-day-legal-limit-extend-inside-black-box. Accessed 26 Mar 2024

Cooper DKC, Ekser B, Tector AJ (2015) A brief history of clinical xenotransplantation. Int J Surg 23:205–210

Correia AS, Anisimov SV, Li J-Y, Brundin P (2005) Stem cell-based therapy for Parkinson's disease. Ann Med 37:487

Cross-Najafi AA, Lopez K, Isidan A, Park Y, Zhang W, Li P, Yilmaz S, Akbulut S, Ekser B (2023) Current barriers to clinical liver xenotransplantation. Front Immunol 13:1. Available via Frontiers https://www.frontiersin.org/articles/10.3389/fimmu.2022.827535/full

Cyranoski D (2008) 5 things to know before jumping on the iPS bandwagon: induced pluripotent stem cells look just like embryonic stem cells, but are easier to create and free of the heavy ethics baggage. David Cyranoski separates fact from fiction in a burgeoning field. Nature 452:406–408

Cyranoski D (2013) US scientists chafe at restrictions on new stem-cell lines. Available via Nature. https://www.nature.com/news/us-scientists-chafe-at-restrictions-on-new-stem-cell-lines-1.13114. Accessed 26 Mar 2024

Cyranoski D (2018) How human embryonic stem cells sparked a revolution. Nature 555:428–430

Cyranoski D (2019) The CRISPR-baby scandal: what's next for human gene-editing. Available via Nature. https://www.nature.com/articles/d41586-019-00673-1. Accessed 26 Mar 2024

da Cruz L, Fynes K, Georgiadis O, Kerby J, Luo YH, Ahmado A, Vernon A, Daniels JT, Nommiste B, Hasan SM, Gooljar SB, Carr A-JF, Vugler A, Ramsden CM, Bictash M, Fenster M, Steer J, Harbinson T, Wilbrey A, Tufail A, Feng G, Whitlock M, Robson AG, Holder GE, Sagoo MS, Loudon PT, Whiting P, Coffey PJ (2018) Phase 1 clinical study of an embryonic stem cell–derived retinal pigment epithelium patch in age-related macular degeneration. Nat Biotechnol 36:328

Daughtry B, Mitalipov S (2014) Concise review: Parthenote stem cells for regenerative medicine: genetic, epigenetic, and developmental features. Stem Cells Transl Med 3:290–298

Davis TS, Wark HAC, Hutchinson DT, Warren DJ, O'Neill K, Scheinblum T, Clark GA, Normann RA, Greger B (2016) Restoring motor control and sensory feedback in people with upper extremity amputations using arrays of 96 microelectrodes implanted in the median and ulnar nerves. J Neural Eng 13:15. Available via NCBI. https://pubmed.ncbi.nlm.nih.gov/27001946/

de Carli GJ, Campos Pereira T (2017) On human parthenogenesis. Med Hypo 106:57–60

de Jesus Medina Arellano M (2012) The rise of stem cell therapies in Mexico: inadequate regulation or unsuccessful oversight? Redbiotica Mag 3:63–78

de Lucas B, Pérez LM, Gálvez BG (2017) Importance and regulation of adult stem cell migration. J Cell Mol Med 22:746–754

de Wert G, Mummery C (2003) Human embryonic stem cells: research, ethics, and policy. Hum Reprod 18(672):680

Dehkordi AN, Babaheydari FM, Chehelgerdi M, Dehkordi SR (2019) Skin tissue engineering: wound healing based on stem-cell-based therapeutic strategies. Stem Cell Res Ther 10:1–20, 7. Available via BMC Springer Nature. https://stemcellres.biomedcentral.com/counter/pdf/10.1186/s13287-019-1212-2.pdf. Accessed 26 Mar 2024

Deshpande DM, Kim Y-S, Martinez T, Carmen J, Dike S, Shats I, Rubin LL, Drummond J, Krishnan C, Hoke A, Maragakis N, Shefner J, Rothstein JD, Kerr DA (2006) Recovery from paralysis in adult rats using embryonic stem cells. Ann Neurol 60:32

Devlin H (2023) Synthetic human embryos created in groundbreaking advance. Available via The Guardian. https://www.theguardian.com/science/2023/jun/14/synthetic-human-embryos-created-in-groundbreaking-advance. Accessed 26 Mar 2024

DeWitt N, Knight J (2002) Biologists question adult stem-cell versatility. Nature 416:354

Dickerson E (2014) Stem cells for hair growth and hair transplant. Mod Aesthetics 2014:48–49

Dittrich R, Beckmann MW, Würfel W (2015) Non-embryo-destructive extraction of pluripotent embryonic stem cells: implications for regenerative medicine and reproductive medicine. Geburtshilfe Frauenheilkd 75:1239–1242

Donne M (2017) Let's be clear: stem cells and popular culture. Available via the Stem Cellar. https://blog.cirm.ca.gov/2017/02/01/lets-be-clear-stem-cells-and-popular-culture. Accessed 26 Mar 2024

Draper JS, Andrews PW (2002) Embryonic stem cells: advances toward potential therapeutic use. Obstet Gynecol 14:309–315

Drukker M, Katchman H, Katz G, Even-Tov Friedman S, Shezen E, Hornstein E, Mandelboim O, Reisner Y, Benvenisty N (2006) Human embryonic stem cells and their differentiated derivatives are less susceptible to immune rejection than adult cells. Stem Cells 24:221–229

Ekser B, Li P, Cooper DKC (2017) Xenotransplantation: past, present, and future. Curr Opin Organ Transplant 22:513–521

Emerich DF, Thanos CG (2005) Nanotechnology and medicine. Expert Opin Biol Ther 3:655–663

Family Guy Fandom (2019) McStroke. Available via Family Guy Fandom. https://familyguy.fandom.com/wiki/McStroke. Accessed 26 Mar 2024

Feltman R (2015) You probably have herpes, but that's really okay. Available via the Washington Post. https://www.washingtonpost.com/news/speaking-of-science/wp/2015/11/02/you-probably-have-herpes-but-thats-really-okay. Accessed 26 Mar 2024

Fieldstadt E (2022) First man to receive a transplanted pig heart died of heart failure, not rejection, encouraging doctors. Available via NBC Universal Media. https://www.nbcnews.com/health/heart-health/first-man-receive-transplanted-pig-heart-died-heart-failure-not-reject-rcna37078. Accessed 26 Mar 2024

Fliesler N (2023) Sickle cell gene therapy and boosting fetal hemoglobin: a 75-year history. Available via Boston Children's Hospital. https://answers.childrenshospital.org/sickle-cell-fetal-hemoglobin-timeline. Accessed 26 Mar 2024

Freeborn D, Dyches T, Olsen Roper S (2017) Lessons learned from a life with type 1 diabetes: adult perspectives. Diabetes Spectr 30:188–194

Freed CR (2002) Will embryonic stem cells be a useful source of dopamine neurons for transplant into patients with Parkinson's disease? PNAS 99:1755–1757

Fu Y, Foden JA, Khayter C, Maeder ML, Reyon D, Joung JK, Sander JD (2013) High frequency off-target mutagenesis induced by CRISPR-Cas nucleases in human cells. Nat Biotechnol 31 (9):822–826. https://doi.org/10.1038/nbt.2623

Fuchs E, Segre JA (2012) Stem cells: a new lease on life. Cell 100:143–155

Gallagher J (2018) Macular degeneration: 'I've been given my sight back'. Available via the British Broadcasting Corporation. https://www.bbc.com/news/health-43458365. Accessed 26 Mar 2024

Gavaghan H (2001) The promise of stem cells. Bull World Health Organ 79:800–801

Geddes L (2016) Pioneering brain implant restores paralysed man's sense of touch. Available via Nature. https://www.nature.com/news/pioneering-brain-implant-restores-paralysed-man-s-sense-of-touch-1.20804. Accessed 26 Mar 2024

Geerts L (2023) Three ways CRISPR is making animal research models more predictive. Available via Genetic Engineering and Biotechnology News. https://www.genengnews.com/insights/three-ways-crispr-is-making-animal-research-models-more-predictive. Accessed 26 Mar 2024

Gentile P, Garcovich S, Bielli A, Scioli MG, Orlandi A, Cervellia V (2015) The effect of platelet-rich plasma in hair regrowth: a randomized placebo-controlled trial. Stem Cells Transl Med 4:1317

Gentile P, Scioli MG, Bielli A, Orlandi A, Cervelli V (2017) Stem cells from human hair follicles: first mechanical isolation for immediate autologous clinical use in androgenetic alopecia and hair loss. Stem Cell Investig 4:1–10

Gladstone Institutes (2018) Researchers create first stem cells using CRISPR genome activation: activating a single gene is sufficient to change skin cells into stem cells. Available via ScienceDaily. https://www.sciencedaily.com/releases/2018/01/180118162449.htm. Accessed 26 Mar 2024

Goldthwaite CA Jr (2018) The promise of induced pluripotent stem cells (iPSCs). Available via the National Institute of Health. https://www.ncbi.nlm.nih.gov/pmc/articles/PMC3652331/. Accessed 26 Mar 2024

Gould Halme D, Kessler DA (2010) FDA regulation of stem-cell–based therapies. N Engl J Med 355:1730–1735

Greene M, Master Z (2018) Ethical issues of using CRISPR technologies for research on military enhancement. Bioeth Inq 15:327–335

Greer T (2023) New study: pig kidneys — for the first time — demonstrate "life-sustaining kidney function" in a human. Available via the University of Alabama at Birmingham. https://www.uab.edu/news/health/item/13712-new-study-pig-kidneys-for-the-first-time-demonstrate-life-sustaining-kidney-function-in-a-human. Accessed 26 Mar 2024

Groeneveld SA, Jongejan N, de Winter BJSAAF, Fiolet ATL (2019) Hacking into a pacemaker; risks of smart healthcare devices. Ned Tijdschr Geneeskd 8:163. Available via NCBI. https://pubmed.ncbi.nlm.nih.gov/30875159/

Gumuskaya G, Srivastava P, Cooper BG, Lesser H, Semegran B, Garnier S, Levin M (2023) Motile living biobots self-construct from adult human somatic progenitor seed cells. Adv Sci 2303575: 12. Available via Wiley Online Library. https://doi.org/10.1002/advs.202303575

Gurevich R (2020) Mock embryo transfer procedure. Available via Very Well Family. https://www.verywellfamily.com/mock-embryo-transfer-procedure-1960216. Accessed 26 Mar 2024

Gutierrez-Aranda I, Ramos-Mejia V, Bueno C et al (2010) Human induced pluripotent stem cells develop teratoma more efficiently and faster than human embryonic stem cells regardless the site of injection. Stem Cells 28:1568–1570

Haapaniemi E, Botla S, Persson J, Schmierer B, Taipale J (2018) CRISPR–Cas9 genome editing induces a P53-mediated DNA damage response. Nat Med 24:927–930

Hajiahmadi Z, Movahedi A, Wei H, Li D, Orooji Y, Ruan H, Zhuge Q (2019) Strategies to increase on-target and reduce off-target effects of the CRISPR/Cas9 system in plants. Int J Mol Sci 20:1–19. Available via Research Gate. https://www.researchgate.net/publication/334957206_Strategies_to_Increase_On-Target_and_Reduce_Off-Target_Effects_of_the_CRISPRCas9_System_in_Plants

Harrison C (2004) Treatment decisions regarding infants, children and adolescents. Paediatr Child Health 9:99–103

He Q, Li J, Bettiol E, Jaconi ME (2003) Embryonic stem cells: new possible therapy for degenerative diseases that affect elderly people. J Gerontol 58:279–287

Hentzea H, Soonga PL, Wanga ST, Phillipsa BW, Puttib TC, Dunna NR (2009) Teratoma formation by human embryonic stem cells evaluation of essential parameters for future safety studies. Stem Cell Res 2:198–210

Hooper J (2017) Does stem cell therapy actually work? Available via Men's Journal. https://www.mensjournal.com/health-fitness/does-stem-cell-therapy-actually-work-w437788/. Accessed 26 Mar 2024

ICCPR, General Comment 36, 124th Sess, adopted 3 Sept. 2019, GE.19-15012(E)

International Movie Database (2010) Futurama "Rebirth" quotes. Available via International Movie Database. https://www.imdb.com/title/tt1620650/characters/nm0921942. Accessed 26 Mar 2024

Izrael M, Slutsky SG, Admoni T, Revel M (2018) Safety and efficacy of human embryonic stem cell-derived astrocytes following intrathecal transplantation in SOD1G93A and NSG animal models. Stem Cell Res Ther 9:152–169

Jehuda RB, Shemer Y, Binah O (2018) Genome editing in induced pluripotent stem cells using CRISPR/Cas9. Stem Cell Rev Rep 14:323–336

Kalomiris M (2022) Using flatworm regeneration to study stem cells. Available via National Institute of Health. https://irp.nih.gov/catalyst/28/5/using-flatworm-regeneration-to-study-stem-cells. Accessed 26 Mar 2024

Kalra K, Tomar PC (2014) Stem cell: basics, classification and applications. Am J Phytomed Clin Ther 2:919–930

Kang H-W, Lee SJ, Ko IK, Kengla C, Yoo JJ, Atala A (2016) A 3D bioprinting system to produce human-scale tissue constructs with structural integrity. Nat Biotech 34:312–319

Kaplan K (2012) Stem cells created without destroying embryos. Available via Los Angeles Times. https://www.latimes.com/health/la-sci-stemcells11jan11-story.html. Accessed 26 Mar 2024

Kaufman D (2004) Embryonic stem cells promising for replacement of blood supply. Available via ScienceBlog.com https://scienceblog.com/4503/embryonic-stem-cells-promising-for-replacement-of-blood-supply. Accessed 26 Mar 2024

Khan FA, Almohazey D, Alomari M, Ameen Almofty S (2018) Isolation, culture, and functional characterization of human embryonic stem cells: current trends and challenges. Stem Cells Int 2018:1–8 (Article ID 1429351). Available via Hindawi. https://downloads.hindawi.com/journals/sci/2018/1429351.pdf?_gl=1*17rivcr*_ga*MTE2MTAwMTc2NC4xNjk3ODA5Mzc0*_ga_NF5QFMJT5V*MTY5NzgwOTM3My4xLjAuMTY5NzgwOTM3My42MC4wLjA.&_ga=2.222582750.477959897.1697809374-1161001764.1697809374

Kirkeby A, Grealish S, Wolf DA, Nelander J, Wood J, Lundblad M, Lindvall O, Parmar M (2012) Generation of regionally specified neural progenitors and functional neurons from human embryonic stem cells under defined conditions. Cell Rep 1:703

Kirkeby A, Nelander J, Hoban DB, Rogelius N, Bjartmarz H, Novo Nordisk Cell Therapy R&D, Storm P, Fiorenzano A, Adler AF, Vale S, Mudannayake J, Zhang Y, Cardoso T, Mattsson B, Landau AM, Glud AN, Sørensen JC, Lillethorup TP, Lowdell M, Carvalho C, Bain O, van Vliet T, Lindvall O, Björklund A, Harry B, Cutting E, Widner H, Paul G, Barker RA, Parmar M (2023) Preclinical quality, safety, and efficacy of a human embryonic stem cell-derived product for the treatment of Parkinson's disease, STEM-PD. Clin Trans Rep 30:1299–1314

Knoepfler P (2018) Does insurance cover stem cell injections at clinics? Available via the Niche. https://ipscell.com/2018/12/does-insurance-cover-stem-cell-injections-at-clinics/. Accessed 26 Mar 2024

Körbling M, Estrov Z (2003) Adult stem cells for tissue repair — a new therapeutic concept? N Engl J Med 349:570–582

Kuriakose D, Xiao Z (2020) Pathophysiology and treatment of stroke: present status and future perspectives. Int J Mol Sci 21:20. Available via NCBI. https://www.ncbi.nlm.nih.gov/pmc/articles/PMC7589849

Lanum N (2023) Hollywood doctor reveals the treatment celebrities are using to reverse aging, Look ten years younger. Available via Fox News. https://www.foxnews.com/media/hollywood-doctor-treatment-celebrities-reverse-aging-look-younger-stem-cell. Accessed 26 Mar 2024

Lee A (2018) Texas' new adult stem cell law underway. Available via Houston Public Media. https://www.houstonpublicmedia.org/articles/news/2017/10/28/244813/texas-new-adult-stem-cell-law-underway-almost. Accessed 26 Mar 2024

Leung A, Crombleholme TM, Keswania SG (2012) Fetal wound healing: implications for minimal scar formation. Curr Opin Pediatr 24:371–378

Lienesch M (1982) Right-wing religion: Christian conservatism as a political movement. Polit Sci Q 97:403–425

Liu X, Wu H, Byrne M, Jaenisch R (1997) Type III collagen is crucial for Collagen I fibrillogenesis and for normal cardiovascular development. Dev Biol 94:1852–1856

Liu H, Reiter S, Zhou X, Chen H, Ou Y, Lenahan C, He Y (2021) Insight into the mechanisms and the challenges on stem cell-based therapies for cerebral ischemic stroke. Front Cell Neurosci 15: 1, 7. Available via Frontiers: https://www.frontiersin.org/articles/10.3389/fncel.2021.637210/full

Lo B, Parham L (2009) Ethical issues in stem cell research. Endocr Rev 3:204–213

Locke JE, Kumar V, Anderson D, Porrett PM (2023) Normal graft function after pig-to-human kidney xenotransplant. JAMA Surg 158:1106–1108. Available via American Medical Association. https://jamanetwork.com/journals/surg/articlepdf/2808483/jamasurgery_locke_2023_ld_230010_1696870110.39561.pdf

Long C, Li H, Tiburcy M, Rodriguez-Caycedo C, Kyrychenko V, Zhou H, Zhang Y, Min Y-L, Shelton JM, Mammen PPA, Liaw NY, Zimmermann W-H, Bassel-Duby R, Schneider JW, Olson EN (2018) Correction of diverse muscular dystrophy mutations in human engineered heart muscle by single-site genome editing. Sci Adv 1:822–826

Lund University (2023) First patient receives milestone stem cell-based transplant for Parkinson's Disease. Available via Lund University. https://www.lunduniversity.lu.se/article/first-patient-receives-milestone-stem-cell-based-transplant-parkinsons-disease. Accessed 26 Mar 2024

Lund RD, Wang S, Klimanskaya I, Holmes T, Ramos-Kelsey R, Lu B, Lanza R (2006) Human embryonic stem cell-derived cells rescue visual function in dystrophic RCS rats. Cloning Stem Cells 8:189

Ma H, Marti-Gutierrez N, Mitalipov S et al (2017) Correction of a pathogenic gene mutation in human embryos. Nature 548:413–419

Madhusoodanan J (2022) After the first pig-to-human heart transplant, scientists look to the future of cardiac xenotransplantation. JAMA 20:1999–2001

Malik K (2020) Science is ever evolving, just like the human race. Available via the Guardian. https://www.theguardian.com/commentisfree/2020/feb/16/how-science-evolves-just-like-human-race. Accessed 4 Mar 2024

Malkoc V (2018) Challenges and the future of 3D bioprinting. J Biomed Imag Bioeng 2:64–65

Marks P (2020) Coronavirus (COVID-19) update: FDA provides updated guidance to address the urgent need for blood during the pandemic, FDA statement. Available via the Food and Drug Administration. https://www.fda.gov/news-events/press-announcements/coronavirus-covid-19-update-fda-provides-updated-guidance-address-urgent-need-blood-during-pandemic. Accessed 26 Mar 2024

Mashambanhaka F (2018) What is 3D bioprinting? – Simply explained. Available via All3dp. https://all3dp.com/2/what-is-3d-bioprinting-simply-explained. Accessed 26 Mar 2024

Maune G (2022) In-Vitro gametogenesis opens new doors for same-sex & infertile, hopeful parents. Available via the Virginia Journal of Bioethics. www.vabioethics.com/content/2022/3/20/in-vitro-gametogenesis-opens-new-doors-for-same-sex-amp-infertile-hopeful-parents. Accessed 26 Mar 2024

Mayo Clinic Staff (2023) Stem cells: what they are and what they do. Available via Mayo Foundation for Medical Education and Research (MFMER). https://www.mayoclinic.org/tests-procedures/bone-marrow-transplant/in-depth/stem-cells/art-20048117. Accessed 26 Mar 2024

McCormack K (2020) Encouraging news for treatment targeting retinitis Pigmentosa. Available via CIRM. https://blog.cirm.ca.gov/tag/retinitis-pigmentosa. Accessed 26 Mar 2024

McDonald JW, Liu XZ, Qu Y, Liu S, Mickey SK, Turetsky D, Gottlieb DI, Choi DW (1999) Transplanted embryonic stem cells survive, differentiate and promote recovery in injured rat spinal cord. Nat Med 5:1410–1412

Medvedev SP, Shevchenko AI, Zakian SM (2010) Induced pluripotent stem cells: problems and advantages when applying them in regenerative medicine. Acta Naturae 2:18–28

Mehrali M, Bagherifard S, Akbari M, Thakur A, Mirani B, Hasany M, Orive G, Das P, Emneus J, Andresen TL, Dolatshahi-Pirouz A (2018) Blending electronics with the human body: a pathway toward a cybernetic future. Adv Sci 5:1–39. Available via Wiley Online Library. https://onlinelibrary.wiley.com/doi/10.1002/advs.201700931

Melton DA (2011) Using stem cells to study and possibly treat type 1 diabetes. Philos Trans R Soc Lond B Biol Sci 366:2307–2311

Menasché P, Vanneaux V, Hagège A, Bel A, Cholley B, Parouchev A, Cacciapuoti I, Al-Daccak R, Benhamouda N, Blons H, Agbulut O, Tosca L, Trouvin J-H, Fabreguettes J-R, Bellamy V,

Charron D, Tartour E, Tachdjian G, Desnos M, Larghero J (2018) Transplantation of human embryonic stem cell–derived cardiovascular progenitors for severe ischemic left ventricular dysfunction. J Am Coll Cardiol 71(429):432
Merienne N, Vachey G, de Longprez L, Meunier C et al (2017) The self-inactivating KamiCas9 system for the editing of CNS disease genes. Cell Rep 12:2980–2991
Mertes H, Pennings G, Van Steirteghem A (2006) An ethical analysis of alternative methods to obtain pluripotent stem cells without destroying embryos. Hum Reprod 21:2749–2755
Microscope Master (2021) Progenitor cells. Available via Microscope Master. https://www.microscopemaster.com/progenitor-cells.html. Accessed 26 Mar 2024
MIT Committee on Assessment of Biohazard & Embryonic Stem Cell Research Oversight (2008) Review and approval for human embryonic stem and induced pluripotent stem cell based research. Available via Massachusetts Institute of Technology. https://facultygovernance.mit.edu/committee/committee-assessment-biohazards-and-embryonic-stem-cell-research-oversight. Accessed 26 Mar 2024
Mitra M (2017) Medical nanobot for cell and tissue repair. Int Rob Auto J 2:218–222
Miyazaki S, Yamato E, Miyazaki J-I (2004) Regulated expression of pdx-1 promotes in vitro differentiation of insulin-producing cells from embryonic stem cells. Diabetes 53:1030. Available via Diabetes Journal. https://diabetes.diabetesjournals.org/content/53/4/1030.short
Moradi S, Jahanian-Najafabadi A, Roudkenar MH (2016) Artificial blood substitutes: first steps on the long route to clinical utility. Clin Med Insights Blood Disord 9:33–41
Murnaghan I (2020) New harvesting techniques, explore stem cells. Available via Explore Stem Cells. www.explorestemcells.co.uk/newharvestingtechniques.html. Accessed 26 Mar 2024
Murugan V (2009) Embryonic stem cell research: a decade of debate from bush to Obama. Yale J Biol Med 82:101–103
Musunuru K (2018) Genome editing: the recent history and perspective in cardiovascular diseases. J Am Coll Cardiol 22:2808–2821
Myers PZ (2016) An abuse of stem cells. Available via Free Thought Blogs. https://freethoughtblogs.com/pharyngula/2016/06/28/an-abuse-of-stem-cells/. Accessed 26 Mar 2024
Narsinh KH, Plews J, Wu JC (2011) Comparison of human induced pluripotent and embryonic stem cells: fraternal or identical twins? Mol Ther 19:635–638
National Cancer Institute (2011) Germline Mutation. Available via National Cancer Institute. https://www.cancer.gov/publications/dictionaries/cancer-terms/def/germline-mutation. Accessed 26 Mar 2024
National Health Service (2018) Pacemaker implantation information. Available via National Health Servies (UK). https://www.nhs.uk/conditions/pacemaker-implantation. Accessed 26 Mar 2024
National Human Genome Research Institute (2017) What are the ethical concerns of genome editing? Available via the National Human Genome Research Institute [hereinafter NHGRI]. https://www.genome.gov/about-genomics/policy-issues/Genome-Editing/ethical-concerns. Accessed 26 Mar 2024
National Human Genome Research Institute (2020) Epigenomes. Available via NHGRI. https://www.genome.gov/27532724/epigenomics-fact-sheet. Accessed 26 Mar 2024
National Institute Of Aging (2023) Parkinson's Disease Information. Available via the National Institute of Aging. https://www.nia.nih.gov/health/parkinsons-disease. Accessed 1 Jan 2024
National Institute of Health (2022) Stem cell basics. Introduction. Available via National Institute of Health. https://stemcells.nih.gov/info/basics. Accessed 4 Jan 2024
National Institute of Health (2023) What is a Stroke? Available via the National Institute of Health. https://www.nhlbi.nih.gov/health/stroke. Accessed 26 Mar 2024
National Kidney Foundation (2024) Understanding the transplant waitlist. Available via the National Kidney Foundation. https://www.kidney.org/content/understanding-transplant-waitlist. Accessed 3 Jan 2024
National Research Council (US) and Institute of Medicine (US) Committee on the Biological and Biomedical Applications of Stem Cell Research (2002) Stem cells and the future of regenerative medicine. National Academies Press (US), Washington (DC). Chapter 3, pp. 4, 10–12,

24, Embryonic Stem Cells. Available via NCBI. https://www.ncbi.nlm.nih.gov/books/NBK223690/

Nedelman M (2018) Human stem cells give monkey hearts a boost after heart attacks, study says. Available via CNN. https://www.cnn.com/2018/07/02/health/monkey-heart-attack-stem-cell-study/index.html. Accessed 26 Mar 2024

New York State Stem Cell Science Dept. of Health (2015) What is the difference between reproductive cloning and therapeutic cloning? Available via New York State Stem Cell Science Dept. of Health. http://stemcell.ny.gov/faqs/what-difference-between-reproductive-and-therapeutic-cloning. Accessed 26 Mar 2024

Newmark P (2006) Flatworms at forefront of regeneration research. Available via the U.S. National Science Foundation. https://www.nsf.gov/news/news_summ.jsp?cntn_id=107916. Accessed 26 Mar 2024

Norris A, Korc M (2010) Chapter 324 - Aberrant signaling pathways in pancreatic cancer: opportunities for targeted therapeutics. In: Handbook of cell signaling, 2nd edn. Academic Press, pp 2783–2798. ISBN 9780123741455. https://doi.org/10.1016/B978-0-12-374145-5.00324-7

Nunes dos Santos RM (2023) Kidney xenotransplantation: are we ready for prime time? Curr Urol Rep 24:287, 296–297

Okano H, Nakamura M, Yoshida K, Okada Y, Tsuji O, Nori S, Ikeda E, Yamanaka S, Miura K (2013) Steps toward safe cell therapy using induced pluripotent stem cells. Circ Res 112:523–533

Ong CS, Yesantharao P, Huang CY, Mattson G, Boktor J, Fukunishi T, Zhang H, Hibino N (2018) 3D bioprinting using stem cells. Pediatr Res 83:223–231

Otsu K, Kumakami-Sakano M, Fujiwara N, Kikuchi K, Keller L, Lesot H, Harada H (2014) Stem cell sources for tooth regeneration: current status and future prospects. Front Physiol 5:1–10. Available via NCBI. https://www.ncbi.nlm.nih.gov/pmc/articles/PMC6344244/

Ouyang L, Yao R, Mao S, Chen X, Na J, Sun W (2016) Three-dimensional bioprinting of embryonic. Biofabrication 7:044101. Available via NCBI. https://pubmed.ncbi.nlm.nih.gov/26531008/

Ovando-Roche P, Georgiadis A, Smith AJ, Pearson RA, Ali RR (2017) Harnessing the potential of human pluripotent stem cells and gene editing for the treatment of retinal degeneration. Curr Stem Cell Rep 3:112–123

Oyamada N, Itoh H, Sone M, Yamahara K, Miyashita K, Park K, Taura D, Inuzuka M, Sonoyama T, Tsujimoto H, Fukunaga Y, Tamura N, Nakao K (2008) Transplantation of vascular cells derived from human embryonic stem cells contributes to vascular regeneration after stroke in mice. J Transl Med 6:7. Available via PubMed. https://pubmed.ncbi.nlm.nih.gov/18823569

Park I-H, Arora N, Huo H, Maherali N, Ahfeldt T, Shimamura A, Lensch MW, Cowan C, Hochedlinger K, Daley GQ (2008) Disease-Specific induced pluripotent stem cells. Cell 134:877–886

Parsa-Parsi RW (2017) The revised Declaration of Geneva: a modern-day physician's pledge. JAMA 318:1971–1972

Passier R, Mummery C (2003) Origin and use of embryonic and adult stem cells in differentiation and tissue repair. Cardiovasc Res 58:324–335

Pati F, Gantelius J, Andersson Svahn H (2016) 3D bioprinting of tissue/organ models. Angew Chem Int Ed Engl 55:4650–4665

Patil M, Singh Mehta D, Guvva S (2008) Future impact of nanotechnology on medicine and dentistry. J Indian Soc Periodontol 12:34–40

PBS News Hour Science (2008) Biotech companies, investors look to adult stem cell research. Available via PBS. https://www.pbs.org/newshour/science/science-july-dec08-stemcell_07-18. Accessed 26 Mar 2024

Penn Medicine (2024) Heart transplant waiting list. Available via Penn Medicine. https://www.pennmedicine.org/for-patients-and-visitors/find-a-program-or-service/transplant-institute/heart-transplant/heart-waiting-list. Accessed 3 Jan 2024

Peters T (2017) What is the difference between stem cell therapy and PRP injections? Available via Todd Peters MD. https://newportspinemd.com/prp-therapy-vs-stem-cell-therapy-whats-difference/. Accessed 26 Mar 2024

Piraino S, Boero F, Aeschbach B, Schmid V (1996) Reversing the life cycle: Medusae transforming into polyps and cell transdifferentiation in *Turritopsis Nutricula* (Cnidaria, Hydrozoa). Biol Bull 190:302

Pires R (2018) What exactly is bioink? – Simply Explained. Available via All3dp. https://all3dp.com/2/for-ricardo-what-is-bioink-simply-explained. Accessed 26 Mar 2024

Pittenge MF, Mackay AM, Beck SC, Jaiswal RK, Douglas R, Mosca JD, Moorman MA, Simonetti DW, Craig S, Marshak DR (1999) Multilineage potential of adult human mesenchymal stem cells. Science 284:143–147. Available via Science.org. https://www.science.org/doi/epdf/10.1126/science.284.5411.143

Powers Maurice J (2023) We'll be living and working to 120 – and it will start within a decade' days doctor to the stars. Available via New York Post. https://nypost.com/2023/09/23/doctor-stem-cells-will-extend-life-to-120-from-this-decade. Accessed 26 Mar 2024

Puiu T (2023) Scientists prove "immortal worms" can regenerate indefinitely and stay forever young. Available via ZME Science. https://www.zmescience.com/feature-post/natural-sciences/animals/invertebrates/immortal-worm-never-ages-forever-young-043294. Accessed 3 Jan 2024

Rabin RC (2022) In a first, man receives a heart from a genetically altered pig. Available via the New York Times Company. https://www.nytimes.com/2022/01/10/health/heart-transplant-pig-bennett.html. Accessed 26 Mar 2024

Ravari H, Hamidi-Almadari D, Salimifar M, Bonakdaran S, Parizadeh MR, Koliakos G (2011) Treatment of non-healing wounds with autologous bone marrow cells, platelets, fibrin glue and collagen matrix. Cytotherapy 13:705–711

Regalado A (2023) A biotech company says it put dopamine-making cells into people's brains. Available via TechnologyReview.com. https://www.technologyreview.com/2023/08/31/1078728/biotech-dopamine-making-cells-into-peoples-brains/. Accessed 26 Mar 2024

Rev. Provost GJ (2016) Catholic Church clear on when human life begins. Available via Eternal Word Television Network. https://www.ewtn.com/catholicism/library/catholic-church-clear-on-when-human-life-begins-3825. Accessed 26 Mar 2024

Revazova ES, Turovets NA, Kochetkova OD, Kindarova LB, Kuzmichev LN, Janus JD, Pryzhkova MV (2007) Patient-specific stem cell lines derived from human parthenogenetic blastocysts. Cloning Stem Cells 9:432–449

Ring K (2018) Stem cell patch restores vision in patients with age-related macular degeneration. Available via California's Stem Cell Agency [hereinafter CIRM]. https://blog.cirm.ca.gov/2020/07/29/encouraging-news-for-treatment-targeting-retinitis-pigmentosa/. Accessed 26 Mar 2024

Rodriguez Fernandez C (2017) CRISPR therapeutics plans first CRISPR/Cas9 clinical trial in Europe for 2018. Available via Labiotech. https://www.labiotech.eu/medical/crispr-therapeutics-clinical-trials. Accessed 26 Mar 2024

Rodriguez Fernandez C (2019) 7 diseases CRISPR technology could cure. Available via Labiotech. https://www.labiotech.eu/crispr/crispr-technology-cure-disease. Accessed 26 Mar 2024

Roget C (2017) The stem cell – a cautionary tale on a global scale. Available vie Medical Tourism Magazine. https://www.magazine.medicaltourism.com/article/stem-cell-cautionary-tale-global-scale. Accessed 26 Mar 2024

Romito A, Cobellis G (2016) Pluripotent stem cells: current understanding and future directions. Stem Cells Int 2016:1. Available via NCBI. https://www.ncbi.nlm.nih.gov/pmc/articles/PMC4699068

Rong Z, Wang M, Hu Z, Stradner M, Zhu S, Kong H, Yi H, Goldrath A, Yang Y-G, Xu Y, Fu X (2014) An effective approach to prevent immune rejection of human ESC-derived allografts. Cell Stem Cell 14:121–130

Roper J, Yilmaz OH (2019) Breakthrough moments: genome editing and organoids. Cell Stem Cell 24:841–842

Rouanet M, Lebrin M, GrossF BB, Cordelier P, Buscail L (2017) Gene therapy for pancreatic cancer: specificity, issues and hopes. Int J Mol Sci 6:1231

Sahara H, Watanabe H, Pomposelli T, Yamadaa K (2019) Lung xenotransplantation. Curr Opin Organ Transplant 22:541–548

Salgado AJ, Oliveira JM, Martins A, Teixeira FG, Silva NA, Neves NM, Sousa N, Reis RL (2013) Tissue engineering and regenerative medicine: past, present, and future. Int Rev Neurobiol 108:1

Sarkar S (2008) Artificial blood. Indian J Crit Care Med 3:140–144

Schapira AH (1999) Medicine, and the future: Parkinson's disease. BMJ 318:311

Serani D (2001) If it bleeds, it leads: understanding fear-based media. Available via Psychology Today. https://www.psychologytoday.com/us/blog/two-takes-depression/201106/if-it-bleeds-it-leads-understanding-fear-based-media. Accessed 26 Mar 2024

Sheldrick G (2018) Wonder stem cell CURE for heart failure: millions could be saved by British trials. Available via the Express. https://www.express.co.uk/life-style/health/972389/heart-attack-stem-cell-treatment-cardiac-heart-failure-cure-arrest-medical-trial. Accessed 26 Mar 2024

Shiroor DA, Wang K-T, Sanketi BD, Tapper JK, Adler CE (2023) Inhibition of ATM kinase rescues planarian regeneration after lethal radiation. EMBO Rep 24:1. Available via European Molecular Biology Organization. https://www.embopress.org/doi/epdf/10.15252/embr.2022 56112

Shroff G (2016) Therapeutic potential of human embryonic stem cells in type 2 diabetes mellitus. World J Stem Cells 8:223–230

Sipherd R (2017) USC stem cell scientists developing a new cure-all for baldness. Available via CNBC. https://www.cnbc.com/2017/10/13/usc-stem-cell-scientists-developing-a-new-cure-all-for-baldness.html. Accessed 26 Mar 2024

Sito S (2017) I've always wondered: does anyone my age have any chance of living for centuries? Available via the Conversation. https://theconversation.com/ive-always-wondered-does-any one-my-age-have-any-chance-of-living-for-centuries-83485. Accessed 26 Mar 2024

Slukvin II (2010) Generation of mature blood cells from pluripotent stem cells. Haematologica 95:1621–1623

Smith WJ (2017) California stem-cell agency wants more voter money. Available via National Review. https://www.nationalreview.com/corner/california-stem-cell-agency-wants-more-voter-money. Accessed 26 Mar 2024

Smolenski J (2015) CRISPR/Cas9 and germline modification: new difficulties in obtaining informed consent. Am J Bioethics 15:35–37

South Park Archive (2019a) Krazy Kripples. Available via South Park Archive. https://southpark.fandom.com/wiki/Krazy_Kripples. Accessed 26 Mar 2024

South Park Archive (2019b) Kenny Dies. Available via South Park Archive. https://southpark.fandom.com/wiki/Kenny_Dies. Accessed 26 Mar 2024

Steck AK, Rewers MJ (2016) Genetics of type 1 diabetes. Clin Chem 57:176

Stein R (2018) Doctors in China lead race to treat cancer by editing genes. Available via National Public Radio [hereinafter NPR]. https://www.npr.org/sections/health-shots/2018/02/21/58533 6506/doctors-in-china-lead-race-to-treat-cancer-by-editing-genes. Accessed 26 Mar 2024

Stein R (2019) First U.S. patients treated with CRISPR As human gene-editing trials get underway. Available via NPR. https://www.npr.org/sections/health-shots/2019/04/16/712402435/first-u-s-patients-treated-with-crispr-as-gene-editing-human-trials-get-underway. Accessed 26 Mar 2024

Steinbeck JA, Choi SJ, Mrejeru A, Ganat Y, Deisseroth K, Sulzer D, Mosharov EV, Studer L (2015) Optogenetics enables functional analysis of human embryonic stem cell–derived grafts in a Parkinson's disease model. Nat Biotechnol 33:204–209

Su S, Hu B, Shao J, Shen B, Du J, Du Y, Zhou J, Yu L, Zhang L, Chen F, Sha H, Cheng L, Meng F, Zou Z, Huang X, Liu B (2016) CRISPR-Cas9 mediated efficient PD-1 disruption on human primary T cells from cancer patients. Sci Rep 6:1–13. Available via Nature https://doi.org/10.1038/srep20070

Suter SM (2015) In vitro gametogenesis: just another way to have a baby? J Law Biosci 3:87–119

Talavera-Adame D, Newman D, Newman N (2017) Conventional and novel stem cell-based therapies for androgenic alopecia. Stem Cells Cloning 10:11

Tangermann V (2018) A CRISPR future: five ways gene editing will transform our world, futurism. Available via Futurism. https://futurism.com/crispr-genetic-engineering-change-world. Accessed 26 Mar 2024

Tarne EC (2018) Trends show more federal funds awarded to non-embryonic stem cell Research. Available via Lozier Institute. https://lozierinstitute.org/trends-show-more-federal-funds-awarded-to-non-embryonic-stem-cell-research/. Accessed 26 Mar 2024

The American Red Cross (2019) Hurricane Dorian creates increased need for blood donations. Available via the Red Cross. https://www.redcross.org/about-us/news-and-events/news/2019/hurricane-dorian-creates-increased-need-for-blood-donations.html. Accessed 26 Mar 2024

The Editors of Encyclopaedia (2023) Cybernetics. Available via Encyclopedia Britannica. https://www.britannica.com/science/cybernetics. Accessed 26 Mar 2024

Thesleff I, Tummers M (2003) Stem cells and tissue engineering: prospects for regenerating tissues in dental practice. Med Princ Pract 12:43

Translational Vision Science & Technology Panel (2014) Chapter 4 - Restoring vision to the blind: stem cells and transplantation. Transl Vis Sci Technol 3:33. Available via NCBI. https://www.ncbi.nlm.nih.gov/pmc/articles/PMC4314983

Tucker V (2007) Shinya Yamanaka reprograms human adult cells into embryonic-like stem cells. Available via University of California San Francisco. https://www.ucsf.edu/news/2007/11/7369/shinya-yamanaka-reprograms-human-adult-cells-embryonic-stem-cells. Accessed 26 Mar 2024

Tucker BA, Mullins RF, Stone EM (2014) Stem cells for investigation and treatment of inherited retinal disease. Hum Mol Genet 23:9

Tyson P (2001) The Hippocratic oath today. Available via Public Broadcasting Service [hereinafter PBS]. https://www.pbs.org/wgbh/nova/article/hippocratic-oath-today. Accessed 26 Mar 2024

U.S. Food & Drug Administration (2009) Information and recommendations for physicians involved in the co-culture of human embryos with non-human animal cells. Available via U.S. Food & Drug Administration [hereinafter FDA]. https://www.fda.gov/BiologicsBloodVaccines/Xenotransplantation/ucm136532.htm. Accessed 12 Jan. 2024

U.S. Food & Drug Administration (2023) FDA approves first gene therapies to treat patients with sickle cell disease (2023). Available via FDA. https://www.fda.gov/news-events/press-announcements/fda-approves-first-gene-therapies-treat-patients-sickle-cell-disease. Accessed 26 Mar 2024

Univ. of Wash. - Health Sciences/UW News, Comm. Relations & Marketing (2014) New cell line should accelerate embryonic stem cell research. Available via ScienceDaily. www.sciencedaily.com/releases/2014/03/140314095104.htm. Accessed 26 Mar 2024

Van der Aa N, Esteki MZ, Vermeesch JR, Voet T (2013) Preimplantation genetic diagnosis guided by single-cell genomics. Genome Med 5:71

Vergara MN, Tsonis PA (2013) 15 - The past, present, and future of tissue regeneration. Section III - Future applications of regenerative pharmacology, 311. Available via Cambridge University Press. https://www.cambridge.org/core/books/regenerative-pharmacology/past-present-and-future-of-tissue-regeneration/229C54EF3E66878085A09702A81C800E. Accessed 26 Mar 2024

Warwick K, Gasson M, Hutt B, Goodhew I, Kyberd P, Andrews B, Teddy P, Shad A (2003) The application of implant technology for cybernetic systems. Arch Neurol 10:1369–1373
Weaver X (2020) Xenotransplantation of the pancreas: a reality check of its historical emergence. Diabetes Clin Res 7:1. Available via ClinMed International Library. https://clinmedjournals.org/articles/ijdcr/international-journal-of-diabetes-and-clinical-research-ijdcr-7-131.pdf?jid=ijdcr
Wei R, Yang J, Hou W, Liu G, Gao M, Zhang L et al (2013) Insulin-Producing cells derived from human embryonic stem cells: comparison of definitive endoderm- and nestin-positive progenitor-based differentiation strategies. PLoS ONE 8:1, 7. Available via NCBI. https://www.ncbi.nlm.nih.gov/pmc/articles/PMC3741181/
Weintraub K (2023a) Beyond IVF: scientists debate ethics of human reproduction without egg and sperm. Available via USA Today. https://www.usatoday.com/story/news/health/2023/04/23/future-treatment-infertility-ethics/11509269002. Accessed 26 Mar 2024
Weintraub K (2023b) With pig-to-human heart transplant, a man hoped to advance medicine. He died weeks later. Available via USA Today. https://www.usatoday.com/story/news/health/2023/10/31/pig-heart-transplant-to-human-update-patient-dies/71397664007. Accessed 26 Mar 2024
Welch A (2018) Patients regain sight with stem cell treatment in U.K. clinical trial. Available via Columbia Broadcasting System News. https://www.cbsnews.com/news/patients-regain-sight-with-stem-cell-treatment-uk-clinical-trial. Accessed 26 Mar 2024
Wen P, Dockser Marcus A (2019) Chinese scientist who gene-edited babies is sent to prison. Via The Wall Street Journal. https://www.wsj.com/articles/chinese-scientist-who-gene-edited-babies-is-sent-to-prison-11577703233. Accessed 26 Mar 2024
Wente M (2010) Autism, vaccines and fear. Available via the Globe and Mail. https://www.theglobeandmail.com/opinion/autism-vaccines-and-fear/article4309831/. Accessed 26 Mar 2024
Williams N (2004) Reeve's stem cell legacy. Curr Biol 14:R907. Available via ScienceDirect https://www.sciencedirect.com/science/article/pii/S0960982204007973
Wilson C (2023) What are 'synthetic embryos' and why are scientists making them? Available via NewScientist. https://www.newscientist.com/article/2378553-what-are-synthetic-embryos-and-why-are-scientists-making-them. Accessed 26 Mar 2024
Wittes B, Chong J (2014) Our cyborg future: law and policy implications. Available via Brookings. https://www.brookings.edu/research/our-cyborg-future-law-and-policy-implications. Accessed 26 Mar 2024
Wooller S (2020) Bionic eye with sharper vision than real eye could be ready in just five years, experts say, New York Post. Available via the New York Post. https://nypost.com/2020/05/21/bionic-eye-with-sharper-vision-than-real-eye-could-be-ready-in-just-five-years-experts-say. Accessed 26 Mar 2024
World Health Organization (2023) Herpes simplex virus. Available via World Health Organization. https://www.who.int/news-room/fact-sheets/detail/herpes-simplex-virus. Accessed 26 Mar 2024
Wu X, Kwong AC, Rice CM (2019) Antiviral resistance of stem cells. Curr Opin Immunol 56:50–59
Yoon CH, Choi HJ, Kim MK (2020) Corneal xenotransplantation: where are we standing? Prog Retin Eye Res 80:100876. Available via Science Direct. https://www.sciencedirect.com/science/article/pii/S1350946220300483
Yu H, Fang D, Kumar SM, Li L, Nguyen TK, Acs G, Herlyn M, Xu X (2006) Isolation of a novel population of multipotent adult stem cells from human hair follicles. Am J Pathol 168:1879–1888
Yu L, Wen H, Liu C, Wang C, Yu H, Zhang K, Han Q, Liu Y, Han Z, Li Z, Liu N (2023) Embryonic stem cell-derived extracellular vesicles rejuvenate senescent cells and antagonize aging in mice. Bioact Mater 29:85–97

Zakrzewski W, Dobrzyński M, Szymonowicz M, Rybak Z (2019) Stem cells: past, present, and future. Stem Cell Res Ther 10:1. Available via National Library of Medicine [hereinafter NCBI]. https://www.ncbi.nlm.nih.gov/pmc/articles/PMC6390367

Zeng A, Li H, Guo L, Gao X, McKinney S, Wang Y, Yu Z, Park J, Semerad C, Ross E, Cheng L-C, Davies E, Lei K, Wang W, Perera A, Hall K, Peak A, Box A, Sánchez Alvarado A (2018) Prospectively isolated tetraspanin+ neoblasts are adult pluripotent stem cells underlying planaria regeneration. Cell 173:1593–1608e.20. Available via NCBI. https://www.ncbi.nlm.nih.gov/pmc/articles/PMC9359418/?report=reader

Zhang M, Yang C, Liu H, Sun Y (2017) Induced pluripotent stem cells are sensitive to DNA damage. Genom Proteom Bioinform 11:320–326

Zheng YL (2016) Some ethical concerns about human induced pluripotent stem cells. Sci Eng Ethics 22:1277–1284

Zivot A, Lipton JM, Narla A, Blanc L (2018) Erythropoiesis: insights into pathophysiology and treatments in 2017. Mol Med 24:1–15

Chapter 3
Claims, Conflicting Claimants, and Their Reasoning

Like many other "hot button" issues, the overall topic of embryonic stem cells brings to mind many arguments from many different perspectives. In the United States, candidates for President have made the topic a part of their campaign platforms in an effort to control the science, and attract voters to their sides. Even the major religions have added to this debate in the most extreme of ways, with an all or nothing attitude. The media, in all its forms have had something to say about this issue, from cartoons to scientific publications, furthering the argument on both sides, but there has been very little work in finding a solution that will enable all sides to be at peace with this area of science. In analyzing the arguments presented below, one must keep in mind to be respectful of all beliefs, and philosophies, and consider the opposing point of views in formulating their own ideas. When dealing with the issue of embryonic stem cells, there are three overall approaches that can be seen, (1) those that are pro-embryonic stem cell research; (2) those that oppose embryonic stem cell research; and (3) those that seek the benefits of embryonic stem cell research without wanting to harm an embryo.

3.1 Pro-Embryonic Stem Cell Research

3.1.1 Currently Suffering Patients and Their Loved Ones

People who are suffering from a variety of diseases, pains, paralysis, different forms of cancer, or a large number of any other medical problems see the use of embryonic stem cells as a potential cure for their ailments.[1] The studies that have already been conducted have made the news, and shown the world that the use of stem cells can vastly improve the lives of those suffering. Obviously, the patients who are suffering

[1] Fuchs and Segre (2012), pp. 143–155.

© The Author(s), under exclusive license to Springer Nature Switzerland AG 2024
J. Weiser, *Embryonic Stem Cells and the Law*,
https://doi.org/10.1007/978-3-031-59750-3_3

would love to have expanded laws so that they too can have a procedure using stem cells, but their families would also feel the same. Due to the fact that many of the more advanced embryonic stem cell studies, and procedures are conducted over seas, being able to undergo any procedure is very costly which limits the amount of people who can access this form of help. Expanding federal funding, and loosening research and stem cell laws would enable many more people to receive treatment in America, which would greatly cut down on costs, such as travel expenses, and hotel costs. Another issue that arises from receiving treatment overseas is that often many insurance providers will not cover the procedures.

Patients tend to all have one thing in common whether they have a new ailment or disability, or they have suffered for years and that is that they are suffering. There are also many patients that medicine has failed to help and they are seeking a "miracle" as a last resort. Knowing that there are both proven and potential cures and treatments to end their suffering, yet they cannot utilize them is perhaps the most egregious and upsetting issue that comes from these outdated laws. When I first told my Doctor that I was planning on tackling the subject of embryonic stem cell laws, he was very excited and immediately introduced me to Dr. Barth Green who is a co-founder and co-head of the Miami Project. Dr. Green was also excited by the ine hearing of my topic choice, and pledged his assistance in any way possible. Dr. Green has published many articles on the potential uses of stem cells in various spinal cord injuries which cause various kinds of paralysis.[2] Virtually on a weekly basis there are studies into all forms of stem cells that are published which lends to offer great hope to all. In fact, just in 2018 alone, especially for some of the most debilitating issues, we are so close to being able to help people who need it the most yet due to the constraining laws, those studying and practicing in this field have their hands effectively tied.[3] Even scientists have published papers based on the fact that neurodegenerative diseases "has received a great deal of scientific and public interest in recent years. This is due to the remarkable pace at which paradigm-changing discoveries have been made regarding the neurogenic potential of embryonic, fetal, and adult cells."[4] There has also been a lot of research conducted using stem cells in areas where there is typically no other method of treatment such as in Sjögren's syndrome which is an autoimmune disease that can harm a vast array of organs in an extremely painful manner.[5] As mentioned above, stem cells have also been instrumental in restoring eye sight to those rendered blind by various disorders.[6] Many of

[2]*See* The Miami Project (2024).

[3]*See Generally* Egawa et al. (2012), p. 145; Lindvall (2004), pp. s42–s50 (describing the ability to use stem cells to potentially cure, and at the least alleviate the symptoms related to various neurodegenerative disorders such as amyotrophic lateral sclerosis ("ALS"), and Parkinson's syndrome).

[4]Le Belle and Svendsen (2002), pp. 389–401.

[5]*Id.* p. 389; Ruan et al. (2017), pp. 2828–2838.

[6]Gater (2018).

these treatments involved the study of and use of embryonic stem cells and had actual, life changing results.[7]

Not all progress comes in the form of a cure or vaccine, in fact, just slowing down certain diseases is often considered a milestone or huge feat.[8] In a 2020 report, the company AstroRX published the results from a study involving patients suffering from amyotrophic lateral sclerosis ("ALS") which is a neurodegenerative disease that essentially degenerates neurons, specifically ones that control a person's muscles so as to make them too weak to use over time, and is considered fatal.[9] AstroRX's study used embryonic stem cells, directly injected into patients' spinal fluid which has proven to slow the progression of the disease by at least three months, even in patients that received the lowest dose.[10] Unfortunately, after only three months the progression of ALS continued, but it is very promising to find a candidate for helping those suffering from a fatal disease, even if it only benefits them for a short amount of time.

It should also be noted that people from other countries come to the United States for medical treatments, so it is very likely that even those from other countries would likely argue for expanded funding and beneficial laws.[11] In fact, approximately 100,000 to 200,000 people from other countries fly into the United States each year for medical treatments.[12] This information specifically does not take into account those travelling from Canada or Mexico which make up a large number as they typically travel "overland" as opposed to flying into the country for their treatments.[13]

3.1.2 *Those Who Believe Access to Medical Benefits Is a Duty Owed by the State*

Many people believe and argue that a person's country of residence owes them the duty of either access to medicine, and/or access to medical treatments.[14] This is not just true for citizens of said countries, but also for those who are there and remain undocumented.[15] The claimed rights do not end with access to medicine and medical treatments, but necessitate that there is a form of equality for all in said access.[16]

[7] Lund et al. (2006), pp. 189–197.

[8] *See generally* Martins (2020).

[9] *See* Zarei et al. (2015), p. 171.

[10] Martins (2020).

[11] *See generally* Chambers (2015).

[12] *Id.*

[13] *Id.*

[14] *See generally* Hestermeyer (2007).

[15] *See* Ruiz-Casares et al. (2010), pp. 329–336.

[16] Bingham et al. (2011), pp. 189–193.

Unfortunately, in practice, courts in the United States have typically shied away from enforcing access to medicine as a constitutional right, as they usually do not attempt to enforce so-called "positive" rights.[17] What this implicitly stands for is the fact that the government cannot preclude those seeking legal treatments from attempting to, or actually receiving them, but there is nothing specific in the Constitution that creates a duty that mandates that the federal government must work to ensure that the person seeking the treatments actually is able to obtain it.[18]

The Universal Declaration in Article 25 specifically states that "Everyone has the right to a standard of living adequate for the health and well-being of himself and of his family, including food, clothing, housing and medical care..."[19] To take this one step further, the ICESCR, includes rights to prevention, treatment, and control of diseases (Art. 12.2 (c)); as well as the right to health facilities, goods and services (Art. 12.2 (d)).[20] However, while it seems as though a positive right was created in this treaty, the United States has not yet ratified this document, even though they are still a state party that has signed the instrument.[21] International "declarations" such as these typically have no binding effect, and instead "represents a very strong commitment by States to its implementation."[22] Some argue that this Declaration should be treated as binding under the principles of customary international law.[23] As discussed at length below, customary international law is viewed as laws that are created through practice due to a State's beliefs that they are required to do so, as opposed to a formal written law.[24] While this argument can be persuasive on an emotional level, there is no universal healthcare in the United States, showing that customary international law is not likely to be applicable over this Article.[25]

3.1.3 *Pharmaceutical & Bio-Tech Companies*

Pharmaceutical companies have already been engaging in stem cell research, and view the potential advancements in medicine and treatments as a lucrative proposition. Many companies have already begun researching effects on paralysis, cancer, diabetes and many other common health problems. Their main problem comes in the form that current U.S. laws prohibit the use of any "new lines" of embryonic stem cells which creates an issue with gaining access here in America to embryos for

[17] Orentlicher (2012), p. 327.
[18] *Id.* 327–328.
[19] *See* United Nations General Assembly (1948).
[20] *See* International Covenant on Civil and Political Rights (1966).
[21] *See generally* Alston (1990), pp. 365–393.
[22] United Nations (2024).
[23] *See generally* Deaton (2011), pp. 1115–1117.
[24] Bodansky (1995), pp. 667–669.
[25] Zieff et al. (2020), p. 580.

studying.[26] These companies have had issues both due to federal laws and even state laws. Do to the legal issue in using and studying embryonic stem cells, pharmaceutical companies have had to mainly focus their studies on the potential for human use, by using animals with similar issues and diseases as humans.[27] Due to the impressive potential using all forms of stem cells, pharmaceutical companies and bio-tech companies have invested a lot of time and money into their research.[28] Studies have been undertaken to see how much money has been donated by pharmacy companies in an attempt to influence voters when stem cell issues have been on various ballots.[29] For example, in just the year 2006 it was found that pharmaceutical companies invested approximately $35 million dollars into a pro-stem cell Missouri election to sway voters to expand the legalities within their state on the use of stem cells.[30] This is not new when it comes to these companies, as the other studies have shown that these investments are commonplace and are on the rise when pharmaceutical companies see a chance to make money.[31] It has been shown that these lobbying expenditures have risen 70% over the last twenty years in just this one area.[32] One of the main benefits of these types of companies using their own funds to make these discoveries is that they need to ensure that their funds are well spent, so there is little waste, and in doing so, many of the problematic or harmful experiments are discovered and removed from further study.[33]

Another, often overlooked aspect of embryonic stem cells is the fact that because they can proliferate at a nearly unlimited rate, scientists will be able to convert them to needed tissue types for study.[34] To take this one step further, scientists can create forms of tissue and test new forms of medication to determine their efficacy and toxicity to the human body.[35] This of course would mark a tremendous benefit for pharmaceutical companies (as well as patients) where so much money has to be spent by the companies in order to test their drug in various stages to obtain U.S. Food and Drug Administration (the "FDA") approval.[36] It has been estimated that on average, pharmaceutical companies must pay in excess of $19 million dollars in order to obtain such approval.[37] Partially due to this high amount of cost, coupled

[26] California's Stem Cell Agency (2014).

[27] Kim et al. (2002), pp. 50–56.

[28] *See* McKernan et al. (2010), pp. 517–520 (stating: "The use of stem cells as research tools has expanded with most of the major pharmaceutical companies using embryonic stem cells (ESCs) or adult stem cells for internal drug discovery programs.").

[29] *See* Wouters (2020), pp. 1–10.

[30] *Id.* p. 5.

[31] Schpero et al. (2022), pp. 1–3.

[32] *Id.* p. 1.

[33] McKernan et al. (2010), pp. 517–520.

[34] Chen (2022), pp. 2–3.

[35] Nature Reports Stem Cells (2007); Doheny (2019).

[36] Johns Hopkins School of Public Health (2018); Wouters et al. (2020), pp. 844–853.

[37] Johns Hopkins School of Public Health (2018).

with that many medicines never even make it to testing, as well as other factors contributes to the bulk of the price that patients have to pay to buy their medicine.[38] Therefore, if scientists can make quicker determinations as to the potential benefits or problems of drugs that they are developing, less money would be lost to the companies, and the entire lengthy FDA approval process could be markedly sped up from the 12–15 year period to something much quicker.[39]

3.1.4 *Those with an Economic Interest in Embryonic Stem Cell Research*

Aside from the pharmaceutical, and biotech companies, there are many others looking to make money from the studying and use of stem cells. One main area of support has come from insurance companies. Insurance providers can be seen as an unlikely supporter of the expansion of stem cell laws, but the fact is that if stem cell transplantation is successful, then the medical costs will fall dramatically, and cause the insurance companies to not have to pay as much as they would if the stem cell procedures were not allowed.[40] This is not to say though that all insurance companies approve or authorize stem cell treatments.[41] As further explained below, insurance providers are usually unwilling to pay for certain stem cell treatments when they are in the research period, and those that are unproven.[42] Further, when the insurance company believes that a stem cell treatment may have success, yet the overall condition of the patient may not change, they are just as likely to deny coverage.[43] Many insurance companies have stated that they would be willing to pay a larger fee over a short amount of time, then to have to cover a person for their entire life. One has to realize that when people suffer from problems that can be helped through the use of stem cells, they often have medical issues that require medical intervention, such as surgeries, hospitalization, and other medical expenses.[44] These expenses tend to be in existence for the person for the remainder of their lives, and facts, an objective everyday observer will understand that furthering studies through expanded research can lead to a host of cures for some of the worst medical problems that are and will be in existence.[45] There are not many people who would turn down

---

[38] Kliff (2018).

[39] van Norman (2016), pp. 171, 178.

[40] Farnia et al. (2014), pp. 66–72.

[41] Hertneky (2018).

[42] *Id.*

[43] *Id.* (discussing a patient who is in need of cancer treatment and sought to benefit from a stem cell-based trial, the insurance company originally denied the request for paying for the trial due to the low chances of success and the fact that their coverage does not typically cover trials).

[44] *Id.*

[45] Fuchs and Segre (2012), pp. 143–155.

medical help when they greatly need it, and out of fear and hope, would likely support a change in American laws to allow for further research so that if they somehow ended up with some ailment, they can be easily cured. National, and international studies have already shown that there are a wide variety of problems that can be cured through the proper use of embryonic stem cells, and that in curing these problems, people will be able to live much longer lives. On top of curing problems, stem cells can help slow down aging, and improve the lives of people in all ages. Even certain religions allow for embryonic stem cell usage when their use would be both "compelling and unreachable by other means."[46]

Some others of these come in the form of stem cell banks which actually create stem cells using other stem cells and then sell them to research labs. Because there are only a small number of hESC lines available for study, these companies have not been able to compete with other larger companies who do the same thing. Another business that revolves around stem cell use are umbilical cord banks. Umbilical cords were found to contain a small amount of stem cells, and new parents have been given the choice to keep and store thee umbilical cords from their children.[47] Doing so can help with curing potential future diseases that the children get, and can even be used for family members.[48] Storing umbilical cords for this purpose is very lucrative, and if the laws were to change to allow for more studying to be allowed, labs and scientist alike would have to depend on some of these banks to store their excess stem cells, and those who purchase stem cells for medical use may find these banks useful as well.[49] As discussed directly below, pharmaceutical companies can generate a fortune for their investors.[50] The fact that stem cells can be changed into virtually any other cell has extreme implications for pharmaceutical companies in that they can choose to research stem cells for many different medical issues and ailments.[51] With the exact benefits unknown, much research is needed in this field, and when there are major successes in trials and experiments, and once these stem cell treatments are approved, the companies making the discoveries can turn a tremendous profit.[52]

In 2019 the company Verified Market Research published their Stem Cell Therapy Market 2020 which forecasted statistics and trends involving the global and United States stem cells markets and investments from 2020 through 2026.[53] It is important to note that Verified Market Research, which is the company that published this report is based on providing information so that the reader will be

---

[46] Abernethy (2011).

[47] *Id.*

[48] *Id.*

[49] Nguyen et al. (2015).

[50] *See generally* Hirsch (2008), pp. 607–636.

[51] Parson (2008), pp. 511–513.

[52] *Id.*

[53] *See generally* Verified Market Research (2019a).

able to focus on "progress and growth."[54] The Report found that as of 2019 the global stem cell market had a value of $117.66 million dollars, and that the global market was projected to hit $255.37 million dollars by 2027 using a formula that a Compound Annual Growth Rate ("CAGR") which shows that the market pertaining to all stem cells will be growing at a steady rate to reach a CAGR of 10.97%.[55] his report specifically states:

> The market for the Stem Cell Therapy industry is extremely competitive, with several major players and small-scale industries. Adoption of advanced technology and development in production are expected to play a vital role in the growth of the industry. The report also covers their mergers and acquisitions, collaborations, joint ventures, partnerships, product launches, and agreements undertaken in order to gain a substantial market size and a global position.[56]

This shows that based on the current state of research, and publications, there are very high hopes for treatments that can be derived from any and all types of stem cells, with the report even further delving into potential diseases that can be cured using stem cell-based treatments.[57] This report specifically does not mention embryonic stem cells by name, and the title of "other cell sources" is all this is used, whereas three separate forms of adult stem cells are explicitly mentioned, and this is likely due to the fact that embryos as a source for stem cells, is not as well studied.[58] One can easily determine that allowing and funding embryonic stem cell research in the United States would easily expand on the potential revenues that these bio-tech and pharmaceutical companies could earn.

For a "real world" example, when President Obama signed Executive Order 13505, reversing the bans implemented by President George W. Bush in 2001, created a wealth of new jobs surrounding embryonic stem cell research.[59] In the state of Michigan alone, when President Obama's Executive Order took effect, it was coupled with the State's pro-embryonic stem cell referendum from the previous year, and it was estimated that approximately 8000 new jobs would be created to benefit the state thanks to funding embryonic stem cell research.[60] This would have acted like a major boost for Michigan's economy because at the time the State had the highest unemployment rate in the United States.[61] Another example can be seen in Amazon's Jeff Bezos investing in an "anti-ageing" group called Altos-Labs which will be advised by one of the most instrumental embryonic stem cell researchers, Doctor Shinya Yamanaka.[62] The newly funded venture has been offering scientists

---

[54] *See* Verified Market Research (2019b).

[55] Verified Market Research (2019a).

[56] *Id.*

[57] *Id.*

[58] *See generally* Henon (2003), pp. 27–45.

[59] *See generally* White House Communications (2009).

[60] *Id.*

[61] *Id.*

[62] Fabiani (2021).

and professors extremely lucrative starting salaries which starts at one million dollars, plus equity.[63]

3.1.5 Those Influenced by News/Media

With the advent of the internet and the ease of access for many people, more and more people have become aware of various studies and advancements in medicine and technology.[64] This has been furthered through the use of social media which allows others to repost websites, news, and opinions instantly in real time.[65] One need only look to the very recent articles from February and March of 2019 relating to not only one, but two patients being cured of H.I.V.[66] The article cited in this paper has been reposted countless times, which provided information showing that a once though incurable disease has been cured thanks to stem cells.[67] The truth of the matter is that stem cells were not the only instrumental aspect of this cure, but the headlines convinced many that they were and that further studies are needed using stem cells to find other benefits.[68] In conducting my interviews, or just speaking with people about stem cells after these articles made the news, I have found many people seemingly more receptive to the use of all forms of stem cells, and more knowledgeable as to the potential benefits and uses. Further, a shocking amount of people were aware of the studies cited in this article that indicate that embryonic stem cells can be harvested without risk to the embryo and even more surprisingly, these people actually know and understand the method and actively compare the procedure to the same procedure that the Japanese scientists claimed to use.[69]

[63] *Id.*

[64] Brooks (2011).

[65] *Id.*

[66] Mandavilli (2019).

[67] *Id.*

[68] *Id.* (discussing that stem cells did not act alone in curing H.I.V. in either of the two cured patients. In fact, the patients first had to undergo a risky form of chemotherapy to kill all of their problematic cells, along with many casualties, made up of non-harmful cells, followed by a bone-marrow transplant with stem cells. The transplanted stem cells effectively replaced the killed problematic cells).

[69] Klimanskaya et al. (2006), pp. 481–485.

3.1.6 *Potential Patients Desiring Positive Law to Prevent Misuse in Obtaining Stem Cells*

An all too often overlooked issue relating to embryonic stem cells is how they are obtained. As explained above, women can choose to donate embryos is most countries for use in medical research. However, this has led to a form of human and organ trafficking by researchers forcing women to donate their embryos.[70] It has also led to other unethical practices such as paying women to "donate" their eggs which has been deemed as completely unethical and illegal by most nations' laws.[71] To better understand the issue and how problems like this can occur, one only need to look to the story of the once very well respected South Korean researcher and veterinarian Hwang Woo-Suk who claimed to have obtained many eggs for use in research.[72] At first, the media gave a lot of positive attention to the researcher, concentrating mainly on the potential cures that could be developed through his using his methods, but after publishing his article, other researchers found problems with his research, which led to his "unmasking" and finding that his claims were likely impossible.[73] For further explanation, it is helpful to look at the famous cloned sheep Dolly, and the process of which created her. In that study, the scientists started out with over 250 eggs, and by the end, only one was viable, which became Dolly.[74] Hwang Woo-Suk claimed to have started with approximately 200 eggs, and ended up with eleven successful stem cell lines.[75] As it turned out, Hwang Woo-Suk not only paid certain women for their eggs, he also coerced two of his own junior researchers to "donate" their eggs.[76] Even more troubling, the women who were paid for the eggs we not paid directly by Hwang Woo-Suk, instead, the researcher contacted various agencies in Malaysia, China, and other Asian countries.[77] The average price paid for the eggs was $1400, and there is no way to know how much the women were actually paid by the third party.[78] It was later found that the women who "donated" their eggs were exploited due to the fact that they were impoverished and taken advantage of.[79] To make matters worse, the methods used in extracting the eggs from these women were potentially deadly, and at the least very dangerous.[80]

[70] Widdows (2009), pp. 5–24.
[71] *Id.* p. 12.
[72] *Id.* p. 11.
[73] *Id.*
[74] *Id.*
[75] *Id.*
[76] *Id.* p. 13.
[77] *Id.*
[78] *Id.*
[79] *Id.* p. 14.
[80] van der Heyden et al. (2009), pp. 25–29.

Given the fact that money is paid for egg and embryo donations, it is very easy to see how people will be able to exploit the vulnerable for their own gains.

This horrible situation is not just relegated to Asia, as other countries that lack or have weak laws allows for the exploitation of vulnerable women.[81] In fact, the paper cited above sought to interview women whom participated in the donor program relating to Newcastle University's egg sharing for research scheme which partnered with various local fertility clinics to obtain embryos to determine if its practices could be deemed exploitive or ethical.[82] Even though it was determined that the school and fertility clinics in this program strictly adhered to England's laws and ethical guidelines, many felt that they were exploited or saw where exploitation could occur in their situation.[83] The critics who saw the potential for exploitation stated that there was no regard given towards the woman's emotional state, or her financial vulnerability, which could lead to women making decisions that they would later regret.[84] The researchers who wrote the cited article seem to have an understandable bias in favor of the program, but have still shared their negative findings which can best be summed up by this quote: "[the] inducement, exploiting the vulnerable, desperate and poor and compromising the patient's capacity to give her informed consent."[85]

3.1.7 Those Who Seek to Increase Scientific Enlightenment

Stem cells of all kinds can be used to test different drug interactions on specific cells, as well as show scientists the earliest stages of growth in organisms.[86] Studying stem cells has even allowed scientists to determine how certain diseases occur, and/or how they can be treated.[87] To view these different interactions, scientists generally coax stem cells into a desired cell type and can then watch how they grow and change, or, test the toxicity of medications and treatments on the cells themselves without having to test them on living human patients.[88] It is easy to see how using embryonic stem cells will provide insight into human development and growth, and at the same time can show scientists and researches how diseases are developed and how they change over time.[89]

---

[81] Haimes et al. (2012), pp. 1199–1214.

[82] *Id.* pp. 1200–1202.

[83] *Id.*

[84] *Id.* p. 1207.

[85] *Id.*

[86] *See generally* Mayo Clinic Staff (2019).

[87] *Id.*

[88] Liu et al. (2017), pp. 1528–1529.

[89] Betts (2010), pp. A432–A437.

3.1.8 *Religious Based Support for Embryonic Stem Cell Research and Use*

While many people know that there are religious groups that are very against embryonic, stem cell and indeed any fetal tissue research, there are other religions that have found themselves more comfortable with this topic, and have championed its cause.[90] It would seem that virtually all of Judaism is pro-embryonic stem cell research, so long as the research is being used "medical or therapeutic purposes."[91] Specific Jewish bodies from each denomination have even issued their positive opinions on this subject, such as The Religious Action Center for Reform Judaism who has stated that it would be both "immoral and unethical" to deny funding and experimentation with stem cells when someone's life could improve through their use.[92] The United Synagogue of Conservative Judaism issued a statement in support of embryonic stem cell research with their own time limit, saying: "Support of stem cell research evolves from the view in Jewish law that an embryo does not have 'full capacity or status' until it is 40 days old."[93] The Reconstructionist Rabbinical Association issued their approval in their "Resolution on Educational and Political Support for Stem Cell Research," which was approved in 2005 stated their support in four parts:

1) "The RRA endorses both adult and embryonic stem cell research as long as new embryos are not created specifically for the purpose of research;"
2) "In recognition of its many potentially life-saving applications, the RRA urges substantial federal funding for both kinds of research; not limited to the original cell lines approved by NIH;"
3) "We call upon RRA rabbis to educate Reconstructionist congregations about the issue, and to support the donation for medical research of embryos that are about to be discarded by fertility clinics and of aborted fetuses;" and
4) We will publicize our stance on this issue by release of this resolution to general and Jewish news media, our congregations, and to local clergy associations."[94]

Clearly, not only was this above stated resolution very pro-stem cells of all kinds, it actually calls for others to be educated on the subject.[95] Even one of the strictest forms of Judaism issued their support for embryonic stem cells in 2005 through the Union of Orthodox Jewish Congregations of America.[96] The Rabbinical Assembly has also exclaimed their support for these types of studies in their "Resolution in

[90] Pew Research Center (2008).

[91] *Id.*

[92] *Id.*

[93] Bennett (2007).

[94] Reconstructionist Rabbinical Association (2005).

[95] *Id.*

[96] Orthodox Union Advocacy Center (2005).

Support of Stem Cell Research and Education" which has two holdings, the first being "the Rabbinical Assembly publicly advocate for the use of human embryonic germ and stem cells for research in all appropriate ways and urge the United States government to again permit and properly fund, stem cell research;" and the second: the Rabbinical Assembly embark on an informational effort to educate our lay leaders and membership regarding the importance of this initiative, which can lead to healing and the mitzvah of *pika nefesh*."[97] The words of choice are very interesting in this latter resolution as the term "*pikuah nefesh*" is used which essentially translates to Jewish law.[98]

Much of the Islamic faith has also been said to be "very compatible with the modern sciences" which was part of a discussion surrounding stem cell laws, yet differs from Judaism as there are multiple areas of disagreement instead of just blanket support.[99] This has been found to stem from the fact that from the point of view of Islam, God has provided humans with many gifts, one of them being health which actually is considered to be one of the most important gifts from God, and as a result humans have what is considered to be a "healthy responsibility for one's own well-being."[100] In Islam, an embryo is not considered to be alive until the time of "ensoulment" which is said to be the time when the soul attaches to the embryo, thus giving it the status of a living being, and this time has been interpreted to be at around day 40 of the pregnancy, yet others believe it does not occur until later on.[101] Most Muslim scholars have based the time of ensoulment on particular verses in the Quran, as well as a Hadith, the Qur'anic verses state:

> And verily We created mankind from a quintessence (of clay). Then We placed him in a place of rest (the womb), firmly fixed (into the uterine lining). Then We made the sperm/egg into a clot of congealed blood. Then of that clot We made a (embryo) lump. Then We made out of that embryo/lump bones and clothed the bones with flesh. Then (finally) We developed out of it another creature (by breathing a human soul into it). So blessed be Allah, the most marvelous Creator[102]

While there is clearly no hard and fast 40-day rule stated in the above quoted text, the issue was further addressed in a Hadith, which does not carry as much weight as the Qur'an, but is extremely important as it considered to be the written sayings and traditions of Muslim Prophet Muhammad.[103] In a particular Hadith, it specifically states that: "Each one of you is constituted in the womb of the mother for forty days, and then he becomes a clot of thick blood for a similar period, and then a piece of flesh for a similar period..."[104] So as the fetus is viewed as without a soul, at least

[97] The Rabbinical Assembly (2005).

[98] *See* Lopatin (2019).

[99] Washington Times (2009).

[100] Ilkilic and Ertin (2010), pp. 151–152.

[101] *Id.* pp. 152–154.

[102] *See* The Qur'an (2019); Maller (2019).

[103] The British Library (2024).

[104] al-Bukhari, Hadith 3036; Ylenfest (2017).

until the 40-day mark, and with the inherent duty to protect one's health, it is very easy to see how the supposed promise of embryonic stem cells would be supported in the Islamic faith.[105]

3.1.9 Global Health Bodies and Non-Governmental Organizations

Due to the nature of embryonic stem cells, there are many groups and individuals who are arguing for the need to expand the study in this area for the benefit of all. A number of independent advocate groups have joined together to form an international coalition of supporters known as the International Consortium of Stem Cell Networks ("ICSCN").[106] Members of the ICSCN meet annually at forums to discuss the new scientific discoveries and theories, potential issues relating to ongoing trials, and the current state of embryonic stem cell laws throughout the world.[107] This is far from the only group of its kind, but all have been working to lobby various countries to alter their laws to benefit those in need.[108] These groups are working to establish large world wide databases so that all can learn and understand the abilities known to these special type of stem cells.[109] There are also a number of health organizations, committees, and councils that comprise of members from many countries who are specialists in their respective fields. These fields comprise of scientists, researchers, doctors, attorneys, ethicists, theologists, and policy makers. When individual countries are creating or amending their laws pertaining to health issues, including embryonic stem cells, the legislative branch of the country looks to reports, conclusions, and recommendations of these groups.

3.1.9.1 World Health Organization

Even leading international scientific bodies such as the World Health Organization (the "WHO") have published guidelines, opinions and bulletins that seem to see the potential benefits of embryonic stem cell therapies, yet caution that there is a desperate need for an international framework, especially when it comes to the ethical aspects.[110] The primary role of the WHO is to act as a coordinator of health within the United Nations system, and as a result, WHO has promulgated their "Guiding Principles on Transplantation" which has been used to influence

---

[105] Ilkilic and Ertin (2010), pp. 151–152.

[106] Chaddah et al. (2011), pp. 315–317.

[107] *Id.*

[108] *See generally* International Stem Cell Forum (2010a).

[109] International Stem Cell Forum (2010b).

[110] *See* Pirnay et al. (2010), pp. 870–872.

legislation as well as professional codes.[111] This set of guidelines was endorsed in 1991 by the General Assembly, and sought to "provide an orderly, ethical and acceptable framework for the acquisition and transplantation of human cells, tissues and organs for therapeutic purposes," and has been revised in 2010.[112] The guidelines stipulate that each jurisdiction will be charged with enacting legislation and implementing the guidelines.[113] Specifically, Guiding Principle 1 requires informed consent be given by the deceased donor, and it must not be found that the deceased would have objected to such a donation.[114] The second Guiding Principle adds that the doctor who was charged with declaring the person as deceased, should not have anything to do with the removal of the deceased's organs, and should not have anything to do with the medical care of the intended recipient of the deceased's organs.[115] Guiding Principle 3 states that the donated material must be used to the fullest extent possible, and has a portion that deals with live donors that does not apply to embryonic stem cells.[116] The fourth Guiding Principle states that measures should be undertaken to protect minors from improper harvesting of their organs and tissues, and only allows for narrow exceptions to this rule which would be determined by the nation's own laws.[117] The fifth Guiding Principle establishes that compensation should not be provided to donors and that any form of tissue should be banned from being sold, further, the Guiding Principle states that it would be permissible to repay the donor for their actual costs that can be verified, in making their donation.[118] The sixth Guiding Principle allows for advertisements that request donations of human tissues, and requires that the local jurisdictions create regulations to make a determination as to the legality of the advertisement.[119] Guiding Principle seven states that costs should not be covered for any form of human tissue obtained through "exploitation or coercion."[120] Guiding Principle eight states that those who engage in the transplantation procedures dealing with human tissues should not be eligible to charge a fee that "exceeds the justifiable fee for the services rendered."[121] The ninth Guiding Principle recommends that the allocation of human tissues "should be guided by clinical criteria and ethical norms," and not based on financial or similar reasoning. The tenth Guiding Principle requires that the procedures for transplantation must be maintained in the safest, most effective, means

[111] *See generally* Association for the Advancement of Blood & Biotherapies (2024).

[112] *See* World Health Organization (2010).

[113] *Id.*

[114] *Id.*

[115] *Id.*

[116] *Id.*

[117] *Id.*

[118] *Id.*

[119] *Id.*

[120] *Id.*

[121] *Id.*

possible, and kept up-to-date for the protection of the donors and the recipients.[122] This Guiding Principle further states that there should be some sort of database that can be used to report issues regarding transplantations, and that this should work on a national in global level.[123] The final Guiding Principle, which is number eleven states that at all levels of the donation and transplantation must be open to scrutiny while protecting the donor's anonymity.[124]

There are a number of Resolutions that the WHO has created pertaining to human tissues that apply to embryonic stem cells, such as WHA 40.13 and WHA 42.5 which both pertain to the commercial trade of organs, state that there is a need for global standards when dealing with situation that can easily be abused. WHA 57.18 also applies in that it sought to have the General-Director collect and review as much global data pertaining to human transplantations so as to be able to update the guidelines as new information becomes available.[125] As of now, the WHO has created Resolution WHA 63.22 which seeks to create a database to be used by its member states, where data can be collected and shared between them.[126] These resolutions, as well as others are adopted by the World Health Assembly which operates as the "decision-making body" for the WHO, and they are charged with creating and enforcing WHO's policies. Through World Health Assembly Resolution WHA 57.18, the WHO effectively urged the member states to establish criteria and provide oversight for the procurement, processing, distribution and transplantation of human cells and tissues.[127] Unfortunately, even member states are not forced to adopt the recommendations of WHO, but should take them very seriously, as they have been created by people who are in a position to know and understand the issues at hand, and tend to work in the targeted field.[128] During the Second Global Consultation on Regulatory Requirements for Human Cells and Tissues for Transplantation: Towards Global Harmonization through Graduated Standards, WHO further stated that it should be up to the more developed countries to aid those who are in need of development with information and technology to assist their people with cutting edge technology and procedures.[129] Creating access to those in need in undeveloped or underdeveloped countries could be life saving for so many people and further emphasizes the need for global cooperation.[130] To add to WHO

Most recently, WHO's Expert Advisory Committee on Developing Global Standards for Governance and Oversight of Human Genome Editing issued two companion reports in relation to genome editing to benefit the public's health.[131] In

[122] *Id.*

[123] *Id.*

[124] *Id.*

[125] *See generally* World Health Organization (2004).

[126] *See* World Health Organization (2010).

[127] World Health Organization (2004).

[128] *See generally* World Health Organization (2004).

[129] *See*: *Id.*

[130] *Id.*

[131] World Health Organization (2021).

discussing the potential benefits of editing genes, the WHO's Expert Advisory Committee recommended that:

> WHO should support members of the scientific community to develop an additional basic and preclinical research Registry for research using human genome editing technologies on (i) human embryos and (ii) germline cells or their progenitors when gametes derived from these will be used to create embryos that will not be used to establish a pregnancy[132]

The potential importance behind editing the genes of an embryo drives this recommendation as doing so can allow an embryo to not inherit something dangerous that the mother would otherwise pass to them, as well as other diseases or debilitating issues.[133] The above statement would seem to further indicate WHO's belief that embryonic stem cells should be further studied for the benefit of all, and in this particular recommendation specifically stated that there is a "need for equity of access to and benefit from human genome editing," as the overall goal is to promote global health and reduce "human suffering."[134] WHO has also published many bulletins pertaining to embryonic stem cell research, which too seeks to create a means to regulate the entire stem cell industry.[135]

3.1.9.2 The World Medical Association

The World Medical Association ("WMA"), a global association made up of physicians, "serve humanity by endeavoring to achieve the highest international standards in Medical Education, Medical Science, Medical Art and Medical Ethics, and Health Care for all people in the world," has issued a statement in regards to embryonic stem cell research, that tries to take into account all of the moral and ethical considerations.[136] In their October 2009 convention, the WMA adopted a recommendation that recognizes the issues that inherently have arisen from the use of embryonic stem cells, and essentially state that only when embryonic stem cells are the only option, then they should be utilized, that local and federal governments should enact legislation that does not prohibit embryonic stem cell research, so long as they conform with the moral and ethical obligations, and finally these moral and ethical issues should follow some form of international law, so as not to be applied differently, or arbitrarily.[137]

---

[132] World Health Organization Expert Advisory Committee on Developing Global Standards for Governance and Oversight of Human Genome Editing, p. 10 (2019) (explaining that currently there are no such requirements for the registries to include certain early data).

[133] *See generally* Kubikova et al. (2023), pp. 737–744; Baylis et al. (2020), pp. 365–377.

[134] *Id.* p. 5.

[135] *See* Lee et al. (2017), pp. 663–664.

[136] World Medical Association (2009).

[137] *See id.*

3.1.9.3 The International Society for Stem Cell Research

The International Society for Stem Cell Research ("ISSCR") is the single largest organization of stem cell researchers in the world, and according to their website, is made up of thousands of scientists, doctors, educators, ethicists and business leaders with the membership spread over 80 countries.[138] The ISSCR has also promulgated their own set of guidelines which are meant to be updated as new information allows.[139] These guidelines establish that scientists and researchers that engage in the scientific study of human embryonic stem cells have a responsibility to maintain rigorous standards to uphold moral and ethical obligations, and further that a uniform agreement or convention should be made on a global scale to establish what these standards are and should be.[140] Further, the ISSCR found that with our current limited knowledge in regards to reproductive cloning, scientists should be completely barred from any experiments attempting to create an embryo.[141] The ISSCR also has a strong belief that there should be some form of monitoring body, to ensure that the methods used, and that the studies themselves do not create and ethical implications.[142] In fact, the ISSCR recommends that monitoring bodies should be established from a local level, all the way to a global level, and should include a means to enforce such rules, and crack down on those in violation of said rules.[143] The ISSCR has also created an outline for how to attempt to ensure compliance with these standards.[144] These methods seek to have authors seeking publication to include a statement, assuring that they have followed the stated guidelines, or similar guidelines; applications for grants must meet stringent criteria, and prove that the proposed study conforms to the guidelines, and even creates a body to monitor compliance.[145] The ISSR also decided to change their guidelines in 2021 in accordance with current scientific understanding, which now removes their 14-day prohibition and instead decided to review each situation on an individual basis.[146] This decision was based partially on the fact that this area of science is evolving through new inventions and new techniques that could allow for greater discoveries if more time was given.[147] Yet, instead of creating a new cut-off, the ISSR has come up with a way to try to find a balance.[148] In this new set of guidelines, the ISSR "recommends that 'national academies of science, academic societies,

---

[138]The International Society for Stem Cell Research (2023).

[139]The International Society for Stem Cell Research (2021a).

[140]*Id.*

[141]*Id.* pp. 6–7.

[142]*Id.*

[143]*Id.* pp. 6, 11, 18, 29, 33–42.

[144]*Id.* pp. 19–20.

[145]*Id.*

[146]*See* The International Society for Stem Cell Research (2021b); Subbaraman (2021).

[147]*See* Lovell-Badge et al. (2021), *p. 1398.*

[148]Baylis (2021).

funders and regulators' engage the public in conversation about the scientific, societal and ethical issues associated with the 14-day limit, and whether this should be extended depending on the research objectives."[149]

3.1.9.4 The International Stem Cell Banking Initiative

The International Stem Cell Banking Initiative (the "ISCBI") is an international body made up of experts from 23 countries with the determination to spread scientific information and guide policy making.[150] One of the loftier goals of the ISCBI is standardize the language used and registration of stem cell lines which will help the scientific community by having biobanks work together to form a global network to make obtaining specific stem cells safer and easier and more efficient, while giving the researchers as much relevant information as possible on the cells and the cell lines.[151] The ISCBI holds workshops to bring the international stem cell community together to try to determine the most ethical, and safest ways for stem cell research and banking to proceed in light of the most recent developments to aid in their guidance of local and national legislation.[152] In 2009, the ISCBI published their Consensus Guidance for Banking and Supply of Human Embryonic Stem Cell Lines for Research Purposes, which effectively discussed how to reach and implement their goals in this field.[153]

3.1.9.5 The International Stem Cell Forum

The International Stem Cell Forum (the "ISCF") is an international body that is made up of 21 funders of embryonic stem cell research who seek to promote "international collaboration and funding support for stem cell research, with the overall aim of promoting global good practice and accelerating progress in this vitally important area of biomedical science."[154] Even with such progressive aims, the ISCF still agrees that banning human cloning, the use of embryonic stem cells should be kept to a minimum, and seeks "[i]international harmonisation of ethical and intellectual property right issues."[155] To accomplish these tasks, the ISCF has created a number of committees who has the purpose of making specific determinations in the areas of ethics, legislation, how to implement their goals.[156]

---

[149] *See Id;* The International Society for Stem Cell Research (2021b); Subbaraman (2021).

[150] *See* International Stem Cell Banking Initiative (2019).

[151] *See generally* Crook et al. (2010), pp. 169–172.

[152] Kim et al. (2019), pp. 1130–1135.

[153] *See generally* The International Stem Cell Banking Initiative (2009), pp. 301–314.

[154] *See* The International Stem Cell Forum (2010a).

[155] *Id.*

[156] The International Stem Cell Forum (2010b).

3.1.9.6 Médecins Sans Frontières/Doctors Without Borders

Doctors Without Borders (the "DWB") is an international humanitarian group that is wholly independent from any government, and mainly works in areas of conflict or disease.[157] This group has even refused funds from the United States and the European Union based on the core values of the DWB which are stated as: "Independence, medical ethics, neutrality and impartiality."[158] While there has been very little information concerning the DWB's stance on embryonic stem cell treatments, policies, or law, they have taken an active stance that can be described as pro-choice, or pro-abortions, when the mother's health is at risk or to prevent unwanted pregnancies or to prevent dangerous forms of abortions from being performed.[159] While this stance on abortions can be deemed as an indicator as to the DWB's view on an embryo's right to life, the DWB has routinely explained that their policy in regards to providing abortions is one meant to protect the life of the mother, and cites examples of mortality rates in countries that ban or overly control abortions.[160]

3.1.9.7 Human Rights Watch

One of the most important non-governmental organizations that works to protect and report abuses in every part of the world is known as Human Rights Watch (the "HRW").[161] The HRW is made up of experts from many varying areas such as lawyers, journalists and others, and they work to change government policy or seek to have governments actively enforce their laws, as well as targeting armed groups and even businesses that harm others.[162] It is also very important to note the extent of their impartiality extends to refusing any funds coming from governments.[163] The HRW is similar to DWB in that they have not taken an actual stance regarding embryonic stem cell research, yet have dealt with the rights of a woman to obtain an abortion.[164] In the HRW discussion involving abortions, they determined that when countries act in a manner to "unreasonably" restrict access to abortions, they are violating the human rights of the person seeking said abortion.[165] In fact, the most important rights are directly implicated and according the HRW include the:

---

[157] *See generally* Medecins Sans Frontieres/Doctors Without Borders, Principles (2023).

[158] de Deus and Reis (2017), p. 263.

[159] Schulte-Hillen et al. (2016), pp. 1–4.

[160] *Id.*

[161] Human Rights Watch (2024).

[162] *Id.*

[163] *Id.*

[164] Human Rights Watch (2017).

[165] *Id.*

> Right to life; rights to health and health care; right to information; rights to nondiscrimination and equality; right to be free from cruel, inhuman, or degrading treatment; right to privacy; right to decide the number and spacing of children; right to security of person; right to liberty; right to enjoy the benefits of scientific progress; right to freedom of conscience and religion.[166]

Even with only a cursory understanding of what embryonic stem cells implicate, it is easy for one to see that the right to life, health and health care, information, liberty, security of person and the right to enjoy the benefits of scientific progress is directly impacted by laws impacting a person's ability to understand, and decide for themselves if a treatment is appropriate for them.[167] HRW also looks to the ability of all to enjoy the same rights, and oftentimes when there are new treatments, the wealthy are the first to obtain them based on the treatments' rarity, or price, and the HRW seeks to protect those who are the most vulnerable and prevent this from happening.[168]

3.1.9.8 The American Association for the Advancement of Science

In its mission statement, the American Association for the Advancement of Science (the "AAAS") states that it were created to "promote and defend the integrity of science...; and promote the responsible use of science in public policy..."; among other goals.[169] In their publications, the AAAS has made recommendations for policy, and challenged current laws and executive decisions pertaining to embryonic stem cell research, including a rather scathing rebuke of former President Bush's veto of beneficial embryonic stem cell research laws which is discussed at length below.[170]

3.2 Against Embryonic Stem Cell Research

Many of the anti-embryonic stem cell groups argue that the destruction of embryos is both morally and ethically wrong, even when there can be great benefits for doing so. This argument still exists, and even still have sway on the laws, even though scientists now know that it is unnecessary to destroy a life in order to save others.[171] There are also group whose arguments stem more from both rational and irrational fear than any personal beliefs, yet are still necessary to overcome so that a successful

---

[166] *Id.*

[167] *Id.*

[168] *Id.*

[169] *See* The American Association of the Advancement of Sciences (2024).

[170] *See generally* Chapman et al. (1999); American Association for the Advancement of Science (2001, 2007).

[171] Lovgren (2006).

framework can be found and alleviate the fears. It is of the upmost importance to mention that those who are against embryonic research are not automatically evil people who are against helping others. In fact, most argue that there should be further protections for all people, including embryos.

3.2.1 *Religious Organizations*

Many people look to their religion for guidance on various topics and issues which also include some of the toughest questions that people will face in their lives.[172] It is also true that many people believe that religion teaches us morality, and ethics.[173] With this in mind, there is no doubt that a subject that necessarily implicate the questions of both when life begins; and after a determination of when life begins, whether it is ever appropriate to end life to benefit those already living. Knowing this, it is imperative to understand that those who have an ethical or moral objection stemming from religious beliefs do not object to this area of research for lack of care, or because they do not believe that science should cross bounds that could, in many people's minds, end life, the fact of the matter is, that these people are arguing on moral and ethical grounds that should be respected.

3.2.1.1 The Catholic Church

Possibly the most vigorous and fundamental opposition against the use of embryonic stem cells, fight against federal funding and loosening restrictions on research comes from religious organizations. It is based on their considered stance as to when life begins. According to the teachings of the Catholic Church, life begins at conception and human life is to be considered sacred.[174] A life is considered taken when an embryo is used for stem cell research because until recently, the only method known to obtain embryonic stem cells was through the donor embryo's destruction.[175] The *Catechism of the Catholic Church,* which was promulgated by Pope John Paul II, states the following:

> Human life must be respected and protected absolutely from the moment of conception. From the first moment of his existence, a human being must be recognized as having the rights of a person—among which is the inviolable right of every innocent being to life.[176]

---

[172] *See* Pargament and Park (2019).

[173] *See generally* Mannion (2023).

[174] *See* Butler (2008).

[175] Klimanskaya et al. (2006), pp. 481–485; Catholic Church (2000).

[176] *See generally* Catholic Church (2000), pp. 547–548.

This above statement found in the *Catechism of the Catholic Church* is very telling further explains that human life is to be protected from the "moment of conception".[177] It is taught that once conception occurs, that all of the rights of a person attach to the unborn.[178] The *Catechism of the Catholic Church* further discusses the fact that the Church has always been against abortions and that this particular view is something that has not changed and cannot be changed.[179] Yet, the strongest statement against abortion comes in section 2273 where it states that people who formally "cooperate" in an abortion are guilty of committing a "grave offense" where the Church will seek to penalize the perpetrators with excommunication.[180] The reason given for such a strong punishment is two-fold, first, the Church views abortion as a crime, in this situation, the taking of an innocent life of another; and second "the irreparable harm done."[181] This particular harm is referring to the taking of the life of the embryo as well as harming the lives of the biological parents of the now destroyed embryo, and even discusses the harm to society as a whole from this taking.[182] The next two statements on abortion in the *Catechism of the Catholic Church* are also very telling as to the position and reasoning for taking said position by the Catholic Church, it states that unborn must be treated with the upmost care as they are still considered "people" even though they have not yet been birthed, that the only procedures that should be carried out on an embryo are the ones that seek to medically help the embryo and nothing that will allow for any "unnecessary" risks to it.[183] Finally, the Catholic Church issues their stance on medical research that directly impacts an embryo by stating: "It is immoral to produce human embryos intended for exploitation as disposable biological material."[184]

One should also consider that the idea of dignity attaching to all life is intrinsic in the Catholic Churches teachings, and the *Catechism of the Catholic Church* explains that the dignity that is inherent in all people comes from that all people are viewed as being "in the image and likeness of God."[185] The fact that all people are born with an inherent dignity that must be respected is not just true for those viewed as clean of sin, and in fact even attaches to those who have been found guilty of the most heinous acts who have received the death penalty.[186] In fact, this was enshrined as part of the *Catechism of the Catholic Church* through Pope Francis's revisions that capital punishment should never be used as a person's dignity itself is not lost during or even after the commission of a serious crime, and that "the death penalty is

---

[177] *Id.*

[178] *Id.* p. 547.

[179] *Id.* p. 548.

[180] *Id.* p. 548.

[181] *Id.*

[182] *Id.*

[183] *Id.* p. 549.

[184] *Id.*

[185] *Id.* p. 424.

[186] Pentin (2018).

inadmissible because it is an attack on the inviolability and dignity of the person, and she works with determination for its abolition worldwide."[187]

Destroying the embryo was never meant to be the goal of the research, but scientists were required to remove a mass of cells from an embryo which was actually only a small amount, but because the human embryo is only made up of roughly eight to ten cells at this time, removing any would more than likely lead to the destruction of the embryo.[188] The destruction of the embryo would appear to violate this religious tenet.[189] Even if the embryonic stem cells could safely be removed without any harm coming to the embryo, it is highly likely that National Catholic Bioethics Center (the "NCBC"), who has the stated mission of upholding the "dignity of the human person in health care and biomedical research" which includes the dignity of the unborn would find that any such scientific manipulation would violate the unborn's dignity. One can draw this inference from the NCBC based on their writings and overall stance, in fact, in 2022 when *Roe v. Wade* was overturned, the NCBC summed up their belief on the unborn... "from the very moment of conception, an irreplaceable and unique human being exists, who not only deserves our protection, but justly has a right to such protection."[190]

Another issue that religious organizations have with this form of research, comes from the potential of cloning humans. For instance, the Catholic Church is very much opposed to the prospect of human cloning, and has stated that "...[C]loning is an affront to the dignity of the individuals who result from cloning. Cloned individuals are deprived of the dignity of sonship, the dignity of being the fruit of the self-giving of a father and a mother."[191] There have also been views expressed by some Buddhist scholars who believe that if the means of procuring stem cells involves the destruction of life, such as embryos, then the religion does not believe that it should be allowed.[192] Interestingly, *Donum Vitae* no. 4 actually allows for some embryonic research, but the embryo must be deemed deceased, and informed consent must still be given.[193] This document which was created in 1987 by the Congregation for the Doctrine of the Faith, is also known as the "Instruction on Respect for Human Life in Its Origin and on the Dignity of Procreation" and also specifies that no harm should be allowed to befall an embryo and that experiments should only be permitted if the embryos are treated as a full human being with respect and dignity, and that there should be no experimentation of any of the embryos for purposes other than those to help others.[194] The document even went on to make it clear that experimentation that

[187] Pope Francis (2017).

[188] Slukvin (2010), pp. 1621–1623.

[189] Amaris and Sivaraman (2019), pp. 419–420.

[190] National Catholic Bioethics Center (2022).

[191] Catholics United for the Faith (2005).

[192] Keown (2016) (explaining that using an embryo to help those who are already born, should be seen as helpful to mankind, especially those in need of medical help).

[193] *See generally* Card Ratzinger and Bovone (1987).

[194] *Id.* p. 29.

is used for the purpose of benefitting science should not be deemed sufficient for research purposes, even if the embryo in question is no longer viable.[195] Because this document was created before many advances in science were made, including in advances in *IVF* treatments which the original 1987 document was against, a follow up was promulgated in 2008 which is known as the *Instruction on Dignitas Personae*.[196] *Instruction on Dignitas Personae* can best be summed up as a set of instructions that were drafted by the Congregation for the Doctrine of Faith which provide the directives regarding embryonic technologies and research which is believed by this group as potentially unethical.[197] This document was created in an attempt to combine "...the formation of conscience" (n. 10) and to encourage biomedical research respectful of the dignity of every human being and of procreation."[198] The opening of this document immediately explains the position of its drafters in regards to an embryo and how they are to be viewed, by stating that all human life, beginning with conception should be afforded the same rights as a living person and requires that this belief should be the cornerstone of every moral and ethical question specifically dealing with when should human dignity attach.[199] Another level was also added in to this framework which was created to balance the "human and the divine" and argues that because God created life, humans are automatically imbued with "unassailable value," which should not be tampered with.[200] The next issue that was discussed was that of *IVF* treatments which the original 1987 text was against, but now states that due to new scientific developments, *IVF* should now be permitted, so long as: "[M]edical intervention respects the dignity of persons when it seeks to assist the conjugal act either in order to facilitate its performance or in order to enable it to achieve its objective once it has been normally performed."[201] It is important to note though that this message is directed to married couples, and still would prohibit the removal of "illicit obstacles to natural fertilization."[202] Another interesting portion of the document states that *IVF* creates so many surplus embryos that approximately 80% end up being destroyed when they eventually go unused.[203] These two abovementioned documents are very important to the Church's overall stance, as they were created to promulgate and defend Catholic Doctrine.[204] It must also be pointed out that Pope Francis stated: "Human life is sacred and inviolable and the use of prenatal diagnosis for selective purposes should be discouraged with strength...." and asked: "Is it legitimate to take out a

---

[195] *id.*

[196] Congregation for the Doctrine of Faith (2010), pp. 51–59; Guevin (2010).

[197] *Id.*

[198] Card Ratzinger and Bovone (1987).

[199] *Id.*

[200] *Id.* p. 5.

[201] *Id.* p. 7.

[202] *Id.*

[203] *Id.* p. 8.

[204] *See generally* Card Ratzinger and Bovone (1987).

human life to solve a problem?... "Is it permissible to contract a hitman to solve a problem?"[205] This highlights the position of that all life is sacred and that one should not take precedent over another, and following the abovementioned dogma, this attaches at the time of conception.

With the unprecedented outbreak of Coronavirus 19 ("Covid-19") and the rush to find treatments and, or cures, fetal tissue and cells have been viewed as a potential source for relief.[206] As a result of studies showing promise in the use of fetal cells in combating Covid-19 there have been many reports of vaccines being developed and used on people which utilized embryonic cells which was met with both hope by some and dismay by others such as the Church.[207] However, due to the urgent need to help as much of the global population as possible in fighting this virus, the Congregation for the Doctrine of the Faith which is the office of the Vatican charged with "promoting and defending church morals and traditions...," specifically stated that: "when ethically irreproachable Covid-19 vaccines are not available ... it is morally acceptable to receive Covid-19 vaccines that have used cell lines from aborted fetuses in their research and production process."[208] This statement is a huge change in stance from the Vatican, and was further bolstered by Pope Francis's approval of the above quoted text by the Congregation for the Doctrine of the Faith.[209] For further understanding of how this decision was made, at least according to the Archdiocese of New Orleans, who explained that in vaccines where scientists did not use cells directly derived from fetal cells in their manufacturing process, it is morally acceptable, as opposed to when it "is morally compromised as it uses the abortion-derived cell line in development and production of the vaccine as well as the testing."[210]

Other arguments against the study of embryonic stem cells is due to the fear of humans "playing God," and this view had widespread support in early evolutionary biology.[211] A hurdle for this area of science stems from the fact that many religious groups believe that humans are a "perfect divine creation" which used to find support from scientists in the earliest days of evolutionary science due to the fact that these scientists misinterpreted evolutionary adaptation with evolutionary perfection.[212] Scientists believed that humans evolved directly to express the most needed traits and characteristics, yet soon after, science recognized the fact that evolution is successful, where those who do not do not survive to spread their genes as successfully as those with the more useful, beneficial traits.[213] This idea has further been

---

[205] Horowitz (2019).

[206] *See generally* Afarid and Sanie-Jahromi (2021), pp. 1–12.

[207] *See:* Rev. Fr. Baker (2021); Rev. Fr. Baker (2020).

[208] Wamsley (2020).

[209] *See*: Vatican News (2020).

[210] *See* Guterl (2021).

[211] Lachmann (2001), pp. 165–168.

[212] *Id.*

[213] *Id.*

perpetuated by some of the world's leaders after evolution was first recognized, yet it stems from a complete misunderstanding.[214] For example, both Karl Marx, and Adolf Hitler believed in this theory and the false notion that there is an "optimal" phenotype, and incorporated it into their future works.[215]

Others who fear humans "playing God" argue that embryos are created just to be destroyed for embryonic stem cells to be obtained. While this argument raises very troubling moral and ethical issues, it fails to realize that embryos do not have to be created and destroyed just to obtain the cells.[216] Even if this were true, there should be some form of balance in protecting those that are already alive and in need of medical help, and those which are only arguably alive. Finally, there is also the fear that using embryonic stem cells to "play God" will lead to people misusing the technology to create so called "designer babies" which would have both the physical and medical attributes that the parents choose for the unborn child.[217]

It should also be noted that the Catholic Church has universally condemned the taking of any human life, even those that were ordered to be put to death under a death penalty law.[218] On August 2, 2018 Pope Francis issued a change to the Catechism of the Roman Catholic Church, which stated that in the past, the Church believed in the use of the death penalty when it was "properly" adjudicated, but now sees that in today's times, that there is a person still is entitled to their dignity even after the commission of a serious crime, and further that there are other options a State has in punishing a convicted person.[219] Pope Francis took this one step further in October of 2020 when he sent an Encyclical which is a letter to all bishops in the Catholic Church titled "*Fratelli Tutti*", "*All Brothers*" that stated the Church's stance that they are for the complete abolishment of the death penalty, even for the most serious cases.[220] Specifically, the letter clearly took a stance, stating: "Today we state clearly that 'the death penalty is inadmissible' and the Church is firmly committed to calling for its abolition worldwide."[221] Even though this issue does not speak to embryonic dignity, it does show very well that the Catholic Church is pro-human life, and seeks to ensure its sanctity under all circumstances.

3.2.1.2 The Southern Baptist Convention

Another religious group that has voiced their opposition to embryonic stem cell research is the Southern Baptist Convention. At a 1999 meeting of the Southern

[214] *Id.*

[215] *Id.*

[216] Dittrich et al. (2015), p. 1241.

[217] Lachmann (2001), p. 167.

[218] *See* Russell (2020).

[219] *Id.*

[220] *Id.*

[221] *Id.*

Baptist Convention, the group passed their "Resolution On Human Embryonic And Stem Cell Research" which states this religious denomination stands in the strongest opposition to federal funding of embryonic stem cells, as it would destroy the embryos to obtain the desired cells.[222] This stance is explained that no human life, including that of the unborn should be destroyed, and the Southern Baptist Convention seeks to have even laboratories that are privately funded to refrain from any research that ends with the destruction of life.[223]

Throughout this document there are areas that reference the future of science, and that during the time of this document's drafting, the drafters believed that adult stem cells could be likely be used in the near future.[224] This Resolution clearly specifies this groups reason for objecting to this area of science which include what they view as the "destruction of embryos," and "[t]he use of human embryos in research would likely lead to an increase in the number of abortions and create a market for aborted embryos and other fetal tissues" including their desire for the federal government to abstain from providing funding for such research.[225]

3.2.1.3 The National Association of Evangelicals

Another religious organization of staunch objectors is the National Association of Evangelicals.[226] In 2005 this group passed their Bioethics & Stem Cell Research Resolution which specifically pertains to regenerative medicine using embryos, and stated that it is their belief that:

> Embryos constitute human life. . .[r]espect for human dignity is paramount in the development of biotechnologies. . .[c]loning human embryos, whether for research or reproductive purposes, must be prohibited. . .[p]atent law must not allow for human embryos, genes, cells, and other tissues to become commodities. . .[f]undamental changes in human physiological nature using biotechnology, genetics, nanotechnology, artificial intelligence, and other means must be prohibited. . .[g]overnment funding for research on the ethical, legal, and social issues raised by these new biotechnology developments is essential and must include vigorous oversight and dissemination.[227]

This resolution speaks to most areas of research that can be done using embryos which is considered morally or ethically wrong by other religious groups, but goes a step further and speaks to an issue of a privacy right in our own genetic make-up, and as a privacy issue, it cannot be used as a "basis for discrimination."[228]

---

[222] The Southern Baptist Convention (1999).

[223] *Id.*

[224] *Id.* (explaining that there is a hope that the potential of adult stem cells will eventually overtake the potential that is believed to be found in embryonic stem cells).

[225] *See id.*

[226] *See generally* National Association of Evangelicals (2005).

[227] *Id.*

[228] *Id.*

3.2.1.4 Lutheran Church-Missouri Synod

The Lutheran Church-Missouri Synod are also major opponents of relaxing or easing laws pertaining to embryonic stem cell research but also promote adult stem cell research and therapies like many other religious groups that are too opposed to embryonic stem cell studies.[229] In the Resolution passed by this organization in 2007, the Church calls on its members to "…speak up for those who cannot speak for themselves."[230] The Resolution also explains the reasoning behind such strong opposition which is that their belief is that life begins at conception and as such, no embryo should be harmed or destroyed as it already has reached the status of a "human life."[231]

3.2.2 Political Conservatives in the United States

Many political conservatives have very similar argument as the religious groups and organizations who also oppose stem cell research and funding in all forms. This can be seen through the laws that were created while conservatives held a majority in Congress. In fact, most if not all of the current, problematic stem cell laws and regulations were proposed and accepted during this time. Another clear example comes in the form of former President George W. Bush's (a political conservative) vetoing of a law that was proposed that would expand federal funding, and reduce limits on stem cell research.[232] It has largely been noted that President Bush won his elections with the assistance and votes of Evangelicals and other religious organizations with similar beliefs, and in doing so, President Bush seemed to have granted these people a great deal of influence over policy and laws that were brought before him.[233] It is also known that President Bush himself was brought up as a "mainstream Methodist" and around his 40th birthday was born again as an Evangelical Christian, thus further cementing his reasoning behind the veto regarding embryonic stem cell research in 2006.[234] Interestingly, this veto was President Bush's first in his five years in office.[235] In July of 2016, the Republican party issued a platform on their principles and policies which specifically called for a total ban of embryonic stem cell research, and would make it a crime to transfer in anyway any fetal tissue

---

[229] *See* The Lutheran Church (2007).

[230] *Id.*

[231] *See generally id.*

[232] Cooper (2008), pp. 153–157.

[233] *Id.* p. 155.

[234] *Id.*

[235] Babington (2006).

obtained from elective abortions.[236] Vice President Mike Pence has vocally opposed the study of embryonic stem cells, and wrongfully believes that adult stem cells are so similar that they make studying embryonic stem cells unnecessary.[237]

3.2.3 People Fearful of Future Experiments Using Stem Cells

A vocal group of people are fearful of what expanding the research into embryonic stem cells will lead to. Many fear that scientists and the medical community can perform practically sci-fi experiments that will lead to future problems. One of the major concerns tends to be in the form of the possibility for human cloning.[238] Another, and much more fictional problem comes in the form of fear of cross-species hybridization, otherwise known as creating a "chimera." A chimera is can best be understood as a single organism that is made up of cells from two or more organisms, and maintains the two separate unique DNA sequences.[239] Even though the term chimera is more often than not used to describe cross-specifies situations, the term can also apply to situations that occur naturally.[240] In fact, the creation of a chimera happens at times when a woman is pregnant with twins, and one twin ends up passing away and being absorbed by the surviving twin.[241] Another example occurs when a patient receives a bone marrow transplant because the recipient already has their own DNA, and when the donor's bone marrow is transplanted, it destroys the recipient's bone marrow to replace it, and then creates blood cells, as bone marrow typically does, and these blood cells contain stem cells which themselves contain the donor's DNA.[242] The third most common natural form of chimeraism occurs when a woman is pregnant, and a small number the fetus's cells get distributed through the mother's blood, mixing with her blood, and forming what is known as "microchimerism."[243] It is important to note that scientists have found that this form of microchimerism occurs in most pregnancies.[244]

It is important to mention though that this fear comes from real experiments, but not on the level of what is feared. Those who have heard of scientists implanting stem cells into different animals, such as sheep, do not realize that it was not to create a new hybrid species, but was done to see the effects of stem cells on problems and

[236] Servick (2016) (discussing the 2016 Republican Platform specifically desires to "make it a crime to acquire, transfer, or sell fetal tissues from elective abortions for research.").

[237] *Id.*

[238] de Magalhães (2012).

[239] *See* Rettner (2016).

[240] *Id.*

[241] *See* Martin (2007), pp. 205–206.

[242] Yam et al. (1987), pp. 399–407.

[243] Boddy et al. (2015), pp. 1106–1118.

[244] *Id.* pp. 1114–1115.

afflictions of animals.[245] There has been a lot of controversy as of late due to new articles discussing the cross hybridization of species' cells for scientific purposes.[246] The experiments sound like something out of a bad science fiction movie. In fact, the headlines specifically mention that there are human cells used in another species, but many people take the headline at face value and fail to understand the potential benefits offered by these experiments.[247] Scientists have often tested potential treatments on animals which use human stem cells to determine their potential effectiveness.[248]

One need only look to a study conducted in China which made global headlines with eye catching titles such as: "Report: Spanish Biologist Working on 'Human-Monkey Hybrids' in China;" "Wait, What? The First Human-Monkey Hybrid Embryo Was Just Created in China;" and "First Human–Monkey Chimeras Developed in China."[249] Based on these headlines, it would make sense for a person who only reads the headlines to believe that a monster creature is being created, yet the end goal of the scientists involved was to create a means to grow human organs, inside the body of another primate, so that in the future, these organs could be harvested and used in a transplantation procedure to help a human being with a problematic organ.[250] However, even if one can find that the scientists were experimenting on such creations purely to benefit humanity, one has to review the opinions of other leading scientists and researchers, many of whom actually disagreed with the core concept of this experiment.[251] Interestingly, those who were against this study do not argue against creating creatures that would otherwise be impossible in nature, but argue that a primate is the wrong organism to use to grow human organs due to the fact that primates are slow growers, and growing human organs would take too long to fully develop.[252] Specifically, veterinarian and researcher, Dr. Pablo Ross, who actually worked on pig-human chimeras in a study stated that it is his belief that the scientists working on this project were actually more interested in answering "questions of evolutionary distance and interspecies barriers..."[253] This can also be seen in the very recent experiment where scientists from Germany and Japan worked together to inject a human gene into seven marmoset fetuses (a species of monkey with a very small brain).[254] This experiment was chosen in order to help determine what role the particular human gene played in

---

[245] Associated Press (2005) (explaining how animal DNA can be mixed with human DNA in stem cell experiments, and highlights and causes more fears as opposed to giving real scientific facts).

[246] Bob Yirka (2018).

[247] *See* Robert (2006), pp. 838–845.

[248] Kriks et al. (2011), pp. 547–551.

[249] *See* Haywayd (2019); Fan (2019); Lanese (2019).

[250] *See* Lanese (2019).

[251] *Id.*

[252] *See id.*

[253] Regalado (2019).

[254] Heide et al. (2020), pp. 546–550.

expanding our brain size and capacity, which was seemingly proven by this study.[255] Scientists essentially prevented the monkeys from being born alive and performed autopsies which showed that the gene did in fact alter the marmoset brains by altering the normal, smooth brain of the monkey into something with more folds and wrinkles, more akin to a human brain, among other discoveries.[256]

When I was first made aware of this argument, I was in disbelief that this could be a fear, yet as discussed at length below, many countries have instituted laws that make such experimental procedures illegal, and punishable. Despite its seemingly sci-fi sounding name, and attributes, this is truly something that is valid for people to fear. It is just important to remember that this one small aspect of what can be, can and should remain illegal in all frameworks except for rare cases where there can be a true medical and/or scientific benefit, otherwise it is just playing God because it is possible to.

As discussed above, there is the practice of xenotransplantation which transplants some living tissue or organ from a non-human into a human.[257] While there are many potential benefits as discussed above, many are fearful of what this area of science can end up creating. There is actually a vast historical background behind this fear which can be seen in many cultures and societies for many centuries, especially in their various mythologies and legends.[258] For example, Dr. Keith Reemtsma, the first person to successfully transplant a kidney from a chimpanzee to a human believes that the legend of Icarus in Greek mythology was the first recorded physical cross of animal attributes to humans in the form of bird wings attached to Icarus so that he could fly.[259] Many other stories come from Hindu mythology, ancient Greek mythology, as well as many others, which too crosses humans with animal attributes, and there are indeed countless more.[260] More recently, a very scary, "mad scientist" type story went public in the form of television specials and documentaries pertaining to Soviet leader Joseph Stalin's attempts to create a half-ape, half-man army, an actual attempt to cross the two species under Soviet biologist and zoologist, Ilya Ivanovich Ivanov.[261] Thankfully, nothing viable came from this, and even though no stem cells were used (stem cells were not even discovered until 1981) but it is this exact form of science fiction that permeates into the public conscious and brings about the fearful response.[262] It is also very important to note that the advocation of using stem cells in this field is only meant to benefit society in helping to find cures and treatments, and not for use in any other field. In sticking to their fears, many people fail to separate the science fiction, or

---

[255] *Id.* pp. 546–547.

[256] *Id.*

[257] *See generally* Schicktanz (2012), pp. 565–574.

[258] Cooper (2012), pp. 49–57.

[259] *Id.* pp. 49–50.

[260] *Id.* p. 49.

[261] Pain (2008); Pacunas and Bosch (2016); and Science Channel (2011).

[262] *Id.*

insane attempts, with the very real potential benefits of such science. However, while there are many potential benefits using this process, there are so many potential risks in the process of xenotransplantation, therefore it is very highly regulated by the FDA due to risks of infections or diseases spreading from animal tissue to humans that humans would otherwise not be susceptible to and then through the further transmission of human to human.[263] Due to the severity of issues that can arise from this process, the U.S. Public Health Service has published guidelines on infectious disease issues in xenotransplantation.[264] While this too seems like something out of a movie, the medical benefits can be tremendous, and the results have been promising, especially when embryonic stem cells are used in the co-culturing process.[265] It has been found that the benefits derived from using embryonic stem cells leads to a better chance of the host (person receiving the transplant) not rejecting the graft and avoiding any host vs. graft reactions which is believed to be due to the "...[L]ow expression of immunogenic antigens by stem cells."[266] Essentially, because stem cells are practically blank, there is no real coding etched into the cells from the original donor.[267]

Again, the primary problem in overcoming an argument such as these is that it is initially based in misunderstanding and fear. There are also various ethical and moral issues pertaining to this area of science, which implicate religion and various "rights groups" organizations. As far as those that are fearful of what experiments can be conducted if regulations on stem cells are reduced, and more studies are allowed, it is important to understand that in the United States there are already various laws in place to protect against horrific and unnatural experiments. Yet as discussed above, once someone has a fear that they believe is rational, it is very hard to overcome it.[268] However, with enough time, and enough proof, it is likely that the benefits will be recognized and the science will prevail, especially once the benefits become more widely known.

---

[263] *Id.* (crafting of the following guidelines on infectious disease transmission after xenotransplantation, the FDA recommend that: (1) You should inform recipients of xenotransplantation products that they and their intimate contacts should defer from donation of blood and other tissues. (2) You should inform patients that they have been treated with a xenotransplantation product and of the risks involved. (3) You should archive patient samples, such as blood, to allow future monitoring for potential infections. (4) You should follow patients for their lifetimes and counsel them to be alert to any unusual symptoms. (5) You should archive samples of the xenotransplantation product. In this case, the nonhuman animal cells used for the co-culture process should be archived).

[264] *Id.*

[265] Behfar et al. (2004), pp. 471–479.

[266] *Id.* p. 478.

[267] *Id.*

[268] *See generally* Shin (2009), pp. 169–191.

3.2.4 Those Who Fear Overpopulation on Earth

There are small groups who believe that due to the limited amounts of natural resources, expanding American lives will inherently cause overpopulation concerns. These concerns lie in the fact that the use of stem cells will likely lead to a major life expectancy expansion.[269] There are differing views in this regard, but often the arguments center around food supplies, fuel reserves, and even the limited amount of land.[270] This is a very small group, but tend to be one of the most vocal on the internet.

3.2.5 Insurance Companies with Patients Seeking to Benefit from Trials with a Zero to Low Chance of Success

As discussed above, there are times when insurance companies tend to deny coverage for patients for stem cell based treatments when the chances of survival from the overall affliction are extremely low.[271] There is no set percentage of success that is given, yet when a man suffering from cancer had only a 50% chance of survival without stem cell treatment, had his chance of survival bumped up to 80% through his enrollment in the stem cell trial, the insurance company decided to allow the patient to begin treatment, yet only authorized one round of the trial.[272] This was largely based on the trials extreme expense and unless the trial proved successful in part, the insurance company believed they would be wasting vast amount of resources on a losing fight.[273] This situation is all too common do to many of the misconceptions relating to stem cells, and people who are seeking a last chance cure. This issue is especially true in other countries around the world that have universal healthcare and as discussed in detail below, many of the countries' courts have had to deal with these exact situations.

3.2.6 Those Against Stem Cell Treatments Based on Negative Representation in the Media

Just as there are many people who watch and understand that the near limitless potential of embryonic stem cells based on what they have seen, read about or have

[269] *Id.* pp. 172–179.

[270] Fight Aging (2006).

[271] Butcher (2018).

[272] *Id.*

[273] *Id.*

been told about, there are certain people who only see the opposite. Factually, these arguments are typically valid, just not in the way these people know. Due to the constrained laws and studies that are conducted in the United States, many unscrupulous people have started businesses and so called "stem cell clinics" which only use adult stem cells, yet due to the fact that these too are unproven, and that the problem "clinics" are created and ran by those strictly seeking profit, these cannot be considered on the same level as what the laws should allow. Just as there are these unscrupulous clinics, there are also valid ones who actually seek to help those in need, yet they too seem to suffer the same fate because they are all lumped in the publics' mind together.[274] Even scarier, many of these "clinics" are actually unlicensed to begin with, or employ "doctors" who are also unlicensed with little to no medical training.[275]

It is very important to understand this argument in its proper context because it may be the hardest to overcome in the publics' mind. First, due to the overly restrictive laws in the United States, many Americans have sought to obtain medical benefits using stem cells abroad. This has actually led to a thriving industry known as "stem cell tourism."[276] Oftentimes, people seeking these benefits like the family discussed above, conduct research into where to go, by looking at which locations have had published successes, and what these locations supposedly offer, yet because these places create hope for those in need, the research typically stops at this point.[277] There are even step-by-step guides to help people understand the so-called truth as to the potential benefits and problems in visiting these clinics.[278] Second, it is important to understand that when these patients seek help from these clinics, they are usually in a stage where there are no other viable options here in the United States, so to many, their only hope remains in these unproven, and potentially unsafe locations.[279] There are also those in the same position yet due to their health circumstances, take the approach of "what do I have to lose" and attempt to gain treatment knowing that it too may not work.[280] When people do not conduct sufficient research they typically fly to other countries with extremely loose, or non-existent stem cell laws, than when tragedy strikes and the treatments cause more harm than help, the news picks up the stories and further publishes them.[281]

As for those who believes that these stem cell clinics offer no help and only harm, they have a sufficient basis based on their knowledge, yet oftentimes fail to realize that these clinics that cause the problems operate in countries that have little to no

[274] *See* Gorn (2018); Cara (2018); Zhang (2017)

[275] Neal (2018).

[276] Kolata (2016).

[277] See *id.* (discussing the manner in which people seek help through the use of stem cell therapies).

[278] *See* Putton (2023); Murdoch and Scott (2010), pp. 16–23.

[279] *Id.* pp. 20–22.

[280] Rachul (2011), p. 8.

[281] Neal (2018).

regulations for stem cells.[282] This problem exists in countries such as Mexico, China, Costa Rica, India, and many others, and occurs because these countries host for-profit businesses, and either have laws that they do not enforce, or fail to enact laws to prevent these problematic clinics from operating, and they offer their supposed "expertise" for a much lower price, than in locations where there are laws in place to protect people.[283] Another issue that fails to be reported on in these instances, is that there are also patients that attempt these treatments, knowing full well that they may not succeed, or could cause further harm, yet due to their various conditions are willing to take the risk.[284]

Depending on the country where these clinics operate, different measures have been taken, and at times, no measures at all. In 2010, the government of Costa Rica's Health Ministry had to close their largest stem cell clinic due to the fact that the offered "treatments" did not seem to provide any benefits.[285] The clinic in question had obtained permission from the Costa Rican government to allow them to store adult stem cells in the form of frozen umbilical cords, or obtaining them from fat tissue, or bone marrow, but had no authority to use the stem cells for treatments.[286] Due to the volume of Westerners and even locals attending the clinic, the government seemed to have overlooked the authority issue until the complaints of lack of success by the patients became too rampant to ignore.[287] Yet, even with this knowledge and the fact that the stem cell clinic was closed down, many patients who were fortunate to have obtained success from this clinic had run to defend the actions and work that the clinic does, and believe that the risk was well worth the reward.[288]

Notwithstanding the facts above, these unproven stem cell treatments, and problematic stem cell "clinics" have created an argument that works against the expanding of stem cell policies and laws. Instead of providing benefits, they only create fear based on their horrific results or lack thereof.[289] Yet with all of these negative articles, there are many wonderful stem cell clinics throughout the world who seek to benefit those in need and have had much success.[290] In fact, my own doctor in Boca Raton, Florida has opened up his own stem cell clinic in South Florida which goes out of their way to fully explain the potential benefits and harms that can arise due to stem cell treatments. These doctors actually offer seminars for their patients where they can ask questions and be taught about everything relating to

---

[282] de Jesús Medina Arellano (2012).

[283] *See* Medical Tourism Magazine (2008).

[284] Levine (2010), pp. 274–281.

[285] Paxman (2010).

[286] *Id.*

[287] *Id.*

[288] *Id.*

[289] *See* LaGrone (2018).

[290] Bowden (2018).

their potential treatments.[291] It is important to note that these doctors do not just rush into using stem cells, and instead, first attempt other methods of treatment, unless they know that the stem cells have a better chance of helping instead of trying other methods that would effectually be a waste of the patients time and end up causing longer and more drawn out issues. My own doctor gave me a print-out discussing the various potentials and issues in undergoing a stem cell procedure, and then had me wait two weeks before I could "accept" the trial procedure. Before agreeing to undergo the procedure, I believe I was aware of virtually all of the risks, and the potential benefits by the time the two weeks had passed.

Another popular argument that those against embryonic stem cell treatments make is that these problems would not occur if adult stem cells were used because they work just as well and should be less regulated.[292] This too is an untrue statement, and clinics that pop up all around the world which only use adult stem cells have just as many problems as embryonic stem cell clinics.[293] Many of these stem cell clinics have opened their doors for unproven treatments which claim to aid in major diseases and devastating conditions, or for treatments that are just to enhance one's appearance.[294] The fact that so many new "clinics" started using stem cells in unproven treatments have led many people to suffer physical harm, as well as mental anguish when the so called "treatments" fail to work has led the FDA to issue a serious warning to stem cell clinics around the country.[295] The FDA even had to publish a new set of guidelines which would limit the scope and abilities of the stem cell clinics to treat various issues.[296]

Additionally, people argue that stem cell sciences have been debunked based on something they have seen or read about problems or issues with stem cell use.[297] Much of this false view comes from the newsworthy situation that took place in 2014 when Japanese researcher, Dr Haruko Obokata claimed that she and her team had discovered a new way to obtain embryonic like stem cells was found in normal adult cells.[298] The so-called discovery was published in two separate articles both in the journal *Nature*, and stated that dipping adult cells into an acidic bath of sorts for 30 min, and named these new cells "STAP" cells which was an acronym for stimulus-triggered acquisition of pluripotency.[299] Unfortunately, no one was able to replicate Dr. Obokata's supposed findings, and an independent investigation took

[291] *See generally* Pur-Form (2024).

[292] Murnaghan (2018).

[293] Lee McFarling (2016).

[294] *Id.* (approximating 200 adult stem cell clinics have opened at the time of this article's publication).

[295] *Id.*

[296] *Id.*

[297] *See generally* Marcon et al. (2017), pp. 465–775.

[298] Gallagher (2014).

[299] *Id.*

place that found Dr. Obokata guilty of misconduct for her false publications.[300] Because this "discovery" was so groundbreaking, the news of the retractions and misconduct has led to many disbelieving in the potential benefits that stem cells of all kinds seem to offer.[301]

As discussed above, there are many young people in the United States who only know of stem cells from television shows which mislead the audience for the purpose of being funny, or adding drama to their stories. When I first told some friends that I too was going to have a stem cell procedure they immediately joked about the source, and referenced Christopher Reeve from the South Park episode "Krazy Kripples." Even though this was clearly only meant to be funny, the problem exists because the subject is highly debated, and people do not conduct their own research into the topic to better understand it. Instead, they heavily rely on the misinformation as portrayed on the TV shows. While there is a benefit in spreading some useful knowledge there is also a harm that cannot be avoided in these situations.[302] This can best be illustrated by realizing that many people have not even heard of stem cells until these TV shows aired, but at the same time, the misinformation meant to be funny scares people into believing the worst most outrageous depictions and leads them to believe that there is some truth to it.

The inverse of this argument is also used fairly often and unfortunately it seems most prevalently used by doctors who are unaware of the potential that stem cells offer.[303] Dr. Mark Berman of Beverly Hills uses adult stem cells in his plastic surgery practice and has seen much success, but even so, Dr. Berman was recently interviewed and stated that "Doctors say it's quackery, nonsensical, no better than placebo," "Everyone on the planet thinks I'm a charlatan and we're just in it for the money."[304] I, myself have even encountered such statements and issues from many trusted doctors and even family members who are in the medical field. For instance, my sister was just recently diagnosed with Multiple Sclerosis, and when my father asked his brother (a doctor in Boston, Massachusetts), stated that he believed stem cells to be more like a myth, yet when asked what he has read on the subject he honestly stated that he has read only "a little" because he has no interest in the subject. This was also the attitude of many other doctors that I went to before I found a forward thinking and interested doctor. In discussions with Dr. Barsell-Edelman who is a fairly new doctor in the field of dermatology, she explained that stem cells are not only the future, but they are so much more.

300 *Id.*

301 Master and Resnik (2013), pp. 321–335.

302 Sweeney (2023).

303 Murnaghan (2018).

304 *Id.*

3.2.7 *People Fearful and Against Neo-Eugenics*

As discussed above, using embryonic stem cells to bring about certain conditions or protections in embryos could likely lead to "designer babies" who would have the most beneficial traits and resistances.[305] This argument takes on two forms, first, that only the wealthy will be able to afford such "designer babies", and second, that there will be a form of divide will occur between those with "designer" genes and those without.[306] While the first argument may start out as true, it is likely that once this is proven, insurance companies will gladly pay for these treatments and the overall price will go down, much like what happens with virtually all medical treatments.

3.3 Those Who Want the Benefits of Embryonic Stem Cells Research Without the Destruction of Embryos

There is a middle ground in this debate who views embryos as more than just a grouping of cells, yet not an actual being with life, yet has the potential for life.[307] Their belief is that embryos should not be created to be destroyed in order to study stem cells, but agree that using already existing embryos should allow for federal funding in order further explore the potential benefits of their use.[308]

References

Abernethy B (2011) Religious views on stem cell research. Available via Public Broadcasting Service [hereinafter PBS]. http://www.pbs.org/wnet/religionandethics/2001/07/27/july-27-2001-religious-views-on-stem-cell-research/15307. Accessed 26 Mar 2024

Afarid M, Sanie-Jahromi F (2021) Mesenchymal stem cells and COVID-19: cure, prevention, and vaccination. Stem Cell Int 2021:1–12. Available via Hindawi. https://downloads.hindawi.com/journals/sci/2021/6666370.pdf

Alston P (1990) U.S. ratification of the covenant on economic, social and cultural rights: the need for an entirely new strategy. Am J Int Law 84:365–393

Amaris M, Sivaraman F (2019) Ethical guiding principles of "do no harm" and the "intention to save lives" in relation to human embryonic stem cell research: finding common ground between religious views and principles of medical ethics. Asian Bioeth Rev 4:409–435

American Association for the Advancement of Science (2001) AAAS statement on President Bush's stem cell policy. Available via American Association for the Advancement of Science. https://www.aaas.org/news/stem-cell-resources. Accessed 26 Mar 2024

---

[305] Lachmann (2001), pp. 165–168.

[306] *Id.*

[307] Caplan and Patrizio (2009), pp. 1074–1075; Devolder (2006), p. 45; Glazer (2019).

[308] *See generally* Brock (2010), pp. 229–237.

American Association for the Advancement of Science (2007) AAAS statement regarding the President's veto of the Stem Cell Research Enhancement Act and the new executive order. Available via American Association for the Advancement of Science. https://www.aaas.org/news/stem-cell-resources. Accessed 26 Mar 2024

Associated Press (2005) Scientists create animals that are part-human. Available via NBC Universal Media. https://www.nbcnews.com/health/health-news/scientists-create-animals-are-part-human-flna1c9443214. Accessed 26 Mar 2024

Association for the Advancement of Blood & Biotherapies (2024) How are stem cells regulated? Available via Association for the Advancement of Blood & Biotherapies. www.aabb.org/aabbcct/therapyfacts/Pages/regulated.aspx. Accessed 3 Jan 2024

Babington C (2006) Stem cell bill gets Bush's first veto. Available via the Washington Post. https://www.washingtonpost.com/archive/politics/2006/07/20/stem-cell-bill-gets-bushs-first-veto/d6343d03-516a-4042-8b1e-2c7bbfff5957. Accessed 26 Mar 2024

Baylis F (2021) Stem cell research community drops 14-day limit on human embryo research. Available via the Conversation. https://theconversation.com/stem-cell-research-community-drops-14-day-limit-on-human-embryo-research-161616. Accessed 26 Mar 2024

Baylis F, Darnovsky M, Hasson K, Krahn TM (2020) Human germline and heritable genome editing: the global policy landscape. CRISPR J 3:365–377

Behfar A, Kane GC, Perez-Terzic C, Alekseev AR (2004) Stable benefit of embryonic stem cell therapy in myocardial infarction. Am J Physiol Heart Circ Physiol 287:471–479

Bennett D (2007) Tackling the moral questions. Available via Center for American Progress Action Fun. https://www.americanprogress.org/article/tackling-the-moral-questions. Accessed 26 Mar 2024

Betts KS (2010) Growing knowledge: using stem cells to study developmental neurotoxicity. Environ Health Perspect 118:A432–A437

Bingham D, Strauss N, Coeytaux F (2011) Maternal mortality in the United States: a human rights failure. Contraception 83:189–193

Bodansky DM (1995) The concept of customary international law. Mich J Int Law 16:667–669

Boddy AM, Fortunato A, Wilson Sayres M, Aktipis A (2015) Fetal microchimerism and maternal health: a review and evolutionary analysis of cooperation and conflict beyond the womb. BioEssays 37:1106–1118

Bowden T (2018) 'Buying hope': unproven stem cell treatments facing tighter regulation. Available via Australian Broadcasting Corporation. www.abc.net.au/news/2018-06-26/unproven-stem-cell-treatments-facing-tighter-regulation/9787050. Accessed 26 Mar 2024

Brock DW (2010) Creating embryos for use in stem cell research. J Law Med Ethics 38:229–237

Brooks R (2011) How social media helps journalists break news. Available Via Social Media Examiner. https://www.socialmediaexaminer.com/how-social-media-helps-journalists-break-news. Accessed 26 Mar 2024

Butcher K (2018) Family gives update on metro man fighting insurance company. Available via Nexstar Media. https://kfor.com/news/family-gives-update-on-metro-man-fighting-insurance-company. Accessed 26 Mar 2024

Butler D (2008) Vatican toughens stance on embryo research. Available via Nature. https://www.nature.com/articles/news.2008.1306. Accessed 26 Mar 2024

California's Stem Cell Agency (2014) World's largest pharmaceutical company signs agreement with ViaCyte for a stem cell treatment for type 1 diabetes. Available via the Vertex. https://investors.vrtx.com/news-releases/news-release-details/vertex-acquire-viacyte-goal-accelerating-its-potentially. Accessed 26 Mar 2024

Caplan AL, Patrizio P (2009) The beginning of the end of the embryo wars. Lancet 373:1074–1075

Cara E (2018) The FDA is taking two shady stem cell clinics to court. Available via G/O Media. https://gizmodo.com/the-fda-is-taking-two-shady-stem-cell-clinics-to-court-1825919369. Accessed Sept 26 Mar 2024

Card Ratzinger J, Bovone A (1987) Congregation for the Doctrine of the Faith: instruction on respect for human life in its origin and on the dignity of procreation replies to certain questions of the day Vatican City 1987. Linacre Q 54:24–49
Catholic Church (2000) Catechism of the Catholic Church, 2nd edn. Our Sunday Visitor, Huntingdon, PA
Catholics United for the Faith (2005) Cloning: an affront to human dignity. In Lay Witness Magazine
Chaddah MR, Dickie BG, Lyall D, Marshall CJ, Sykes JB, Bruijn LI (2011) Meeting report of the International Consortium of Stem Cell Networks' Workshop Towards Clinical Trials Using Stem Cells for Amyotrophic Lateral Sclerosis/Motor Neuron Disease. Amyotroph Lateral Scler 12:315–317
Chambers A (2015) Trends in U.S. health travel service trade: USITC Executive briefing on trade. Available via available via United States International Trade Commission. https://www.google.com/url?sa=t&rct=j&q=&esrc=s&source=web&cd=&cad=rja&uact=8&ved=2ahUKEwjbl7rAl8P-AhW7ZzABHatqBc4QFnoECBkQAw&url=https%3A%2F%2Fwww.usitc.gov%2Fpublications%2F332%2Fexecutive_briefings%2Fchambers_health-related_travel_final.pdf&usg=AOvVaw3cpoUY2iXmsaniJRY1Vwtr. Accessed 26 Mar 2024
Chapman AR, Frankel MS, Garfinkel MS (1999) Stem cell research and applications: monitoring the frontiers of biomedical research. Produced by the American Association for the Advancement of Science and Institute for Civil Society
Chen G, Yin S, Zeng H, Li H, Wan X (2022) Regulation of embryonic stem cell self-renewal. Life (Basel) 12:1–17. Available via the National Library of Medicine [hereinafter NLM]. https://www.ncbi.nlm.nih.gov/pmc/articles/PMC9410528/
Cooper ME (2008) Life as surplus: biotechnology and capitalism in the neoliberal era. University of Washington Press, pp 153–157
Cooper DKC (2012) A brief history of cross-species organ transplantation. Proc Bayl Univ Med Cent 25:49–57
Crook JM, Hei D, Stacey G (2010) The international stem cell banking initiative: raising standards to bank on. In Vitro Cell Dev Biol Anim 46:169–172
Dark Matters: Twisted but True, season 1, episode 1, "The Philadelphia Experiment, ape-man army, zapped to death," Science Channel (original air date Aug. 31, 2011)
de Deus S, Reis R (2017) Doctors without borders: coherent principles. SUR Int J Hum Rights 14: 259–264
de Jesús Medina Arellano M (2012) The quest for stem cell science regulation in Mexico: ethical, legal and religious controversies. In the Faculty of Humanities. University of Manchester. Available via the University of Manchester. https://www.escholar.manchester.ac.uk/api/datastream?publicationPid=uk-ac-man-scw:165929&datastreamId=FULL-TEXT.PDF. Accessed 26 Mar 2024
de Magalhães JP (2012) Cloning and stem cells. Available via Senescene. https://jp.senescence.info/cloning-and-stem-cells. Accessed 26 Mar 2024
Deaton M (2011) Universal declaration of human rights. In: Chatterjee DK (ed) Encyclopedia of global justice. Available via Springer, Dordrecht, pp 1115–1117. https://doi.org/10.1007/978-1-4020-9160-5_664. Accessed 22 Jan 2024
Devolder K (2006) What's in a name? Embryos, entities, and ANTities in the stem cell debate. J Med Ethics 32:43–48
Dittrich R, Beckmann MW, Würfel W (2015) Non-embryo-destructive extraction of pluripotent embryonic stem cells: implications for regenerative medicine and reproductive medicine. Geburtshilfe Frauenheilkd 75:1239–1242
Doheny K (2019) Stem cells and health advances: where are we now? Available via WebMD. https://www.webmd.com/brain/news/20190812/stem-cells-and-health-advances-where-are-we-now. Accessed 26 Mar 2024
Egawa N, Kitaoka S, Tsukita K, Naitoh M, Takahashi K, Yamamoto T, Adachi F, Kondo T, Okita K, Asaka I, Aoi T, Watanabe A, Yamada Y, Morizane A, Takahashi J, Ayaki T, Ito H,

Yoshikawa K, Yamawaki S, Suzuki S, Watanabe D, Hioki H, Kaneko T, Makioka K, Okamoto K, Takuma H, Tamaoka A, Hasegawa K, Nonaka T, Hasegawa M, Kawata A, Yoshida M, Nakahata T, Takahashi R, Marchetto MCN, Gage FH, Yamanaka S, Inoue H (2012) Response to comment on "Drug screening for ALS using patient-specific induced pluripotent stem cells". Sci Transl Med 4:145

Fabiani A (2021) Jeff Bezos is reportedly funding Altos Labs, a new anti-ageing venture aiming to cheat death. Available via Screenshot. https://screenshot-media.com/the-future/science/altos-labs-anti-ageing. Accessed 26 Mar 2024

Fan S (2019) Wait, what? The first human-monkey hybrid embryo was just created in China. Available via SingularityHub. https://singularityhub.com/2019/08/13/wait-what-the-first-human-monkey-hybrid-embryo-was-just-created-in-china. Accessed 26 Mar 2024

Farnia S, Gedan A, Boo M (2014) Impact of the Affordable Care Act on stem cell transplantation. Curr Hematol Malig Rep 9:66–72

Fight Aging (2006) Overpopulation, not a problem now, and never will be. Available via Fight Aging! https://www.fightaging.org/archives/2006/09/overpopulation.php. Accessed 26 Mar 2024

Fuchs E, Segre JA (2012) Stem cells: a new lease on life. Cell 100:143–155

Gallagher J (2014) Stem cell scandal scientist Haruko Obokata resigns. Available via the British Broadcasting Corporation. https://www.bbc.com/news/health-30534674. Accessed 26 Mar 2024

Gater R (2018) Curing blindness with stem cells – here's the latest science. Available via the Conversation. https://theconversation.com/curing-blindness-with-stem-cells-heres-the-latest-science-92412. Accessed 26 Mar 2024

Glazer ES (2019) Infertility: extra embryos — too much of a good thing? Available via Harvard Health Publishing. https://www.health.harvard.edu/blog/infertility-extra-embryos-too-much-of-a-good-thing-2019042216476. Accessed 26 Mar 2024

Gorn D (2018) FDA injunctions send clear message: stem cell clinics beware. Available via KQED. https://www.kqed.org/futureofyou/441482/fda-takes-action-against-california-stem-cell-clinics. Accessed 26 Mar 2024

Guevin BM (2010) Responses to Dignitas personae - Part I of II. Natl Cathol Bioethics Q 10:51–59

Guterl F (2021) COVID-19 vaccines and fetal tissue: the science and controversy explained. Available via News Week. https://www.newsweek.com/covid-19-vaccines-fetal-tissue-science-controversy-explained-1575863. Accessed 26 Mar 2024

Haimes E, Taylor K, Turkmendag I (2012) Eggs, ethics and exploitation? Investigating women's experiences of an egg sharing scheme. Sociol Health Illness 34:1199–1214

Haywayd J (2019) Report: Spanish biologist working on 'human-monkey hybrids' in China. Available via Breitbart News Network. https://www.the-scientist.com/news-opinion/first-humanmonkey-chimeras-developed-in-china%2D%2D66231. Accessed 26 Mar 2024

Heide M, Haffner C, Murayama A, Kurotaki Y, Shinohara H, Okano H, Sasaki E, Huttner WB (2020) Human specific ARHGAP11B increases size and folding of primate neocortex in the fetal marmoset. Science 369:546–550

Henon PR (2003) Human embryonic or adult stem cells: an overview on ethics and perspectives for tissue engineering. In: Elçin YM (ed) Tissue engineering, stem cells, and gene therapies. Advances in Experimental Medicine and Biology, vol 534, pp 27–45

Hertneky D (2018) Oklahoma man fighting with insurance company over cancer trial. Available via Griffin Media. https://www.newson6.com/story/5e348ccd527dcf49dad7b97c/oklahoma-man-fighting-with-insurance-company-over-cancer-trial. Accessed 26 Mar 2024

Hestermeyer HP (2007) Human rights and the WTO: the case of patents and access to medicines. OUP Catalogue, Oxford University Press

Hirsch ML (2008) Side effects of corporate greed: pharmaceutical companies need a dose of corporate social responsibility. Minn J Law Sci Technol 9:607–636

Horowitz J (2019) Pope Francis says abortion, even of a sick fetus, is like hiring a 'hitman'. Available via the New York Times. https://www.nytimes.com/2019/05/25/world/europe/pope-abortion-sick-fetus.html. Accessed 26 Mar 2024
How Can Stem Cells Advance Medicine? (2007) Nature Reports Stem Cells. Available via Nature. https://www.nature.com/articles/stemcells.2007.23. Accessed 26 Mar 2024
Human Rights Watch (2017) Q&A: human rights law and access to abortion. Available via Human Rights Watch. https://www.hrw.org/news/2017/07/24/qa-human-rights-law-and-access-abortion. Accessed 26 Mar 2024
Human Rights Watch (2024) About Us. Available via Human Rights Watch. https://www.hrw.org/about/about-us. Accessed 3 Jan 2024
Ilkilic I, Ertin H (2010) Ethical aspects of human embryonic stem cell research in the Islamic world: positions and reflection. Stem Cell Rev Rep 6:151–161
Instruction on Dignitas Personae (2010) Congregation for the Doctrine of Faith, 1987
International Covenant on Civil and Political Rights, Dec. 16, 1966, S. Exec. Rep. 102-123, 103 999 U.N.T.S. 171
International Stem Cell Banking Initiative (2019) About ISCBI. Available via International Stem Cell Banking Initiative. https://www.iscbi.org/about. Accessed 26 Mar 2024
International Stem Cell Forum (2010a) About. Available via International Stem Cell Forum. https://stemcellforum.org/about_stem_cell_research.cfm. Accessed 26 Mar 2024
International Stem Cell Forum (2010b) Initiatives. Available via International Stem Cell Forum. https://stemcellforum.org/forum_initiatives.cfm. Accessed 26 Mar 2024
Johns Hopkins School of Public Health (2018) Cost of clinical trials for new drug FDA approval are fraction of total tab. Available via Johns Hopkins School of Public Health. https://www.jhsph.edu/news/news-releases/2018/cost-of-clinical-trials-for-new-drug-FDA-approval-are-fraction-of-total-tab.html. Accessed 26 Mar 2024
Keown D (2016) No harm applies to stem cell embryos, one Buddhist's view. Available via Beliefnet. http://www.beliefnet.com/News/Science-Religion/2004/04/No-Harm-Applies-To-Stem-Cell-Embryos-One-Buddhists-View.aspx. Accessed 26 Mar 2024
Kim J-H, Auerbach JM, Rodríguez-Gómez JA, Velasco I, Gavin D, Lumelsky N, Lee S-H, Nguyen J, Sánchez-Pernaute R, Bankiewicz K, McKay R (2002) Dopamine neurons derived from embryonic stem cells function in an animal model of Parkinson's disease. Nature Pub Group 418:50–56
Kim JH, Alderton A, Crook JM, Benvenisty N, Brandsten C, Firpo M, Harrison PW, Kawamata S, Kawase E, Kurtz A, Loring JF, Ludwig T, Man J, Mountford JC, Turner ML, Oh S, da Veiga Pereira L, Pranke P, Sheldon M, Steeg R, Sullivan S, Yaffe M, Zhou Q, Stacey GN (2019) A report from a workshop of the International Stem Cell Banking Initiative, (held in collaboration of global alliance for iPSC therapies and the Harvard Stem Cell Institute, Boston, 2017). Stem Cells 37:1130–1135
Kliff S (2018) The true story of America's sky-high prescription drug prices. Available via Vox. https://www.vox.com/science-and-health/2016/11/30/12945756/prescription-drug-prices-explained. Accessed 26 Mar 2024
Klimanskaya I, Chung Y, Becker S, Lu S-J, Lanza R (2006) Human embryonic stem cell lines derived from single blastomeres. Nature Pub Group 444:481–485
Kolata G (2016) A cautionary tale of stem cell tourism. Available via the New York Times Company. https://www.nytimes.com/2016/06/23/health/a-cautionary-tale-of-stem-cell-tourism.html. Accessed 26 Mar 2024
Kriks S, Shim J-W, Piao J, Ganat YM, Wakeman DR, Xie Z, Carrillo-Reid L, Auyeung G, Antonacci C, Buch A, Yang L, Flint Beal M, Surmeier DJ, Kordower JH, Tabar V, Studer L (2011) Dopamine neurons derived from human ES cells efficiently engraft in animal models of Parkinson's disease. Nature 480:547–551
Kubikova N, Phil D, Keefe DL, Wells D, Oktay KH, Feinberg EC (2023) Should we use CRISPR gene editing in human embryos? Fertile Battle 120:737–744. Available via https://www.fertstert.org/article/S0015-0282(23)00766-5/fulltext

Lachmann P (2001) Stem cell research—why is it regarded as a threat? EMBO Rep 2:165–168
LaGrone K (2018) New lawsuit against US stem cell after patient loses vision; family ties called into question. Available via Scripps Media. https://www.abcactionnews.com/news/local-news/i-team-investigates/new-lawsuit-against-us-stem-cell-after-patient-loses-vision-family-ties-called-into-question. Accessed 26 Mar 2024
Lanese N (2019) First human–monkey chimeras developed in China. Available via the Scientist. https://www.the-scientist.com/news-opinion/first-humanmonkey-chimeras-developed-in-china%2D%2D66231. Accessed 26 Mar 2024
Le Belle JE, Svendsen CN (2002) Stem cells for neurodegenerative disorders 'Where do we go from here?'. BioDrugs 16:389–401
Lee T-L, Lysaght T, Lipworth W, Hendl T, Kerridge I, Munsie M, Stewart C (2017) Regulating the stem cell industry: needs and responsibilities. Bull World Health Organ 95:663–664
Levine AD (2010) Stem cell tourism: assessing the state of knowledge. A J Law Tech Soc 7:274–282
Lindvall O, Kokaia Z, Martinez-Serrano A (2004) Stem cell therapy for human neurodegenerative disorders–how to make it work. Nat Med 10:S42–S50
Liu S, Yin N, Faiola F (2017) Prospects and frontiers of stem cell toxicology. Stem Cells Dev 26: 1528–1539
Lopatin A (2019) Pikuach nefesh: the Jewish value of saving a life. Available via My Jewish Learning. https://www.myjewishlearning.com/article/pikuach-nefesh-the-overriding-jewish-value-of-human-life. Accessed 26 Mar 2024
Lovell-Badge R et al (2021) ISSCR guidelines for stem cell research and clinical translation: the 2021 update. Stem Cell Rep 16:1398
Lovgren S (2006) Stem cells can be collected without destroying embryos, scientists show. In National Geographic News issue No. 28
Lund RD, Wang S, Klimanskaya I, Holmes T, Ramos-Kelsey R, Lu B, Girman S, Bischoff N, Sauvé Y, Lanza R (2006) Human embryonic stem cell–derived cells rescue visual function in dystrophic RCS rats. Cloning Stem Cells 8:189–199
Maller AS (2019) When a human soul enters the fetus? Available via IslamiCity. https://www.islamicity.org/19988/when-a-human-soul-enters-the-fetus. Accessed 26 Mar 2024
Mandavilli A (2019) H.I.V. is reported cured in a second patient, a milestone in the global AIDS epidemic. Available via the New York Times Company. https://www.nytimes.com/2019/03/04/health/aids-cure-london-patient.html. Accessed 26 Mar 2024
Mannion G (2023) The humble path to ethics schopenhauer, religion and morality. Routledge. 9780754608233
Marcon AR, Murdoch B, Caulfield T (2017) Fake news portrayals of stem cells and stem cell research. Regen Med 12:765–775
Martin A (2007) The chimera of liberal individualism: how cells became selves in human clinical genetics. Hist Sci Soc 22:205–222
Martins I (2020) AstroRx slows ALS progression at all doses tested, trial shows. Available via ALS News Today. https://alsnewstoday.com/2020/08/05/astrorx-slows-als-progression-at-all-doses-tested-trial-shows. Accessed 26 Mar 2024
Master Z, Resnik DB (2013) Hype and public trust in science. Sci Eng Ethics 19:321–335
Mayo Clinic Staff (2019) Stem cells: "What they are and what they do?" Available via Mayo Foundation for Medical Education and Research (MFMER). https://www.mayoclinic.org/tests-procedures/bone-marrow-transplant/in-depth/stem-cells/art-20048117. Accessed 26 Mar 2024
McFarling UL (2016) FDA moves to crack down on unproven stem cell therapies. Available via STAT News. https://www.statnews.com/2016/02/08/fda-crackdown-stem-cell-clinics. Accessed 26 Mar 2024
McKernan R, McNeish J, Smith D (2010) Pharma's developing interest in stem cells. Cell Stem Cell 6:517–520

Medecins Sans Frontieres/Doctors Without Borders (2023) Principles. Available via Médecins Sans Frontières. https://www.doctorswithoutborders.org/who-we-are/principles. Accessed 3 Jan 2024
Medical Tourism Magazine (2008) Mexico: a medical tourism mecca across the border. Available via Medical Tourism. https://www.magazine.medicaltourism.com/article/mexico-a-medical-tourism-mecca-across-the-border. Accessed 26 Mar 2024
Murdoch CE, Scott CT (2010) Stem cell tourism and the power of hope. Am J Bioethics 10:16–23. Available via Taylor & Francis Online. https://www.tandfonline.com/doi/full/10.1080/15265161003728860
Murnaghan I (2018) Adult vs. embryonic stem cells. Available via Explore Stem Cells. www.explorestemcells.co.uk/adultvsembryonicstemcells.html. Accessed 26 Mar 2024
National Association of Evangelicals (2005) Bioethics & Stem Cell Research Resolution. Available via the National Association of Evangelicals. https://www.nae.org/bioethics-and-stem-cell-research. Accessed 26 Mar 2024
National Catholic Bioethics Center (2022) The NCBC praises Supreme Court's decision to protect the right to life on the feast of the sacred heart. Available via National Catholic Bioethics Center. https://www.ncbcenter.org/resources-and-statements-cms/dobbs. Accessed 26 Mar 2024
Neal DJ (2018) 'Dr. Compassionate 420' gave 'vampire face lifts', cops say she had no medical license. Available via the Miami Herald. https://www.miamiherald.com/news/state/florida/article213382644.html#storylink=cpy. Accessed 26 Mar 2024
Nguyen V, Meak L, Carroll J (2015) Is umbilical cord blood banking worth the cost? Available via NBC Universal Media. http://www.nbcbayarea.com/investigations/Is-Umbilical-Cord-Blood-Banking-Worth-the-Cost-301242771.html. Accessed 26 Mar 2024
Orentlicher D (2012) Rights to health care in the United States: inherently unstable. Scholarly Works 1054:325–347
Orthodox Union Advocacy Center (2005) OU welcomes U.S. House passage of Stem Cell Research Enhancement Act. Available via Union of Orthodox Jewish Congregations of America. https://advocacy.ou.org/ou_welcomes_us_house_passage_of_stem_cell_research_enhancement_act. Accessed 26 Mar 2024
Pacunas B, Bosch D (2016) Joseph Stalin's humanzee experiments. Bosch Media, LLC
Pain S (2008) Blasts from the past: the Soviet ape-man scandal. Available via NewScientist. https://www.newscientist.com/article/mg19926701-000-blasts-from-the-past-the-soviet-ape-man-scandal. Accessed 26 Mar 2024
Pargament KI, Park CL (2019) The psychology of religion. In times of stress: the religion-coping connection. https://doi.org/10.4324/9780429495915-4
Parson AB (2008) Stem cell biotech: seeking a piece of the action. Sci Direct Cell Press 132:511–513
Paxman R (2010) Health Ministry closes Costa Rica's largest stem cell clinic. Progress Educational Trust. https://www.bionews.org.uk/page_92391. Accessed 4 Jan 2024
Pentin E (2018) Pope Francis changes Catechism to say death penalty 'inadmissible'. Available via National Catholic Register. https://www.ncregister.com/blog/pope-francis-changes-catechism-to-say-death-penalty-inadmissible#.W2MLVlMvwW8. Accessed 26 Mar 2024
Pew Research Center (2008) Religious groups' official positions on stem cell research. Available via Pew Research Center. https://www.pewresearch.org/religion/2008/07/17/religious-groups-official-positions-on-stem-cell-research. Accessed 26 Mar 2024
Pirnay J-P, Vanderkelen A, Zizi M, De Vos D, Rose T, Laire G, Ectors N, Verbeken G (2010) Human cells and tissues: the need for a global ethical framework. Bull World Health Organ 88: 870–872
Pope Francis (2017) Address to participants in the meeting organized by the Pontifical Council for the Promotion of the New Evangelization. 11 Oct. 2017. L'Osservatore Romano
Pur-Form (2024) Stem cell therapy 'trail blazers'. Available via Pur-Form. https://stemcellorthopedic.com/learn-more. Accessed 4 Jan 2024

Putton K (2023) Nine things to know about stem cell treatments. Available via International Society for Stem Cell Research. https://www.aboutstemcells.org/info/nine-things-to-know-about-stem-cell-treatments. Accessed Sept 26 Mar 2024

Rachul C (2011) What have I got to lose? An analysis of stem cell therapy patients' blogs. Health Law Rev 20:5

Reconstructionist Rabbinical Association (2005) Resolution on educational and political support for stem cell research. Available via Reconstructionist Rabbinical Association. https://therra.org/resolutions/rra-stem-cell-resolution-edited.pdf. Accessed 26 Mar 2024

Regalado A (2019) Scientists are making human-monkey hybrids in China. Available at MIT Technology Review. https://www.technologyreview.com/2019/08/01/652/scientists-are-making-human-monkey-hybrids-in-china. Accessed 26 Mar 2024

Rettner R (2016) 3 human chimeras that already exist. Available via Scientific American. https://www.scientificamerican.com/article/3-human-chimeras-that-already-exist. Accessed 26 Mar 2024

Rev. Fr. Baker J (2020) COVID-19 vaccines: moral evaluation. Available via Catholic Diocese of Madison. https://madisoncatholicherald.org/covid-vaccines-morality. Accessed 26 Mar 2024

Rev. Fr. Joseph Baker's Letter (2021) Covid-19 vaccines: moral evaluation-updated. Available via Catholic Diocese of Madison. https://madisondiocese.org/documents/2021/3/Executive%20Summary%20COVID-19%20Vaccines%20aker%20210306.pdf. Accessed 26 Mar 2024

Robert JS (2006) The science and ethics of making part-human animals in stem cell biology. FASEB J 20:838–845

Ruan GF, Zheng L, Huang J-S, Huang W-X, Gong B-D, Fang X-X, Zhang X-Y, Tang J-P (2017) Effect of mesenchymal stem cells on Sjögren-like mice and the microRNA expression profiles of splenic CD4+ T cells. Exp Ther Med 13:2828–2838

Ruiz-Casares M, Rousseaua C, Derluyn I, Watters C, Crépeau F (2010) Right and access to healthcare for undocumented children: addressing the gap between international conventions and disparate implementations in North America and Europe. Soc Sci Med 70:329–336

Russell A (2020) Pope Francis releases encyclical cementing Catholic Church's opposition to the death penalty. Available via the American Bar Association. https://www.americanbar.org/groups/committees/death_penalty_representation/publications/project_blog/pope-francis-encyclical-death-penalty-opposition. Accessed 26 Mar 2024

Schicktanz S (2012) Xenotransplantation, encyclopedia of applied ethics, 2nd edn. Academic Press, pp 565–574. https://doi.org/10.1016/B978-0-12-373932-2.00052-1

Schpero WL, Wiener T, Carter S, Chatterjee P (2022) Lobbying expenditures in the US health care sector, 2000-2020. JAMA Health Forum 3(10):e223801. Available via the Journal of the American Medical Association. https://jamanetwork.com/journals/jama-health-forum/fullarticle/2797734

Schulte-Hillen C, Staderini N, Saint-Sauveur J-F (2016) Why médecins sans frontières (MSF) provides safe abortion care and what that involves. Conflict Health 10:1–4

Servick K (2016) Embryonic stem cells and fetal tissue research—will Trump intervene? Available via American Association for the Advancement of Science. https://www.sciencemag.org/news/2016/11/embryonic-stem-cells-and-fetal-tissue-research-will-trump-intervene. Accessed 26 Mar 2024

Shin LM (2009) Israel liberzon, the neurocircuitry of fear, stress, and anxiety disorders. Neuropsychopharmacology 35:169–191

Slukvin I (2010) Generation of mature blood cells from pluripotent stem cells. Haematologica 95:1621–1623

Subbaraman N (2021) Limit on lab-grown human embryos dropped by stem cell body. Available via Scientific American. https://www.scientificamerican.com/article/limit-on-lab-grown-human-embryos-dropped-by-stem-cell-body. Accessed 26 Mar 2024

Sweeney A (2023) Ode to stem cells… and South Park? New study highlights how stem cells could restore retina function. Available via Pie Magazine. https://piemagazine.org/new-study-highlights-how-stem-cells-could-restore-retina-function. Accessed 26 Mar 2024

The American Association of the Advancement of Sciences (2024) Mission and History. Available via the American Association of the Advancement of Sciences. https://www.aaas.org/mission. Accessed 3 Jan 2024

The British Library (2024) Hadith collection: what is a hadith? Available via the British Library. https://www.bl.uk/collection-items/hadith-collection. Accessed 3 Jan 2024

The Hadith, Sahih al-Bukhari: 3036, Book 56, Hadith 242

The International Society for Stem Cell Research (2021a) Guidelines for the conduct of human embryonic stem cell research version I. Available via the International Society for Stem Cell Research. https://www.isscr.org/guidelines. Accessed 26 Mar 2024

The International Society for Stem Cell Research (2021b) Guidelines for stem cell research and clinical translation. Available via the International Society for Stem Cell Research. http://isscr.org/guidelines. Accessed 26 Mar 2024

The International Society for Stem Cell Research (2023) About available via the International Society for Stem Cell Research. https://www.isscr.org/about. Accessed 26 Mar 2024

The International Stem Cell Banking Initiative (2009) Consensus guidance for banking and supply of human embryonic stem cell lines for research purposes. Stem Cell Rev Rep 5:301–314

The International Stem Cell Forum (2010a) About the ISCF. Available via the International Stem Cell Forum. https://stemcellforum.org/about_the_iscf.cfm. Accessed 26 Mar 2024

The International Stem Cell Forum (2010b) Forum Initiatives Available via the International Stem Cell Forum. https://stemcellforum.org/forum_initiatives.cfm. Accessed 26 Mar 2024

The Lutheran Church (2007) 63rd regular convention of the Lutheran Church—Missouri Synod, 2007 Resolution 6-02. Available via the Lutheran Church—Missouri Synod. https://reporter.lcms.org/2007/convention-sets-lcms-position-on-stem-cell-research. Accessed 26 Mar 2024

The Miami Project (2024) Barth A. Green, M.D. Available via the Miami Project. https://www.themiamiproject.org/faculty/barth-green. Accessed 2 Jan 2024

The Qur'an (2019), 23:12–14

The Rabbinical Assembly (2005) Resolution in support of stem cell research and education. Available via The Rabbinical Assembly. https://www.rabbinicalassembly.org/resolution-support-stem-cell-research-and-education. Accessed 26 Mar 2024

The Southern Baptist Convention (1999) Resolution on human embryonic and stem cell research. Available via the Southern Baptist Church. https://www.sbc.net/resource-library/resolutions/resolution-on-human-embryonic-and-stem-cell-research. Accessed 26 Mar 2024

United Nations (2024) Declaration on Human Rights Defenders. Available via: the Office of the High Commissioner for Human Rights https://www.ohchr.org/en/special-procedures/sr-human-rights-defenders/declaration-human-rights-defenders. Accessed: 2 Jan 2024

United Nations General Assembly (1948) Universal Declaration of Human Rights, G.A. Res. 217 (III) A, U.N. Doc. A/RES/217(III) (Dec. 10, 1948), Art. 25

van der Heyden MAG, van de Ven T, Opthof T (2009) Fraud and misconduct in science: the stem cell deduction. Neth Heart J 17:25–29

van Norman G (2016) Drugs, devices, and the FDA: Part 1: an overview of approval processes for drugs. JACC Basic Transl Sci 3:170–179

Vatican News (2020) Vatican CDF says use of anti-Covid vaccines 'morally acceptable'. Available via Vatican News. https://www.vaticannews.va/en/vatican-city/news/2020-12/vatican-cdf-note-covid-vaccine-morality-abortion.html. Accessed 26 Mar 2024

Verified Market Research (2019a) Global Stem Cell Therapy Market Size by Cell Source, By Therapeutic Application, By Type, By Region and Forecast, Report ID: 24113. Available via Verified Market Research. https://www.verifiedmarketresearch.com/product/stem-cell-therapy-market. Accessed 26 Mar 2024

Verified Market Research (2019b) About Us. Available via Verified Market Research. https://www.verifiedmarketresearch.com/about-us. Accessed 26 Mar 2024

Wamsley L (2020) Vatican oks receiving COVID-19 vaccines, even if research involved fetal tissue. Available via Available via National Public Radio. https://www.npr.org/sections/

coronavirus-live-updates/2020/12/21/948806643/vatican-oks-receiving-covid-19-vaccines-even-if-research-involved-fetal-tissue. Accessed 26 Mar 2024
Washington Times (2009) Iran at the forefront of stem cell research. Available via the Washington Times. https://www.washingtontimes.com/news/2009/apr/15/iran-at-forefront-of-stem-cell-research. Accessed 26 Mar 2024
White House Communications (2009) The Obama-Biden economic plan: creating jobs, strengthening the economy for Michigan families. Available via the White House President Barack Obama Archives. https://obamawhitehouse.archives.gov/progressreports/Michigan. Accessed 26 Mar 2024
WHO Expert Advisory Committee on Developing Global Standards for Governance and Oversight of Human Genome Editing (2021) Human Genome Editing Recommendations. Available via World Health Organization. https://www.who.int/publications/i/item/9789240030381. Accessed 4 Jan 2024
Widdows H (2009) Border disputes across bodies: exploitation in trafficking for prostitution and egg Sale for stem cell research. Int J Fem Approaches Bioeth 2:5–24
World Health Organization (2005) First global consultation on regulatory requirements for human cells and tissues for transplantation (Ottawa, Nov. 29, 2004 - Dec. 1, 2004). Available via the World Health Organization. https://www.who.int/publications/i/item/9241593296. Accessed 26 Mar 2024
World Health Organization (2010) WHO guiding principles on human cell, tissue and organ transplantation. Available via the World Health Organization. https://iris.who.int/bitstream/handle/10665/341814/WHO-HTP-EHT-CPR-2010.01-eng.pdf?sequence=1. Accessed 26 Mar 2024
World Health Organization Expert Advisory Committee on Developing Global Standards for Governance and Oversight of Human Genome Editing (2019) Human genome editing recommendations, 10. Available via World Health Organization. https://www.who.int/publications/i/item/WHO-SCI-RFH-2019-02. Accessed 4 Jan 2024
World Medical Association (2009) Statement on embryonic stem cell research. Adopted by the 60th WMA General Assembly. New Delhi, India. October 2009. Available via the World Medical Association. https://www.wma.net/policies-post/wma-statement-on-embryonic-stem-cell-research. Accessed 26 Mar 2024
Wouters OJ (2020) Lobbying expenditures and campaign contributions by the pharmaceutical and health product industry in the United States, 1999-2018. JAMA Intern Med 5:1–10
Wouters OJ, McKee M, Luyten J (2020) Estimated research and development investment needed to bring a new medicine to market, 2009-2018. JAMA 323:844–853
Yam P, Petz LD, Knowlton RG, Wallace BR, Bruce R, Stock AD, de Lange G, Brown VA, Donis-Keller H, Blume KG (1987) Use of DNA restriction fragment length polymorphisms to document marrow engraftment and mixed hematopoietic chimerism following bone marrow transplantation. Transplantation 43:399–407
Yirka B (2018) Second successful human-animal hybrid: sheep embryo with human cells. Available via Science X Network. https://phys.org/news/2018-02-successful-human-animal-hybrid-sheep-embryo.html. Accessed 26 Mar 2024
Ylenfest (2017) Islam and the beginning of human life. Available via Petrie-Flom Center at Harvard Law School. https://blog.petrieflom.law.harvard.edu/2017/12/08/islam-and-the-beginning-of-human-life. Accessed 26 Mar 2024
Zarei S, Carr L, Reiley L, Diaz K, Guerra O, Fernandez Altamirano P, Pagani W, Lodin D, Orozco G, Chinea A (2015) A comprehensive review of amyotrophic lateral sclerosis. Surg Neurol Int 6:171

Zhang S (2017) A woman went blind after stem cells were injected in her eyes, and she isn't the only one. Available via the Atlantic. https://www.theatlantic.com/science/archive/2017/10/stem-cell-eye/541299. Accessed 26 Mar 2024

Zieff G, Kerr ZY, Moore JB, Stoner L (2020) Universal healthcare in the United States of America: a healthy debate. Medicina (Kaunas) 11:580. Available via NLM. https://www.ncbi.nlm.nih.gov/pmc/articles/PMC7692272/

Chapter 4
Past Trends in Decisions and Conditioning Factors

One of the most important results that this book seeks is to find a solution to the problem while taking into account all of those with different arguments discussed at length above, and find a workable solution that benefits the majority while minimizing any harm to those who outright oppose the issue on unique grounds. As discussed below, many countries have their own laws regarding this topic, and have expressed their reasoning for such laws in the discussions that led to the laws being passed, through public comments, through the laws themselves, and through case law that has emerged due to issues regarding the laws in question. It is extremely important to view the laws both beneficial and harmful to embryonic stem cell research, so that a workable scenario can emerge which will allow for some embryonic stem cell research which can be deemed beneficial, and banning, or placing more stringent requirements on research which may not be deemed as beneficial. It is imperative to view the laws of many other countries to find a sustainable and agreeable solution because even though there is no "perfect" set of laws in any country, looking to each can help find useful, and important aspects in order to draft a lasting law that encompasses the best from each, without the problematic limitations which we can look to, in order to learn from.

4.1 International Law

Before looking to the current laws of other countries, it is imperative to discuss relevant international laws which are or can be binding on state parties. International law is typically only binding if in the form of a treaty, executive agreement or as customary international law.[1] The United States Constitution provides for how a treaty is to become law by stating that the President: "shall have Power, by and with

[1] *See generally*: Mulligan (2023).

© The Author(s), under exclusive license to Springer Nature Switzerland AG 2024

J. Weiser, *Embryonic Stem Cells and the Law*,
https://doi.org/10.1007/978-3-031-59750-3_4

the Advice and Consent of the Senate, to make Treaties, provided two-thirds of the Senators present concur."[2] Internationally, the Vienna Convention on the Law of Treaties (the "Vienna Convention") was entered into as a treaty itself that specifies what must be included in an international agreement for it to be binding upon its members that complete both the signatory process, showing the desire to enter into the treaty and the ratification process whereby the potentially agreeing country accepts the agreement as law.[3] It is necessary to mention though that the United States has signed the Vienna Convention but has not yet ratified it, meaning it is not yet binding on the country.[4] Executive agreements are something that only the acting President of the United States can enter into.[5] It essentially is an agreement between the United States President, and a foreign government, and does not require senate approval.[6] The President's authority to bind the nation through an executive agreements has been challenged in the past, and each time that Court has held that because the President has the power and control over foreign affairs, issuing such agreements is within the scope of office.[7] Finally, there is something known as customary international law, which "arises from a 'general and consistent practice of states followed by them from a sense of legal obligation (*opinio juris sive necessitatis*).'"[8] This "general and consistent" practice must be both widespread and representative of the participation of states, including those whose rights are directly affected by this practice, and the amount of time that this practice has been in place for an appreciable amount of time.[9] While this type of law is not actually physically agreed upon by the United States through ratification or an executive agreement, the Supreme Court has held that customary international law is binding on it.[10] Courts in the United States have routinely applied customary international law and that a review of "federal court opinions demonstrates that customary international law has long been incorporated by the federal judiciary for federal decision-making..."[11] Essentially, customary international law is viewed and treated as federal common law, as it has been described as self-executing, meaning that it does not require any act of Congress to implement it, and it has the legal authority to preempt inconsistent state laws.[12] This is due to the fact that as a state party to various international documents, certain countries have a duty to adhere to the international laws which may require

[2]U.S. Const. amend. Art. II, §2.

[3]United Nations (1969), p. 331; Paust (1990), pp. 59–60.

[4]*See generally:* Fon and Parisi (2008), pp. 383–406.

[5]Knoll (1970), p. 94.

[6]*Id.*

[7]*Id.; U.S. v. Belmont*, 301 U.S. 324 (1937); *U.S. v. Pink*, 315 U.S. 203 (1942).

[8]Scoville (2016), p. 1895; Restatement (Third) of the Foreign Relations Law of the United States (1987).

[9]Baker (2010), p. 179.

[10]*Paquete Habana v. U.S.* 175 U.S. 677 (1900).

[11]Paust (1999), pp. 310.

[12]Bradley and Goldsmith (1997), pp. 860–861; Henkin (1984), p. 1561.

the allowance of embryonic stem cell research, or in less specific terms, research that can benefit the health of citizens of said state party. The international laws come in the form of the 1948 Universal Declaration of Human Rights, Article 25 mentions the right to health as a part of an adequate standard of living, among others, and the ICESCR which again, provides a right to health.[13] There have been various international agreements which are discussed at length below that address certain aspects of embryonic stem cell research, which come in the form of treaties or other agreements and localized laws. Aside from the potential criticism facing the effectiveness of the United Nations, there has only been a small group of laws or recommendations that have been produced by the international body which can possibly deemed to be some form of tacit acceptance.

4.1.1 Universal Instruments

4.1.1.1 Universal Declaration of Human Rights

The Universal Declaration of Human Rights is another non-binding, but widely agreed upon Declaration.[14] The fact that this Declaration was adopted by the United Nation's General Assembly is what would normally cause the rights recognized within it to be non-binding on State Parties, but Courts consider it "part and parcel of customary international law," which means that the rights recognized within its text are enforceable.[15] It has the main focus of assuring the rights of dignity to all humans.[16] That being said, there is no actual time limit given for when life actually is considered to have begun.[17] Article 1 treats "All human beings are born free and equal in dignity and rights.[18] The meaning behind the idea of being "born free" is that everyone, no matter their nationality, race, sexuality, gender, or anything else that can make someone different from others are to be treated equally.[19] Some would argue that this statement "explicitly premises human rights on birth" which would effectively mean that the unborn has no rights, and that the Declaration applies only to those who have actually been born and not to embryos.[20] They are endowed with reason and conscience and should act towards one another in a spirit of

[13] *See generally*: Office of the United Nations High Commissioner for Human Rights (2008a).

[14] Universal Declaration of Human Rights (1948a).

[15] Chan (2007), pp. 235–236; Higgins (1978), pp. 11–16; Hannum (1998), pp. 144–158; United Nations (2023).

[16] Resnik (2007), pp. 215–220.

[17] *Id.* pp. 215–216.

[18] Universal Declaration of Human Rights (1948b), Art. 1.

[19] Office of the United Nations High Commissioner for Human Rights (2019), p. 9.

[20] Copelon et al. (2005), pp. 120–129.

brotherhood"[21] Yet, others would argue that Article 3's right to life would counter this theory, including the author of General Comment No. 36 to the International Covenant on Civil and Political Rights.[22]

Article 3 discusses the right to life, yet only states that "[e]veryone has the right to life, liberty and security of person" yet falls short of discussing what is meant by "everyone".[23] As mentioned above, this Declaration is as such, non-binding but there are very persuasive arguments that at least some of this Declaration should be interpreted as having a binding effect as it has become customary international law, and customary international law essentially functions similarly to common law, meaning that it is "general practice accepted as law," and as a result it is binding.[24] Therefore, if some of the provisions of the Declaration are to be treated as customary international law, those provisions could then be interpreted as binding on the United States.[25] Many of the rights recognized in the Universal Declaration of Human Rights are protected in international treaties such as the International Covenant on Civil and Political Rights, and the International Covenant on Economic, Social and Cultural Rights.[26]

4.1.1.2 International Covenant on Civil and Political Rights

The ICCPR as discussed above, is considered a multilateral treaty, or a treaty between more than two countries, and was adopted by the United Nationals General Assembly in Resolution 2200A (XXI) in 1966 but took effect in 1976.[27] State parties to the ICCPR are promising that they will guarantee certain human rights which include many important areas, such as the right for all people to self-determination, or the ability of all people to essentially control their own lives, absent illegal intrusions by governments.[28] The treaty also lists certain rights that guarantees people their right to life, and the right to be free from harms such as torture, procedural fairness; freedom from various forms of discrimination including political, racial. religious, speech, association, among others.[29] Of specific importance to the topic at hand is Article 6 of the ICCPR deals specifically with the right to life.[30] This Article states that "Every human being has the inherent right to life. This right

[21] Office of the United Nations High Commissioner for Human Rights (2018).

[22] Human Rights Committee, General Comment No. 36 (2017).

[23] Office of the United Nations High Commissioner for Human Rights (2018); Nugraha (2016), p. 238.

[24] Zamfir (2018); Hannum (1996), p. 289; *See generally* Kunz (2017), p. 662.

[25] *Id; Paquete Habana v. U.S.* 175 U.S. 677 (1900), p. 700.

[26] *See* Human Rights Watch (2022).

[27] *See generally* United Nations Treaty Series § 999 U.N.T.S. 171.

[28] *Id.*

[29] *Id.* pp. 174–179.

[30] *Id.* p. 174.

shall be protected by law. No one shall be arbitrarily deprived of his life."[31] This document also provides for a legal remedy for those claiming that their rights under the ICCPR have been violated.[32] In order to do so, and to monitor compliance and issue directives, the ICCPR also established a body known as the Human Rights Committee to do so.[33]

In reading Article 6's statement in regards to a right to life, General Comment No. 36 issued by the Human Rights Committee in September of 2019 specifically states that:

> [T]he International Covenant on Civil and Political Rights recognizes and protects the right to life of all human beings. The right to life is the supreme right from which no derogation is permitted, even in situations of armed conflict and other public emergencies that threaten the life of the nation. The right to life has crucial importance both for individuals and for society as a whole.[34]

While it would seem that this statement would hold that the right to life is absolute and requires an active approach by the state parties, Article 6 has typically been viewed as a negative right, meaning that the state party cannot act in a manner that would violate the provisions of Article 6.[35] Yet, General Comment No. 36 of 2019 paragraph 26 seems to create a positive right in that it states that:

> The duty to protect life also implies that States parties should take appropriate measures to address the general conditions in society that may give rise to direct threats to life or prevent individuals from enjoying their right to life with dignity... These general conditions may include...the prevalence of life-threatening diseases...[36]

The same paragraph further suggests that state parties must meet in order to be in compliance with the ICCPR in dealing with the right to life including the potential need for state parties to make an active effort to work on efforts to help citizens with certain diseases and disabilities holding:

> States parties should also develop strategic plans for advancing the enjoyment of the right to life, which may comprise measures to fight the stigmatization associated with disabilities and diseases, including sexually transmitted diseases, which hamper access to medical care...and for improving access to medical examinations and treatments designed to reduce maternal and infant mortality.[37]

While it would seem that this General comment not only requires state parties to take certain actions instead of merely refraining from doing so, it also particularly mentions improving treatments for particular medical issues, as well as the duty

---

[31] *Id.*

[32] *Id.* pp. 173–174.

[33] *Id.* p. 176.

[34] *See* Human Rights Committee (2019).

[35] *See* Farha (2016).

[36] *See* Human Rights Committee (2019), p. 6.

[37] *Id.*

for state parties to take actions in improving the lives of its citizens.[38] From the above quoted statements, it would seem that the ICCPR would not only allow for research to be conducted on embryonic stem cells, but can also be read to imply that state parties are required to investigate their fitness in treating certain conditions and diseases to not only promote the "enjoyment" of life but also to treat certain diseases and medical issues.[39] It is important to note that there are currently many studies being conducted using embryonic stem cells to reduce both infant mortality and maternal mortality from many various causes.[40]

The Human Rights Committee also expanded on its stance on the right to life in General Comment No. 36 from 2017 and 2018.[41] This general comment discusses many areas that pertain to the right to life, and the protections that the state party must follow in order to comply with the articles of the ICCPR, but in relevant part, the right to life is discussed in relation to legal abortions.[42] Where many would argue that an embryo is a life that counts as a life that must be protected under the ICCPR, the ICCPR actually impliedly states otherwise by containing a provision for voluntary abortions, and by discussing the protections that must be in place to ensure that the woman or girl who is having the abortion is protected and her life is not put at risk in seeking or in having the abortion procedure.[43] Yet, there can be an argument that this comment differentiates between abortions for those who need one to protect their own life or health, as opposed to those that are purely voluntary.[44] While this is by no means a statement authorizing or promoting the destruction of embryos, this comment clearly allows for non-medically necessary pregnancy terminations by stating: "Although States parties may adopt measures designed to regulate voluntary terminations of pregnancy, such measures must not result in violation of the right to life of a pregnant woman or girl, or her other rights under the Covenant."[45]

Unfortunately, there is no further guidance in relation to what rights the unborn life has, but the General Comment also states that "…the right to life is not absolute."[46] The General Comment goes on to state that although the ICCPR does not specify when it would be appropriate to deny someone their right to life, but does specifically state that the denial of such right must not be "arbitrary."[47] Again, there

---

[38] *See generally* Human Rights Committee (2019).

[39] *Id.* pp. 4–6, 13.

[40] *See generally* Walker et al. (2018), pp. 97–124 (explaining the use of the "embryonic stem cell test" which is used to determine "embryotoxicity."); Duong et al. (2011), pp. 118–138 (discussing that embryotoxicity is a cause of issues that could lead to infant death); Chen et al. (2020), pp. 761–775 (discussing the beneficial use of embryonic stem cells in fighting maternal diabetes.)

[41] *See generally* Human Rights Committee (2017); Human Rights Committee, General Comment No. 36 (2017).

[42] Human Rights Committee, General Comment No. 36 (2017).

[43] *Id.* p. 2.

[44] *See generally id.*

[45] *Id.*

[46] *Id.* at No. 10.

[47] *Id.*

is nothing specific in regards to the unborn, but arguments can likely be made on both sides as to whether the destruction of an embryo violates this "arbitrary" standard, yet.[48]

Some have even taken to responding to the General Comments on the Right to Life found in Article 6, such as the Society for the Protection of Unborn Children who feared that the ICCPR and its General Comments: "wrongfully excludes the right to life of the most vulnerable human beings, unborn babies"[49] One important issue to note is that the United States does not hold certain provisions as actual law.[50] The United States in their "observation" to a General Comment No. 36 of 2017 on the ICCPR's Article 6 right to life stated that:

> the range of issues the Committee considers to fall within the scope of the inherent right to life and the obligations of States Parties under Article 6 is overly expansive and the Committee provides little or no authoritative legal support or treaty analysis grounded in established rules of treaty interpretation under international law to support many of its positions. The Committee's citations to its own work products, whether in the form of general comments, concluding observations and recommendations, or "views" on Protocol communications, do not in and of themselves provide legal support under international law. They merely represent a collection of the Committee's prior consistent, non-binding views and carry no greater weight or authority than when first published.[51]

The United States also recommended that the Human Rights Committee should include language in their General Comments that they:

> make explicit at the beginning of any final general comment that it reflects the Committee's views, which are not legally binding, and that the purpose of the general comment is to provide recommendations to States Parties with regard to their implementation of the Covenant and in fulfilling their periodic reporting requirements... In keeping with its advisory mandate, the Committee should refrain from providing its recommendations in imperative ("must") or mandatory ("required") terms.[52]

Further, the United States wrote in its observations in regards to the 2017 General Comment No. 36 on the right to life that it was the ICESCR that should be used in relation to human rights health issues, and that Article 6 of the ICCPR does not contain:

> [t]he right to the enjoyment of the highest attainable standard of health is not commensurate with a right to be healthy or a right not to succumb to disease. It is, instead, oriented toward the progressive realization, in accordance with a State's available resources, of the right for an individual to enjoy the highest attainable standard of health.[53]

While the above quoted text is specifically referring to the obligations that were potentially imposed by the ICCPR in regards to access to abortions, one can easily

[48] *Id.* pp. 2–14.
[49] *See* Society for the Protection of Unborn Children (2017).
[50] *See* Human Rights Committee (2017).
[51] *Id.* p. 3.
[52] *Id.*
[53] *Id.* pp. 3–4.

determine the United States' stance on the matter.[54] Even more so, with this information it would seem that the United States would argue to hold the information provided in the General Comments is meant to be merely suggestive as opposed to authoritative.[55] This would seem to indicate that the United States treats only certain provisions of General Comments with any form of authority.[56]

4.1.1.3 International Covenant on Economic, Social and Cultural Rights

The ICESCR is a similar instrument to the ICCPR as it too is a multinational treaty dealing with human right, and was adopted by the United Nations General Assembly in Resolution 2200A (XXI) in 1966 and became effective in 1976.[57] Similarly to the ICCPR, it too begins with the powerful statement that "All peoples have the right of self-determination. By virtue of that right they freely determine their political status and freely pursue their economic, social and cultural development."[58] The differences from the ICCPR begin to appear starting in Article 2, where the idea of "progressive realization" is introduced in relation to the principles enshrined in the ICESCR. .[59] Progressive realization refers to the obligations of state parties to this treaty to take actions in order for the for the rights described in the ICESCR to be fully available to all of its citizens.[60] Further, this idea actually imposes on state parties to act immediately, even if the country is finding their resources "scarce."[61] The rights recognized in the ICESCR's are unique in relation to the ICCPR as they seek to protect labor rights, or the right to work in safe working conditions with equal opportunities, the right to form unions.[62] Next, the ICESCR states the right to social security, which specifically includes the right to social insurance.[63] Article 10 of the ICESCR next discusses the right to family life where it states that: "The widest possible protection and assistance should be accorded to the family, which is the natural and fundamental group unit of society, particularly for its establishment and while it is responsible for the care and education of dependent children."[64] Section 2 further elaborates that special protections should be afforded to a mother both before

---

[54] *Id.*

[55] *Id.*

[56] *Id.*

[57] *See generally*: Equality and Human Rights Commission (2019); Cop and Eymirliolu (2005), pp. 115–146.

[58] United Nations Treaty Series § 999 U.N.T.S. 173.

[59] Office of the United Nations High Commissioner for Human Rights (2008b), p. 13.

[60] *Id.* pp. 13–16.

[61] *Id.* p. 14.

[62] United Nations General Assembly (1966), pp. 3–4.

[63] *Id.* p. 5.

[64] *Id.* p. 1.

and, for a reasonable time after childbirth, during such time the mother should be granted social security benefits.[65] This section also describes the special protections that children are to be afforded such as the right to be free from any kind of discrimination, exploitation, and employment based protections.[66] The next set of protections enshrined in the ICESCPR focus on the right to an adequate standard of living which includes the right to food, housing and clothing.[67] The next articles recognize the right to health while providing guidance on the steps to ensure progressive realization of such; the right to a free education; and the right "[t]o take part in cultural life… [t]o enjoy the benefits of scientific progress and its applications… [t]o benefit from the protection of the moral and material interests resulting from any scientific, literary or artistic production of which he is the author."[68]

As discussed above, Article 12 of the ICESCR recognizes right to health.[69] This Article begins by specifically recognizing that everyone has "the right of everyone to the enjoyment of the highest attainable standard of physical and mental health."[70] It next provides the steps needed by State Parties to "achieve the full realization of this right" which are:

(a) The provision for the reduction of the stillbirth-rate and of infant mortality and for the healthy development of the child;
(b) The improvement of all aspects of environmental and industrial hygiene;
(c) The prevention, treatment and control of epidemic, endemic, occupational and other diseases;
(d) The creation of conditions which would assure to all medical service and medical attention in the event of sickness.[71]

This issue has been ripe with questions as to how what standard of health should be applied and at what times this right should be recognized so there has been no shortage of general comments and general recommendations specifically regarding Article 12's right to health.[72] To make a determination as to how the right to health should be applied, Economic and Social Council issued a General Comment in 2000, and focused on how the right to health should be structured.[73] The main issue that this General Comment sought to clarify is the duties owed under Article 12 as to what constitutes "enjoyment of the highest attainable standard of physical and mental health"?[74] The first aspect the General Comment looked to tackle was the

---

[65] *Id.* p. 2.
[66] *Id.*
[67] *Id.* p. 6.
[68] *Id.* pp. 6–9.
[69] *Id.* pp. 6–7.
[70] *Id.* p. 6.
[71] *Id.* p. 7.
[72] Hunt (2016), p. 114.
[73] *See* Economic and Social Council (2000).
[74] *Id.* pp. 1–2.

origin and the bound of Article 12's right to life.[75] In issuing this General Comment it was found that:

> The right to health is closely related to and dependent upon the realization of other human rights, as contained in the International Bill of Rights, including the rights to food, housing, work, education, human dignity, life, non-discrimination, equality, the prohibition against torture, privacy, access to information, and the freedoms of association, assembly and movement. These and other rights and freedoms address integral components of the right to health.[76]

While this would point to the issue that one's right to health does not exist in a bubble that is able to supersede or negatively impact the rights of others that are deemed equally important.[77] Yet, this was not the end of the discussion because the ICESCR states that the right to health encompasses the highest attainable standard, so the General Comment went onto discuss the fact that the United Nations General Assembly chose not to adopt the World Health Organization's definition of health that they have published in their own WHO Constitution.[78] The WHO's definition of in regards to the right to health is much more wide reaching, definitive and concrete as it encompasses: "a state of complete physical, mental and social well-being and not merely the absence of disease or infirmity."[79] This would seem very telling as to the ICESCR's standing as to a right to health, yet the discussion could not end there because there was still no true understanding of the meaning behind "highest attainable standard" in relation to healthcare, so, a discussion on the history pertaining to this right was further discussed.[80] In looking as to how to appropriately recognize this right, it was understood that one must take into account a person's own "biological and socio-economic preconditions" as well as the State Party's particular resources, and that the right itself contains both "freedoms and entitlements."[81] The freedoms protect "the right to control one's health and body, including sexual and reproductive freedom, and the right to be free from interference, such as the right to be free from torture, non-consensual medical treatment and experimentation."[82] While the entitlements recognize and include "the right to a system of health protection which provides equality of opportunity for people to enjoy the highest attainable level of health."[83]

The General Comment also called for what essentially sounds like a form of cultural relativism which is a view "that ethical and social standards reflect the cultural context from which they are derived" meaning that one cannot judge another

[75] *Id.* p. 2.

[76] *Id.*

[77] *Id.*

[78] *Id.* pp. 2–18; *See:* World Health Organization (2023).

[79] *Id.*

[80] United Nations General Assembly (1966), p. 3.

[81] *Id.* pp. 2–3.

[82] *Id.*

[83] *Id.*

due to their own preconceived notions because what is "normal" to some may not be to others, and one must respect that there are different beliefs, practices and opinions throughout the vast number of cultures on this planet.[84] This view was created in response to those who believe that human rights are universal for all, at all times because this fails to account for cultural beliefs and cultural values that others may not share.[85] For the purpose of the right in question, this means that one must not only look to the cultural beliefs of the individual, but also must look to the availability, accessibility, acceptability, and quality in attaining the highest standard of health.[86] The availability requirements requires the State Party to make available adequate health-care facilities and programs, yet this General Comment states that the level of care that is required must be based in part on the "developmental level" of the country.[87] In dealing with the accessibility requirement, the General Comment states that the health care facilities must not discriminate in allowing physical access, economic access and access to relevant information.[88] The acceptability requirements provides that the health care facilities, goods and services "must be respectful of medical ethics and culturally appropriate, i.e. respectful of the culture of individuals, minorities, peoples and communities, sensitive to gender and life-cycle requirements, as well as being designed to respect confidentiality and improve the health status of those concerned."[89] Finally, the quality requirement in the General Comment relates to the fact that the services provided must be done so based on the cultural factors discussed above and requires that whatever is provided be based in sound science and be "medically appropriate" for the situation at hand.[90]

One must also consider the Second Optional Protocol to the ICESCR which was an instrument that could be signed by a State Party to the ICESCR and recognizes that people should be free from fear of capital punishment.[91] While this is not necessarily indicative of how the ICESCR would react to or treat embryonic research, it is telling as to the overall opinion on how lives should be respected, regardless of their station, situation, or anything else.[92]

General Comment No. 14 leads to an interesting conclusion, as it would seem to indicate that there are times when State Parties would actually be required to provide embryonic stem cell treatments to those in need when the state is advanced enough economically and scientifically, the science shows it would help in alleviating condition, and it can be done in a safe manner.[93] For this to be possible it would

---

[84] *See generally id*; Österman (2021), pp. 124–134.

[85] Binder (1999), pp. 211–221; Agyeman and Momodu (2019), pp. 23–46.

[86] *See* United Nations General Assembly (1966), pp. 3–4.

[87] *Id.*

[88] *Id.*

[89] *Id.*

[90] *Id.*

[91] United Nations General Assembly (1989).

[92] *Id.*

[93] *See:* Nugraha (2016), p. 234.

also have to occur in a country where embryonic research is permitted, whether it be for cultural reasons or other, and while the embryonic stem cell treatment would be required immediately for State Parties that are in a position to be able to do so.[94] And due to the fact that the right to health is meant to be one that is progressively realized, State Parties that eventually become able to adequately provide embryonic stem cell treatment and where it is culturally permitted would be required to begin embryonic stem cell treatments immediately once able to do so within the confines of the ICESCR.[95] In any event, the United States would not have to abide by the ICESCR as have not yet ratified the Treaty which would be needed to give it the weight of law.[96]

4.1.1.4 Universal Declaration on the Human Genome and Human Rights

This Declaration was unanimously adopted at a conference in 1997, and was than endorsed by the United Nation's General Assembly.[97] The goal of the conference was to "set universal standards in the field of bioethics with due regard for human dignity and human rights and freedoms, in the spirit of cultural pluralism inherent in bioethics."[98] This Declaration recognizes that studies involving genetic research which incorporates stem cell research has tremendous potential in improving human lives through the health benefits of these studies, yet states that human rights should always prevail over these studies, and that respecting human dignity is paramount.[99] The only real prohibition in this Declaration that relates to stem cell research comes in the form of a statement that reproductive cloning of human beings will not be permitted.[100] This is in line with the laws from many countries that deal with stem cell research. The only real issue is that there is nothing pertaining to when a being is technically "alive."[101] Even though this is merely a non-binding Declaration, it is still something that the member states should reflect in their own laws.[102]

[94] *Id.*

[95] *Id.* at 235.

[96] *See:* United States Mission Geneva (2017).

[97] United Nations Educational, Scientific and Cultural Organisation (UNESCO) (1997).

[98] United Nations Educational, Scientific and Cultural Organisation (UNESCO) (2005).

[99] Shalev (2002), pp. 137–151.

[100] *Id.* p. 141.

[101] United Nations Educational, Scientific and Cultural Organisation (UNESCO) (2000), p. 5.

[102] *Id.*

4.1.1.5 United Nations Declaration on Human Cloning

This Declaration was adopted by the General Assembly and had the goal of prohibiting different forms of human cloning.[103] This included the cloning of embryos for the use of obtaining stem cells but member states such as the Poland, stated that they agree with the Declaration, but embryonic stem cells "should be permitted only when the stem cells or stem cell lines were obtained from supernumerary cells, or when donors had expressed their willingness to permit it."[104] Due to the fact that this Declaration was adopted through a majority vote and not a unanimous one, there were objections from countries who have strong support for all forms of stem cell research.[105] It is also important to note that just like other instruments adopted by the General Assembly of the United Nation it is considered non-binding on State Parties, meaning that they do not have to adhere to what the General Assembly adopted.[106]

4.1.1.6 UNESCO's International Bioethics Committee

Due to the potential benefits and hazards that come with embryonic stem cell research, many international bodies have felt obligated to discuss them in detail in an attempt to find a workable solution, focusing on the ethical dilemmas that may come up. In November of 2000, UNESCO's International Bioethics Committee ("IBC") met in Quito, Ecuador to discuss what was known about embryonic stem cells and their potential benefits, and all of the potential ethical issues that come with them.[107] This international group discussed the following in great detail: (1) the science behind embryonic stem cell therapies; (2) possible future applications using embryonic stem cells; (3) existing international provisions relating to embryonic stem cells; (4) the philosophical and religious views pertaining to embryonic stem cell research; (5) ethical arguments relating to when an embryo is considered a "human being" or "alive"; (6) options on the potential uses of embryos for this form of research; (7) issues relating to embryos that are specifically created for embryonic stem cell research; (8) the overall ethical restraints when dealing with such a sensitive issue; and finally (9) the conclusions made by this working group.[108]

While the scientific aspects, and the potential benefits have been discussed above, the UNSESCO working group also looked to the potential future benefits in this research area.[109] The working group focused mainly on the issue of rejection of

---

[103] United Nations General Assembly (2005), UN Doc. GA 10333.

[104] *Id.*

[105] *Id.*

[106] Schwebel (1979), pp. 301–303.

[107] International Bioethics Committee (2001).

[108] *Id.*

[109] *Id.* p. 2.

donor or artificial organs, and the fact that those derived from embryonic stem cells do not pose such a threat.[110] Further, the working group also discussed information provided to them by leading scientists and researchers and how using embryonic stem cells can lead to cures and treatments for some of the worst and most pervasive diseases and disorders.[111] With the extreme importance of potential embryonic stem cell applications, the working group looked to international and even more localized provisions, yet found very little guidance due to the extreme differences in laws from throughout the world.[112] The working group next looked to the philosophical and religious issues that relate to this topic and found that most arguments against embryonic stem cell research have roots in the overall issue as to when human life technically begins.[113] This same issue seems to be the foundations for ethical issues as well, however, it is important to note that the working group specifically looks to *IVF* embryos that become surplus when they no longer serve the *IVF* purposes.[114]

Prior to issuing any findings or decisions, the IBC first delved into the argument as to when life begins, by avoiding as much of it as possible, instead, the IBC stated that they were only considering surplus embryos, prior to implantation in the uterus.[115] The IBC further found that due to the obvious "intrinsic" ethical issues, there should be certain rules as to when an embryo should be allowed to be used for stem cell research purposes:

> The IBC took account of the view that what gave the embryo a status in this context was its individual potential to develop into a person. If this potential were not present in a particular case, because there is no possibility of implantation (e.g., for medical reasons or due to parental decision, excluding any commercial incentive), then it is ethically defensible to use the embryo for medical and therapeutic research needs because it has no potential to develop into a person. The only other option is its destruction.[116]

Unfortunately, the above is not the final pronouncement of the issue because there are other methods and means of acquiring embryos, so the IBC, issued guidelines for the four main scenarios that scientists have gone about harvesting embryos for research, all of which have their own potential ethical issues:

> (1) the embryo which is created by in vitro fertilization (*IVF*) in order to be implanted in the uterus and identified as suitable for this purpose; (2) the embryo which has been created in vitro to be implanted as in (1), but which is supernumerary (excess embryos having to be created to insure successful pregnancy in infertility treatments); (3) the embryo which is created by oocyte-sperm fertilization for purposes of research or for purposes of the development of stem cell lines; and (4) the "embryo" which is created through the transfer of the nucleus of a donor cell into the denucleated oocyte.[117]

---

[110] *Id.* p. 3.

[111] *Id.*

[112] *Id.* pp. 3–6.

[113] *Id.* pp. 6–8.

[114] *Id.* pp. 8–9.

[115] *Id.* p. 9.

[116] *Id.* p. 10.

[117] *Id.*

Considering the above scenarios, the IBC adopted the following conclusions which unfortunately do not give much in terms of true guidance, yet offer their thoughtful and important opinion.[118] First, the IBC explained that they recognized the extreme importance of allowing science to progress using embryonic stem cell research because of its almost limitless potential; that a true framework should be created at each national level to take into account each sides' beliefs and ethical opinions; new studies should seek new methods in obtaining embryonic stem cells, and finally that "[i]n all aspects of research involving human embryos, particular importance should be given to respect of human dignity and the principles set out in the Universal Declaration of Human Rights (1948) and the Universal Declaration on the Human Genome and Human Rights (1997)."[119]

Perhaps most interestingly, this working group, which was then followed by the Bioethics Committee discussed that this topic is one of the most difficult they have encountered in terms of ethical issues and considered not issuing a final opinion and just say: "In these circumstances, one might ask whether it is possible for a body such as the IBC to pronounce on the subject, or whether one might simply conclude that the use of embryos for research purposes is a matter of private conviction."[120] Thankfully though, after careful analysis of the leading arguments, the working group proposed that its members should not shy away from such a contemptuous area of science and reflect an attitude of "to each his own", but instead should come up with some sort of regulations to ensure to the best possibility that embryos are not treated as any other lab equipment, or a test animal, and therefore should not just be created for the sole purpose of being destroyed for science.[121]

4.1.2 Regional Instruments

4.1.2.1 Europe

4.1.2.1.1 The Charter of Fundamental Rights of the European Union

In December of 2000, the European Union adopted the Charter of Fundamental Rights of the European Union in Nice, France.[122] This Charter sets out necessary political, social and economic rights that all European institutions must follow when implementing European law, deciding cases in European Courts or using European law or while acting in an official capacity.[123] This Charter's main purpose was to

[118] *Id.* p. 13.

[119] *Id.* p. 16.

[120] *Id.* p. 9.

[121] *Id.*

[122] European Commission (2024).

[123] Johnson (2017).

create a law that focused on European law because the European Union was still working on deciding to become a state party as a group on the European Convention on Human Rights.[124] In effect, this Charter essentially codifies the principles of European Convention on Human Rights but adds new protections as well, which include for the purposes of this book: a guarantee of human dignity, including bioethics, as well as a right to physical and mental integrity.[125]

Laws that apply to the European Union have to first go through a commission known as the Commission of the European Communities (the "CEC"), or the European Commission which serves three main functions which are: "to initiate proposals for legislation, to ensure that the provisions of the various treaties underlying the EU are applied, and to manage and execute EU policy."[126] As a direct result of their charge, the CEC had to review the issue of embryonic stem cells, and to do so, the CEC reviewed the origin of embryonic stem cells, their potential for various treatments, the various proposed sources of embryonic stem cells, how to create and enforce laws pertaining to embryonic stem cells, and of course review the various ethical issues.[127] To accomplish this great task, the CEC looked to many different sources including breaking publications, current laws within individual member states, as well as laws from non-European Union countries.[128] Interestingly, when the laws of the United States that pertain to embryonic stem cell research were brought up, the CEC found that there was a great disparity between the public and private sectors, and that real legislation did not exist which constrained the abilities of scientists, specifically National Institute of Health (the "NIH") scientists.[129] When this issue was taken up by the CEC, the Commission ended up finding that the idea of pluralism, or the idea that different cultures will have different philosophical beliefs and/or ethical guidelines to follow, and that all must be respected, so that it would be up to each member state to allow or disallow the study of embryonic stem cells.[130] Further, in the member countries where this form of research is permitted, the CEC states that "strict public control by a centralised authority" must be in place, even if the research is fully within the private sector.[131] The CEC also created a legal framework for when the study of embryonic stem cells are permitted, which requires special protections for potential donating women to prevent an increased burden on them due to the short supply of embryos; the donor must give free and informed consent; a real risk-benefit analysis which specifically states that the proposed research should not be "exaggerated;" the anonymity of the donor; prohibits the buying and selling of embryonic cells and fetal tissues; and finally, that this

[124] *Id.*

[125] *Id.*

[126] *See* Commission of the European Communities (1999).

[127] Commission of The European Communities (2003).

[128] *Id.*

[129] *Id.* p. 78.

[130] *Id.* p. 80.

[131] *Id.* p. 81.

framework should be revisited from time to time as new information can give rise to a change in any of these requirements.[132]

As mentioned above, the European Union also created the European Group on Ethics in Science and New Technologies ("EGE") which was created to be "[a]n independent, multi-disciplinary body" in Commission Decision (2016/835).[133] The idea behind the creation of the EGE was to form an advisory committee to advise relevant governing bodies on ethical aspects in dealing with new and emerging scientific research and techniques in relation to European Union laws and policies.[134] This Committee in theory, answers directly to the President of the European Commission, but in reality, also works with the European Parliament, Council, and Commission, and it is to always be made up of individuals with varying backgrounds in law, science, policy making, and ethics.[135]

On November 14, 2000, the EGE issued Opinion n°15—14/11/2000—"Ethical Aspects of Human Stem Cell Research and Use", which looks for a way to harmonize all of the relevant European Treaties and laws with the need to advance in the medical arena, while maintaining their ethical standards.[136] To accomplish their task, the EGE looked to the known research and history of all types of stem cells, as well as their potential benefits, and at the same time looked to relevant national law, and international law and to formulate a set of guidelines that they believed would ensure the protection of the dignity of the embryo, and the person donating the embryo.[137] This comprehensive overview of stem cell science delved into aspects that many other international bodies did not consider and even discussed the observed benefits of using embryonic stem cells in patients with Parkinson's disease.[138]

> Based on the positive results of experimentation on rodents and primates, clinical trials in patients with Parkinson's disease have been performed on around 200 patients over the last 10 years especially in Sweden and the USA. They have shown that the transplantation of neural cells derived from the human foetus can have a therapeutic effect, with an important reduction of the symptoms of the disease in the treated patients. The clinical improvement among these patients has been observed for 6-24 months after transplantation and in some cases for 5-10 years. It has recently been shown that 10 years after the transplantation surgery, the transplanted neural cells were still alive and producing dopamine, the compound which is deficient in the brain of patients with Parkinson's disease. However, this therapeutic approach still remains experimental. In addition, the availability of neural foetal tissue is very limited. Five to six aborted foetuses are needed to provide enough neural foetal tissue to treat one Parkinson's patient. That is why new sources of neural cells have been explored in some countries such as the US and Sweden. The aim is to derive neural stem cells from

---

[132] *Id.* pp. 82–83.

[133] *See* European Group on Ethics in Science and New Technologies (2017).

[134] *Id.*

[135] *Id.*

[136] The European Group on Ethics in Science and New Technologies to the European Commission (2000).

[137] *Id.*

[138] *Id.* p. 6.

> foetuses: these stem cells could be induced to proliferate in culture, providing much greater amounts of neural tissue for transplantation.[139]

Before finalizing their recommendation, the EGE discussed what they termed the "[f]undamental ethical principles at stake" which they had to weigh in coming to their conclusion and include: (1) "the principle of respect for human dignity"; (2) "the principle of individual autonomy", which includes informed consent requirements and respect to privacy; (3) "the principle of justice and beneficence", in relation to the "improvement and protection of health"; (4) "the principle of freedom of research", which the EGE states must be weighed against "other fundamental principles"; (5) "the principle of proportionality" which requires the research to be done in a manner that is required to achieve the stated objective when no other reasonable alternatives are available; and (6) the EGE states that a precautionary approach should be taken with the "potential long-term consequences" on the individual and society to be considered.[140] In the end, the EGE promulgated their guidelines which tackles each type of stem cell separately, first, and least ethically questionable, adult stem cells, whether coming from an actual adult or from an umbilical cord merely require informed consent from the donor; second, that "[t]he retrieval of foetal tissues to derive stem cells requires, besides informed consent, that no abortion is induced for the purposes of obtaining the tissue and that the termination timing and the way it is carried out are not influenced by this retrieval;" and finally states that it should be up to each member state to the approving or forbidding of essentially creating embryos for experimentation.[141]

Even with such a strong, multi-national body to help direct and formulate legislation involving emerging bio-technologies and treatments, there are still problems due to the fact that this Committee operates in a "grey" area, because there is no direct mention in any of the European Union's treaties or legislation.[142] Further, the EGE stated that they have to respect the idea of "pluralism" meaning that they have to recognize that there are multiple co-existing sets of laws, as each country has their own, and these other laws must be respected.[143]

While there is very little hard law in relation to embryonic stem cells specifically in the European Union, a Directive was passed which seeks to set standards of "quality, and safety for the donation, procurement, testing, processing, preservation, storage and distribution of human tissues and cells."[144] This Directive also sets the minimum standards that the laboratories that work in this type of field must meet in order to be certified.[145] A "sister" directive was also issued pertaining to how to store

---

[139] *Id.* pp. 6–7.

[140] *Id.* p. 7.

[141] *Id.* p. 14.

[142] Busby et al. (2008), pp. 840–842.

[143] The European Group on Ethics in Science and New Technologies to the European Commission (2000).

[144] *See* European Parliament (2004), pp. 1–11.

[145] Charalambouse and Genakritis (2013), pp. 161–162.

and trace the location of the donations as well as any adverse reactions noted in using the donated samples, known as Directive 2006/17/EC.[146] The main issue with these Directives is that many of the European Union member states have failed to implement them into their laws, except for Cyprus Republic, who incorporated them into their laws through a Regulatory Administrative Act 438/2008, and Greece, who incorporated them into their laws by Presidential Decree 26/2008.[147] It is also important to note that the European Union also has passed Directive 2001/20/EC which applies to clinical trials of medicines which directly affects embryonic stem cell research.[148]

Scholars have found that the European Union tends to have three "legislative tendencies" when dealing with embryonic stem cell research laws which are:

1. The creation of embryos for the sole purpose of conducting research on the created embryos is usually forbidden;
2. Some countries not only allow for the creation of embryos for study, but also allow for therapeutic cloning; and
3. The remaining countries tend to allow embryonic stem cell research to be conducted on unused embryos which were created originally for *IVF* purposes, and at the same time forbid the creation of embryos that are intended to exclusively be used in research.[149]

Even though as discussed above, the European Union has been somewhat silent as to when an embryo is to be treated as a person with a right to life, stating that it should be up to the member nations, a 2011 case that was heard by the Court of Justice of the European Union held contradictorily.[150] This case primarily deals with the patentability of embryonic stem cell methods due to Greenpeace suing Dr. Brustle, who was attempting to patent an application that looked to be beneficial to those suffering from Parkinson's disease through the use of embryonic stem cells.[151] The European Union's leading law on harvesting embryonic stem cells is called Directive 98/44 (the legal protection of biotechnological inventions), and states that patents cannot be obtained for the creation of embryos to than destroy during the removal of the stem cells.[152] Further, the Directive banned its member states from issuing patents for "uses of human embryos for industrial or commercial purposes."[153] German law allows for the assessment of whether an embryo is to be deemed a living being, yet there is no room for this assessment under the European

[146] *Id.* p. 162.

[147] *Id.*

[148] *Id.*

[149] *Id.* pp. 162–164.

[150] *See* Denoon (2011), pp. 85–87 (discussing the case of CJEU: Brüstle Oliver v Greenpeace case, C-34/10, 2011).

[151] *Id.*

[152] *Id.*

[153] *Id.*

Union's standard, so the case was sent to the Court of Justice of the European Union to decide which law to apply.[154] The Advocate General Yves Bot gave a broad definition of a human embryo, stating that "totipotent cells carrying within them the capacity to evolve into a complete human being must be legally classified as human embryos and must therefore be excluded from patentability."[155] The Court held that "'Article 6(2)(c) of Directive 98/44 excludes an invention from patentability where the technical teaching which is the subject-matter of the patent application requires the prior destruction of human embryos or their use as base material, whatever the stage at which that takes place and even if the description of the technical teaching claimed does not refer to the use of human embryos.'"[156]

This left open the ability for researchers to continue with their embryonic stem cell research, yet made patenting of such methods impossible, and even removed from the already patented methods any means to protect or enforce their patent at law.[157] This was troubling for many reasons, most importantly though, this may slow down or halt the progress of vital studies to benefit and protect those in need. Those engaged in the business of this form of research felt as though they would have no other choice but to move their studies elsewhere, and risk losing certain scientists and funding.[158] The Court in this case seemed to decide it under a moral or ethical reasoning, even though the European Union's own laws argue that the member nations should be left to decide when an embryo is to be afforded the protections of a life.[159] At the same time, many doctors and scientists are unfazed where he explained: "Many industries are based on materials which are not patentable; Coca Cola uses sugar as the basic material. Then make your money by the process of manufacturing – in our case making therapies and the associated equipment and infrastructures."

In 2016 The Court of Justice of The European Union partially reversed, or at least further clarified their previous holding regarding the patentability of stem cells from human embryos.[160] In this case, the court lifted their previous ban on patenting the process of creating embryonic stem cells that were derived using a particular method, known as parthenogenesis.[161] This process only uses an unfertilized ovum, that is then coaxed to cause cellular division.[162] The end result is a group of stem cells that will never be able to form into a human being.[163] In the holding,

[154] *Id.*

[155] Sautier (2012).

[156] *Id.*

[157] *Id.*

[158] den Exter (2016), p. 311.

[159] *Id.*

[160] *Id.*; *CJEU: International Stem Cell Corporation v. Comptroller General of Patents*, C-364/13, 2014.

[161] *Id.* pp. 5–7.

[162] *Id.*

[163] *Id.*

Advocate General Cruz Villalón created a test of sorts for determining when a patent can be issued for processes used to obtain embryonic stem cells, which is that so long as the human ovum (should be treated as human embryo) in question has no chance of developing into a live human being, then the process is patentable.[164]

Aside from case law dealing within the realm of embryos, there are a few treaties that were entered into by the members of the European Union that create positive law in different regards for embryonic research. First, there is the Treaty of Amsterdam, Article F which further explains the Treaty of the European Union.[165] Also entered into at the same time as the Treaty of Amsterdam was the Treaty Establishing the European Community, which in Article 152 (4)(a) discusses the origin of organs and tissues derived from human beings and how the member states are able to further restrict their laws in their home countries.[166] The European Union has further published the European Parliament and Council Directive 98/79/EC which involves the ethics of certain *IVF* procedures.[167] Third, fundamental ethics considerations determined by Parliament and Council Decision of 22 December 1998 in the 5th Framework Programme of the European Community for Research, Technological Development and Demonstration Activities Article 7 have played a role in embryonic stem cells policies due to the fact that research of this kind directly impacts ethical issues.[168] Next, The European Union drafted and ratified The Convention on Human Rights and Biomedicine of the Council of Europe, which was technically the first international treaty aimed at protecting human dignity and human rights set ethical, sets scientific standards for transplantations, and other novel medical treatments, as well as adds prohibitions that require informed donor consent.[169] This Treaty also promotes public debate in areas of emerging biotech arenas, and appoints the European Court of Human Rights to give advisory opinions pertaining to issues relating to the interpretation of this Treaty.[170] The Committee that created the Treaty is known as the Committee on Bioethics, and replaced the Steering Committee on Bioethics.[171]

---

[164] *Id.* pp. 13–16.

[165] *See* European Union: Council of the European Union (1997), p. 103.

[166] *See* European Union, Parliament and Council (1998), pp. 74–75.

[167] *See* European Parliament and Council (1998), pp. 5–7.

[168] *Id.* p. 6.

[169] Council of Europe (1988), pp. 2–3; Andorno (2005), pp. 134–138.

[170] Council of Europe (1988), pp. 2–3.

[171] *Id.*

4.1.2.2 Council of Europe

4.1.2.2.1 Convention for the Protection of Human Rights and Dignity of the Human Being with Regard to the Application of Biology and Medicine: Convention on Human Rights and Biomedicine/The Oviedo Convention

This Convention was established by the Council of Europe in 1997, and contains a very similar clause to that of the Universal Declaration on the Human Genome and Human Rights as it related to human cloning.[172] Although the United States is not a member of the Council of Europe and has not ratified this treaty, this Convention reflects a vast majority of the international views regarding cloning and how it can relate to stem cell research.[173]

This is such an important instrument as it is the only "international legally binding instrument on the protection of human rights in the biomedical field."[174] The Oviedo Convention specifically implicates stem cells, future experiments and treatments using stem cells by requiring informed consent of donors, scientific research, transplantations, and those who feel as though their rights have been infringed can seek redress in the European Court of Human Rights.[175] Article 18 even states that "where the law allows research on embryos in vitro, it shall ensure adequate protection of the embryo."[176] Further, Article 18 prohibits the cloning of human embryos for research, and also has an explicit prohibition against human cloning which was codified in the Protocol on the Prohibition of Cloning Human Beings.[177] This Convention explicitly states that there is no clear agreement as to the inherent rights of an embryo among the countries who have signed this agreement: "…the absence of a unanimous agreement on the definition, the member States decided to allow domestic law to provide clarifications for the purposes of the application of that Convention"[178] When situations arise where there is a disagreement as to how a treaty should be applied to specific state parties, a margin of appreciation should be granted to allow the State Party to act in accordance with their held beliefs.[179] Essentially, this idea must look to a state party's cultural, historic and philosophical differences in determining how the Convention should be applied.[180]

---

[172] Oviedo 4.IV.1997.

[173] *See generally* Shalev (2002), pp. 137–151.

[174] *See* Ashford (2021), pp. 907–909; Council of Europe (2024).

[175] *See* Holm (2004), pp. 113–118.

[176] Council of Europe Parliamentary Assembly (2003a).

[177] *See* Council of Europe (2001); Winter (1997), p. 192.

[178] *Vo v. France*, 53924/00 [2004] ECHR 326.

[179] Kratochvil (2011), pp. 327–329.

[180] *Id.* p. 332; Thompson Reuters Practical Law (2024).

4.1.2.2.2 European Convention on Human Rights

The European Convention on Human Rights (the "ECHR") is a treaty between all member states of the Council of Europe which was created to protect the human rights of the citizens of the countries.[181] The ECHR contains 18 articles and 14 protocols, and is enforced though The European Court of Human Rights.[182] Unfortunately, the Convention has not been used in a context relating to embryonic stem cells, but has been used in cases dealing with abortion, which has some similar ethical issues to those that arise from embryonic stem cells.[183] This is helpful because also like stem cells, abortion is not a right that is directly stated in the ECHR.[184] However, the application of the ECHR seems to change depending on if the member state legally allows abortions, or not.[185] It is also important to note that the ECHR does not directly mention/include or directly preclude prenatal life, as to life that is to be protected according to Article 2.[186]

A seminal case coming from France has had to deal with certain issues where embryonic stem cell issues permeate without being the actual root for the matter as a whole. This was especially true in the 2004 case *Vo v. France*, where the European Court of Human Rights had to decide an issue pertaining to when a being is actually considered to be alive.[187] This case posed an interesting situation, the initial plaintiff went to her doctor during her pregnancy, and while there, another woman with the same last name arrived for a medical procedure, accidentally, the doctor got the two women confused and accidentally performed the procedure on the pregnant plaintiff, necessitating a therapeutic abortion.[188] A criminal complaint was filed in 1991 and originally, the court held that the doctor was guilty of unintentional homicide for the accidental killing of the plaintiff's unborn baby, subsequent appeals overturned the case and found for the doctor, and finally the plaintiff brought the case to the European Court of Human Rights.[189] The human rights court held that the issue pertaining on when a being is technically alive should be an issue left for the national level, meaning that France should prescribe the timing as to when an embryo should be deemed a human being.[190]

> It follows that the issue of when the right to life begins comes within the margin of appreciation which the Court generally considers that States should enjoy in this sphere, notwithstanding an evolutive interpretation of the Convention, a "living instrument which

[181] *See* Equality and Human Rights Commission (2019).

[182] *Id.*

[183] *See generally* Puppinck (2013), pp. 142–193.

[184] *Id.* p. 165.

[185] *Id.* pp. 146–147.

[186] Council of Europe (1950), p. 6.

[187] *Vo v. France*, 53924/00 [2004] ECHR 326.

[188] *Id.* pp. 7–8.

[189] *Id.* p. 3.

[190] *Id.* p. 32.

> must be interpreted in the light of present-day conditions" (see Tyrer v. the United Kingdom, judgment of 25 April 1978, Series A no. 26, pp. 15-16, § 31, and subsequent case-law).[191]

Another issue came in the form of the fact as to how to properly apply the European Convention of Human Rights Article 2(1) which requires member states to take "…appropriate steps to safeguard the lives of those within its jurisdiction" meaning that each member state owes a duty to everyone to protect their right to life.[192] In coming to a holding over this issue the Court held:

> At European level, the Court observes that there is no consensus on the nature and status of the embryo and/or foetus…, although they are beginning to receive some protection in the light of scientific progress and the potential consequences of research into genetic engineering, medically assisted procreation or embryo experimentation. At best, it may be regarded as common ground between States that the embryo/foetus belongs to the human race. The potentiality of that being and its capacity to become a person – enjoying protection under the civil law, moreover, in many States, such as France, in the context of inheritance and gifts, and also in the United Kingdom… – require protection in the name of human dignity, without making it a "person" with the "right to life" for the purposes of Article 2. The Oviedo Convention on Human Rights and Biomedicine, indeed, is careful not to give a definition of the term "everyone", and its explanatory report indicates that, in the absence of a unanimous agreement on the definition, the member States decided to allow domestic law to provide clarification for the purposes of the application of that Convention… The same is true of the Additional Protocol on the Prohibition of Cloning Human Beings and the Additional Protocol on Biomedical Research, which do not define the concept of "human being"… It is worth noting that the Court may be requested under Article 29 of the Oviedo Convention to give advisory opinions on the interpretation of that instrument.[193]

As discussed above, the Court found that they were left with no choice as to how or when to apply human rights to that of the unborn, as the Oviedo Convention's stated explanatory report requires the Court to use domestic law when there are disagreements among the member states.[194]

The court further found that at best, the embryo in question was not yet a person, but rather had the potential to be one, and that even under the Charter's Second Article, which protects the human dignity of a person, recognizing that all have a right to life, the embryo was not entitled to the full protections, but agreed that certain protections should be granted to a lesser degree than having the right to life.[195] This case can be deemed extremely crucial when dealing with embryonic stem cell research for obvious reasons. The European Court of Human Rights expressly stated that an embryo is not to be considered a full human being with the right to life.[196] Yet at the same time, the court recognized the importance of protecting the dignity of all, including embryos.

---

[191] *Id.* p. 37.
[192] *Id.* pp. 39–40.
[193] *Id.* p. 38.
[194] *Id.* pp. 16, 38.
[195] *Id.* pp. 39–40.
[196] *Id.* pp. 36–42.

A crucial case was heard by the European Court of Human Rights, in *Evans v United Kingdom*, 43 E.H.R.R. 21 (2004).[197] The facts of the case can best be summarized thusly, the plaintiff, Ms. Evans, was engaged Mr. Johnston (both residents of the United Kingdom), during their engagement Ms. Evans discovered she had ovarian cancer, and as a result had eggs removed and fertilized by Mr. Johnston, resulting in six embryos which were frozen until Ms. Evan's body was ready for *IVF*.[198] Soon thereafter the two split, and Mr. Johnston contacted the embryo holding facility and requested the six embryos be destroyed.[199] As per Human Fertilisation and Embryology Authority's law, Ms. Evans was contacted regarding Mr. Johnston's request because the Human Fertilisation and Embryology Authority law requires both parties consent for *IVF* to continue, and with Mr. Johnston changing his mind, the embryos would have to be destroyed.[200] The High Court held that Mr. Johnston was entitled to remove his consent to disallow Ms. Evan's from having the *IVF* continue because if any of the embryos were successful, Mr. Johnston would have a legal duty over them, and because the couple was no longer together, the Court found that the duty was one which should not be forced upon Mr. Johnston.[201] Ms. Evans then appealed to the House of Lords but her appeal was denied, so she sought to have her case heard by the European Court of Human Rights.[202]

Ms. Evans claimed that the European Court of Human Rights was under an obligation to protect the rights of her frozen embryos, as they are alive and under the Court's jurisdiction.[203] The Court, in a 5-2 opinion found for Mr. Johnston, stating that although they feel for Ms. Evan's plight, and even though Article 8 of the European Convention of Human Rights (the right to a family life), they could not override Mr. Johnston's withdrawal of consent.[204] Most importantly for this book's purposes, the Court further held that an argument as to when embryo should be deemed "alive" "comes within the margin of appreciation which the Court generally considers that States should enjoy...", thus finding that the United Kingdom's law requiring both parents to consent is what controls.[205]

While this case can be deemed beneficial for embryonic stem cell research, the same problems persist as in many other situations. There is no multi-national agreement, or law set up that is easy to define. The constant fight as to when the right to life should attach seems to be a major problem in many cases and philosophical and religious arguments.

---

[197] *Evans v. United Kingdom* (2007), pp. 570–585.

[198] *Id.* p. 574.

[199] *Id.* 572.

[200] *Id.* p. 576.

[201] *Id.* pp. 575–581.

[202] *Id.* pp. 581–582.

[203] *Id.*

[204] *Id.* 582–584.

[205] *Id.;* Council of Europe (1950), p. 6.

Next, the case of *Parrillo v. Italy*, ECHR 27 Aug 2015 dealt with the supposed ownership/control made by the "parent" of an embryo.[206] The applicant, Parrillo sought to invalidate a legislative ban that prevented her from donating her unused embryos (obtained for *IVF* purposes) in respect to her right to have a private life, peaceful enjoyment of her possessions and her right to self-determination.[207] Prior to bringing her cause of action, the applicant sought to have her embryos returned for a number of years but was denied, and by the time she requested that her supernumerous embryos to be donated for research, Italy passed a ban on embryonic research.[208] The Court held against Parrillo due to the fact that the right to donate her embryos to science was found to be an unimportant aspect of her identity, and that because there was essentially a lack of agreement between the member nations of the European Union, the Human Rights Court gave great deference to Italy's own laws regarding embryonic research. The most important and impactful decisions that came from this case were that two exceptions were fleshed out in Italy's laws through dictum as to when one can conduct research on an embryo and are as follows:

1. "therapeutic scientific research that was taken with the aim of protecting the health and development of the embryo when no alternative modes existed;" and
2. Allows "embryonic stem cell research when it was performed exclusively on stem cell lines obtained from embryos destroyed outside of Europe."[209]

4.1.2.3 Americas

4.1.2.3.1 The American Convention on Human Rights

The American Convention on Human Rights was adopted at the Inter-American Specialized Conference on Human Rights, San José, Costa Rica, 22 November 1969 (came into force in 1978), with the purpose of creating a system of liberty and justice based on the respect "for the essential rights of man."[210] The American Convention includes various articles that must be evaluated within the embryonic stem cell context, such as Article 4's right to life which in part recognizes that "Every person has the right to have his life respected. This right shall be protected by law and, in general, from the moment of conception. No one shall be arbitrarily deprived of his life."[211] Article 4 also recognizes that the death penalty should be banned by countries that did not have it in place prior to signing, and that those that do have

[206] *Parrillo v. Italy*, ECHR 27 Aug 2015.
[207] *Id.* p. 10.
[208] *Id.* p. 3.
[209] *Id.* at 48 (Judge Pinto De Albuquerque's concurring opinion).
[210] U.S. Legal (2024).
[211] Organization of American States (1969), p. 123.

it, must act in accordance with the law and not add crimes for which it could be applied.[212] This Convention also recognizes the right to Humane treatment in Article 5 "[e]very person has the right to have his physical, mental, and moral integrity respected."[213] This Article also recognizes that people should be protected from all forms of torture and how criminal proceedings and criminal punishments should be handled.[214] As discussed below, human rights issues in the Americas is handled by the Inter-American Court of Human Rights but it is important to note that the United States has yet to ratify this Convention, meaning that it is not enforceable as law.[215] As this Convention contains a right to life, it is crucial to understand what and who this right protects.[216]

It would seem that those who meet the criteria of "person" would be protected, but the question remains as to what constitutes a person?[217] In order to make this determination, one must look to the history of how the word "person" is interpreted in practice.

The Inter-American Court of Human Rights has dealt with the use of stem cells, only through cases involving in vitro fertilization, yet has stated that when there are options that can save lives, the use of stem cells should not be denied.[218] The Inter-American Court has also had to decide on when life begins in a Costa-Rican case where Costa Rica banned the use of in vitro fertilization.[219] The manner in which the Court decided the starting of actual life came in the form of deciding when embryos should be protected. The Court's ruling had the effect that upon conception, the embryo enjoyed little to no rights or protections at all, and that protections and rights have to occur on a gradual basis.[220] With this ruling, the Inter-American Court has seemingly held that embryos are not technically "alive" or "human beings" until a certain time period has been met, but the time period was not actually given, and there was no guidance as to resolving this issue.[221] In another case that dealt partially dealt with embryos, *Ana Victoria Villalobos et al. v. Costa Rica, Case v. Costa Rica*, the petitioner challenged Costa Rica's ban on access to *IVF* treatments.[222] Costa Rica claimed that the band was necessary in order to protect embryonic life as it was

---

[212] *Id.* pp. 145–146.

[213] *Id.* p. 146.

[214] Paul (2012).

[215] *See* ACLU (2024).

[216] Organization of American States (1969).

[217] *Id.*

[218] *See L.H. A. y otra c/ I.O.M.A. y otra* Abeledo Perrot no. 20090394, slip. op. at 10 (Arg.) (2009) (holding that embryos are not technically "alive" at conception, but that over time the embryo starts to develop into a "being".)

[219] Gennarini (2013).

[220] Center for Reproductive Rights (2012).

[221] *See Ana Victoria Villalobos et al. v. Costa Rica* (2004) (holding that embryos are not to be considered live human beings immediately up conception, and that the previous holding from the lower courts in 2000, which stated otherwise was to be overturned).

[222] *Id.*

their belief that *IVF* procedures available at the time imply a "high loss of embryos."[223] Further, Costa Rica claimed that Article 31 of the Costa Rican Civil Code protects embryos 300 days before birth.[224] While the issue of Article 4 was not at issue in this case, the Court still held that they have long recognized the rights of people to freely decide how and when they want children, and this recognition "includes the right of all to make decisions concerning reproduction free of discrimination, coercion, and violence as expressed in human rights documents."[225] It was cautioned though that Article 32 of the American Convention requires that the rights recognized are only controlling so long as they do not wrongfully limit the rights of others.[226] While this holding clearly does not specify when a being should be considered alive for the proposes of this Convention, it chooses not to offer a blanket protection of embryonic life by through not finding that an embryo's right to life should be protected by banning procedures that can lead to embryo destruction.[227]

In a more recent case, *Xdkmok Kdsek Indigenous Community v. Paraguay*, the Inter-American Court heard a case that involved how Article 4's right to life's definition of "person" applied to the unborn yet chose not to explore how the unborn should be treated.[228] The plaintiffs in this case, the Xtikmok Kisek Indigenous community, argued that due to Paraguay's lack of recognition to this Community's ancestral land, leaving them without adequate resources, which has left them vulnerable to a number of societal issues such as having various health problems.[229] While there were other claims and arguments made in this case in relation to those who were in fact born, meaning that no argument as to the definition of "person" would come into play, but the pertinent issues arise from one unborn child who died due to fetal distress, and the other that died for unknown reasons.[230] In the end, while the Court did find violations by Paraguay for the allegations as to those not "unborn," it held that "the representatives and the Commission have not presented arguments regarding the alleged violation of the right to life of the "unborn," so that, given the absence of grounds, the Court lacks facts on which to form an opinion as to the State's responsibility in these cases."[231] Thereby not actually answering if those who are still yet to be born are in fact protected by Article 4 of the American Convention.[232]

---

[223] *Id.* p. 3.

[224] *Id.* p. 5.

[225] *Id.* p. 10.

[226] *Id.*

[227] *Id.*

[228] *See Xdkmok Kisek Indigenous Cmty. v. Paraguay*, Inter-Am. Ct. H.R. (ser. C) No. 214, 1–79, 30. (2010).

[229] *Id.* p. 5; Paul (2012), pp. 210–212.

[230] Paul (2012).

[231] *Id.*

[232] *Id.* p. 53.

One of the seminal cases involving the right to life that was decided by the Inter-American Commission on Human Rights, the court was asked whether the right to life guaranteed in the Charter was to be applied to embryos.[233] While this case was not in fact heard by the Inter-American Court, it did still retain the Court's understanding as to when the right to life is realized in the Americas.[234] The facts of the case were relatively simple, yet the question presented was a crucial one.[235] The initial case was brought by the petitioners (Christian White, Gary Potter and Catholics for Political Action), stating that the state of Massachusetts and the United States had violated the American Charter by overturning a conviction of manslaughter for a doctor who performed an abortion.[236] The petitioners charged that this ruling went against the Charter's provisions regarding the right to life; and right to health, but the Commission held that because the Charter was not yet in force, they could not hold otherwise.[237] Specifically, the petitioners argued that Article 4 of the Convention protects the unborn as it states that the right to life is to be protected by law from the "moment of conception."[238] The Commission also stated though that even if the Charter was in force, an embryo's right to life is not absolute, in that it is something that occurs gradually.[239] Further, the Commission stated that the right to life can also be restricted if there is an important goal that can be met by not granting this right in an absolute form.[240] In discussing the petitioner's claim as to Article 4 being violated, the Commission held that:

> In the light of this history, it is clear that the petitioners' interpretation of the definition given by the American Convention on the right of life is incorrect. The addition of the phrase 'in general, from the moment of conception' does not mean that the drafters of the Convention intended to modify the concept of the right to life that prevailed in Bogota, when they approved the American Declaration. The legal implications of the clause 'in general, from the moment of conception' are substantially different from the shorter clause 'from the moment of conception' as appears repeatedly in the petitioners' briefs.[241]

At the time the "hot button" issue was regarding abortion, but it is clear how the dicta in this case can lead one to see that this can apply to embryos in a stem cell research capacity. Not only are most of the embryos merely moments old, there is also a prevailing goal that can be met thanks to this form of research which could benefit many more people. It is also worth noting that the goal should be found in stem cell cases more easily when the embryo in question would be subject to destruction if not used.

---

233 *White and Potter (Baby Boy) v. United States* (1981).

234 *Id.*

235 Petersen (2005), p. 457.

236 *Id.*

237 Women's Link Worldwide (2024).

238 *White and Potter (Baby Boy) v. United States* (1981).

239 Petersen (2005).

240 *Id.* pp. 457–458.

241 The Inter-American Commission On Human Rights (1981), pp. 18–19.

4.1.2.4 Africa

4.1.2.4.1 The African Charter on Human and Peoples' Rights

The African Charter on Human and Peoples' Rights of 1981 is one of the most through and comprehensive international conventions in existence and was ratified and put into force in 1986 by a majority of member states of the Organization of African Unity.[242] The Charter encompasses many similar rights as the Inter-American, and European Charters, especially in regards to protections of embryos.[243] Article 4 provides for the "right to life"; and Article 16 provides for the right to enjoy both physical and mental health.[244] Further, Article 16 even states that member nations have a duty protect the mental and physical health of their people.[245] This is impactful on embryonic stem cell research for the same reasons as the other Charter's provisions relating to the right to life and/or health in that the debate usually encompasses the question of how far these rights extend. Does the Charter grant protection to embryos or are they not to be deemed a living being for the sake of these protections? The African Charter is more unique in that it also provides that medical treatments should be freely available to those in need, and presents another similar question: Does the right to health include the right to conduct research to benefit those in need, and do medical treatments that require embryonic stem cells fall into the category of the right to health?[246] To answer these difficult questions, or to provide some sort of insight, the Organization of African Unity established The African Human Rights Court.[247] Unfortunately there is not much in the area of embryonic stem cell issues that have been addressed by this Court, and what has been addressed has not seemed to cover issues similar to those in Europe or the Americas, however, many cases have held that certain areas of law are best left up to the member states, and this Court may act in a similar manner as the European Court to relegate these situations to their member states.[248]

Although this approach has been seen elsewhere, it is more problematic in Africa, because it is believed that the member states do not seem to treat the African Charter or even the African Human Rights Court with the respect it deserves, and instead follows their own laws and customs.[249] Commentators have stated that until this issue changes, there will be much less weight afforded to these institutions.[250]

---

[242] *See generally* Mutua (1993); Gittleman (1982), pp. 698–690.

[243] *Id.*

[244] Organization of African Unity (OAU) (1982).

[245] *Id.*

[246] *Id.* p. 5.

[247] Claiming Human Rights (2010).

[248] Mutua (2000).

[249] *Id.* at 34.

[250] *Id.*

4.1.2.5 Islamic States

4.1.2.5.1 Islamic Fiqh Council

The Islamic Fiqh Council was created to discuss and further understand difficult, and novel issues dealing with "Muslim Ummah" or Muslim community under the Muslim World League.[251] The stated goal of the Council is to take these new issues and issue rulings based on interpretations of the Holy Qur'an and the immaculate Sunnah.[252] Due to the potential that embryonic stem cells hold in healing and helping those in need, the Council met in 2003 to discuss what the boundaries of permissibility are for those wanting to ensure that they are not going against Muslim law, while still finding a way to benefit from stem cells.[253] During this meeting, the issue of when life begins was decided to take place after the ensoulment, or the moment where a human life obtains a soul, and that once this occurs, it was unanimously agreed that once this has taken place, the human life is sacred and cannot be "transgressed against."[254] In the end, the Council agreed that embryonic stem cells were a permissible area of science to study so long as the embryo has not reached the period of ensoulment, but like many other religious bodies comprised of different sects and faiths, there was still those who took issue with the timing as to when life truly begins, and the range argued for was from 40–120 days, showing how far some groups disagree.[255]

4.2 Countries with Favorable Stem Cell Laws

There are many countries who have laws that greatly differ from those in force in other countries, yet all are in place to protect, and to benefit their citizens through stem cell research. They consider their research to be therapeutic, and often times even include the ability to clone embryos for the sake of science. Therefore, it is very important to understand the different legislations from around the world which should be used to help influence some of the outdated American laws.

---

[251] Muslim World League (2023).

[252] *Id.*

[253] *See* Al-Tabba' et al. (2020).

[254] *Id.* pp. 1, 6.

[255] *Id.*

4.2.1 Australia

Australia has one of the most permissive set of laws which directly conflict with that of the United States.[256] In 2002, the Research Involving Human Embryos Act, and the Prohibition of Cloning Act were created and then followed by the Regulation of Human Embryo Research Amendment Bill in 2006.[257] Prior to the creation of these laws, the Catholic Church sought to influence the Prime Minister Howard, the Prime Minister of the country at the time to ban all forms of embryonic stem cell research.[258] This influence came in two forms first, from the Catholic Archbishop of Melbourne, Denis Hart, and second from a Catholic member of Parliament Kevin Andrews, both on the grounds that experimenting on embryos violates their right to life.[259] At the same time surveys of Australian citizens showed widespread support for embryonic stem cell research in the amount of 75%.[260] This has actually led many companies who study all aspects of stem cells to move from the U.S. to Australia where they can conduct research without government interference. The main goal is to create new medical techniques through research, and although they have a similar set of guidelines as in the United States, it is much more loosely construed, and allows for the derivation of new hESC lines so long as a license is sought and granted by the government.[261] Most of the funding for this form of research comes from the government itself, as well as a small amount of money coming from commercial enterprises.[262]

4.2.2 Europe

4.2.2.1 European Union and the Council of Europe

There is still no real consensus coming from the Council of Europe, or the European Union, yet as new advancements and knowledge come about, there are more countries starting to accept and allow for embryonic stem cell research.[263] As of now, the European Group on Ethics in Science and New Technologies has made it illegal to carry out experiments where embryos are created for the sole purpose of

---

[256] Sinclair and Schofield (2007), pp. 221–223.

[257] Australian Stem Cell Centre (2010).

[258] Oakley (2002), pp. 228–231.

[259] *Id.*

[260] *Id.* p. 228.

[261] *See generally* National Stem Cell Foundation of Australia (2019).

[262] *See generally* Sinclair and Schofield (2007), pp. 221–223.

[263] Lenoir (2000), pp. 1425–1427.

research. This has caused some issues with countries who are member of the European Union, so many have created their own laws in this area.[264]

4.2.2.2 United Kingdom

The U.K. is believed to have one of the most permissive forms of stem cell legislation.[265] Through the Human Fertilisation and Embryology Act of 1990, a set of goals and ethical standards have been created.[266] This was one of the earliest forms of legislation with such permissive laws. The ethical guidelines were fully codified in 2001, and allow for embryonic stem cell research to be conducted for reasons dealing with fertility, contraception, research in gene and chromosomal abnormalities, and for the research of potential cures that can be learned through stem cell use.[267] In a manner similar to that of Australia's laws, a license is required and can be sought through the Human Fertilisation and Embryology Authority (the "HFEA"), so long as a the above reasons have been met.[268] Since the creation of the Act, the U.K. has amended it many times, and has added other groups to permit licenses depending on what the reason for the research is.[269] The HFEA also plays a crucial role in monitoring research conducted in the arena of embryo studies.[270] Due to the vast amount of potential research and conflicting issues, the U.K. has created the House of Lords Stem Cell Committee whose primary purpose is to oversee and ensure proper implementation of embryonic stem cell research.[271] Even though this is the current rules regarding embryos, this has long been the position of the United Kingdom.[272] As a result of creating a law that seeks to regulate something so controversial as embryonic stem cells, many challenges to the House of Lords authority over the given subject.[273] For instance, the case of *R (on the application of Quintavalle) v Secretary of State for Health*, the authority of the House of Lords was called into question because the HFEA was attempting to regulate embryos created through scientific methods, and not through traditional intercourse or *IVF* procedures.[274] Using a technique known as somatic cell nuclear transfer ("SCNT"),

---

[264] Council of Europe (2003).

[265] Isasi and Knoppers (2006), pp. 9–25.

[266] *Id.* p. 14.

[267] *Id.*

[268] Human Tissue Authority (2024); Council of Europe Parliamentary Assembly (2003b).

[269] *Id.*

[270] Mahalatchimy (2017).

[271] House of Lords (2002).

[272] Bahadur (2001), pp. 785–787.

[273] Pattinson (2005), pp. 391–401.

[274] *R (on the application of Quintavalle) v Secretary of State for Health*, 2 All Engl. L. Rep. 113–30 (2003); *R (Bruno Quintavalle on behalf of the ProLife Alliance) v Secretary of State for Health* (2003) UKHL 13 (2003).

where and embryo is completely created though science, by taking genetic material which is removed from an egg and replaced by an adult cell or embryo body cell (in this case an embryo) to create an embryo.[275] Essentially the newly created embryo is a clone of the being that supplied the initial DNA.[276] Because the situation being challenged was technically a clone and not an embryo, Pro-Life Alliance, a pro-life group, claimed that because there was no fertilization process as, the law could not authorize the House of Lords to license research for clones because the law only covers fertilized eggs. The House of Lords disagreed based on their interpretation of Parliament's role in creating the law.[277] The holding by the House of Lords' Lord Bingham in this case best sums up their reasoning:

> The court's task, within the permissible bounds of interpretation, is to give effect to Parliament's purpose. So the controversial provisions should be read in the context of the statute as a whole, and the statute as a whole should be read in the historical context of the situation which led to its enactment....While it is impermissible to ask what Parliament would have done if the facts had been before it, there is one important question which may permissibly be asked: it is whether Parliament, faced with the taxing task of enacting a legislative solution to the difficult religious, moral and scientific issues mentioned above, could rationally have intended to leave live human embryos created by CNR outside the scope of regulation had it known of them as a scientific possibility. There is only one possible answer to this question and it is negative.[278]

The holding in this case clearly impacts all areas of embryonic stem cell research, and shows that the House of Lords has full control when it comes to virtually anything that uses or effects embryos.[279] This is not just true for what was known when the law was created and passed, but also future research that could not have been dreamt up by Parliament at the time of this law's drafting.[280]

Further, in English law, a fetus is not considered a person, and has no rights until the time of birth, yet is to be treated with respect.[281] Another relevant agency from the U.K. is the Human Tissue Authority which was created through the passage of the Human Tissue Act of 2004 which provides the country with rules for storing human tissues outside of specific studies.[282]

The United Kingdom also has passed what is known as the Human Rights Act 1998.[283] This Act was created for the purpose of incorporating the rights contained

---

[275] Progress Educational Trust (2017) (explaining that "cell nuclear replacement (CNR) which is the term used in the case is the same as somatic cell nuclear transfer").

[276] *Id.*

[277] *R (on the application of Quintavalle) v Secretary of State for Health*, 2 ALL. ENGL. L. REP. 113–130 (2003).

[278] *R v S of S for Health ex parte Quintavalle (on behalf of Pro-Life Alliance)*, 2 WLR 692, House of Lords (2003).

[279] *Id.*

[280] Pattinson (2005).

[281] Bahadur (2001).

[282] Mahalatchimy (2017).

[283] Human Rights Act (1998).

in the European Convention on Human Rights into United Kingdom law.[284] Another goal of this Act was to remove the necessity for United Kingdom citizens to have to go to the European Court of Human Rights, and instead, now allows their own judges to hear these cases with the need that they implement the laws of the European Court of Human Rights.[285] The Act added to the United Kingdom's laws, certain rights to reproductive issues and practices, such as the right to infertility treatments, the status of a fetus, abortions, as well as the need for consent, as illustrated in *Evans v. United Kingdom*.[286] Article 2 of the Act states that "everyone's right to life shall be protected by law" can lead to issues involving the United Kingdom's overly permissive laws regarding the treatment of embryos, yet this has not been the case, and many are arguing for further laws permitting experimentation on embryos for purposes other than what is already allowed.[287] In fact, people have argued that this decision shows that the European Court of Human Rights may not be fully compatible with European Union law, this is especially problematic because the European Union is bound to respect the European Court of Human Rights.[288] It is also important to note that this decision is not appealable as there is no right to appeal.[289]

4.2.2.3 Sweden

Sweden is arguably the leading country researching embryonic stem cells in all of Europe.[290] This is in part due to the fact that Sweden was the first of all Nordic countries to enact a specific law dealing with embryonic research back in 1991.[291] The Swedish Research Council has found that there are over 100 post-doctorate scientists working in this field.[292] Much of Sweden's success in this field was due to the fact that Sweden has promulgated multiple laws relating to embryonic stem cell research over a number of years which has led to the follow laws being passed and enforced (as well as others discussed below):

> The Activities Involving Human Eggs for Research or Treatment Purposes Act 1991 allows for research on embryos for up to 14 days after fertilization and their destruction afterwards. Subsequent regulation has built on this. The Biobanks in Medical Care Act 2002 set up a national stem cell bank. The 2005 Act on Genetic Integrity allows the creation of human

[284] *Id.*

[285] Bahadur (2001), p. 785.

[286] *Id.* at 786.

[287] *Id.*

[288] Plomer (2011).

[289] *Id.*

[290] Hallermalm (2002).

[291] Nordic Council of Ministers (2007).

[292] *See* Hallermalm (2002).

> embryos for research using somatic cell nuclear transfer (SCNT): 'therapeutic cloning'. This technique can be used only following approval by an ethics committee and by donors.[293]

Further, the government of Sweden relies on Swedish National Council of Bioethics ("Statens Medinsk-Etiska Råd") to provide advice from a scientific point of view in regards to any and all ethical and legal issues that may appear as the field of embryonic stem cell research evolves.[294] It is also important to understand that the 1991 Act has been updated throughout the years, but has always had a few very specific provisions such as the requiring of informed consent of the gamete's donors.[295] The Act also stipulates that these donated embryos cannot be implanted into the uterus of a woman, and that all research cannot be used to develop "hereditary genetic effects…"[296] The Act has also added that research proposals must be evaluated on ethical grounds and was cemented in the Ethics Review of Research Involving Humans Act of 2003.[297]

Sweden was also the first of the Nordic countries to pass a law that allowed the researchers to use somatic cell nuclear transfer techniques with human embryos, and at the same time allowed for embryos to be created for the purpose of research by fertilization.[298] The success of this particular Act can be attributed to the serious and thoughtful discussions pertaining to the ethics of such liberal laws for this area of research.[299] The Swedish government stated that they understood the ethical risks as discussed but believed that the expected outcomes and benefits of the passage of the Act would outweigh the potential risks.[300] Sweden also allows for experimentation and research on aborted fetuses so long as proper informed consent is given by the woman whom carried the fetus in the first place.[301] Sweden further requires that the woman be informed as to what experiments will or would be conducted on the fetus, and finally the National Board of Health and Welfare must also provide specific approvals as to the specific research.[302]

---

[293] *See* Small (2020).

[294] *Id.*

[295] Nordic Council of Ministers (2007).

[296] *Id.*

[297] *Id.*

[298] *See* Swedish Code of Statutes no 2006:351 (2006).

[299] Nordic Council of Ministers (2007).

[300] *See* Regeringens proposition 2003/04:148 (2004), pp. 28–29, 148; Regeringens Proposition (1990), p. 33.

[301] *Id.*

[302] *Id.*

4.2.2.4 Denmark, Netherlands, Bulgaria, Finland, Greece, and Portugal

The above European countries have laws that allow for embryonic stem cell research, and medical purposes.[303] Although the time period allowed for obtaining stem cells from embryos varies among these countries, all afford some degree of latitude for research and grants of government funding.[304] Typically, these countries allow for surplus embryos, created for in vitro fertilization, to be used as a source for obtaining embryonic stem cells, so long as some form of informed consent is provided by the donor.[305] Most of these countries have formed their own versions of advisory committees for the purpose of discussing and recommending how the laws should be amended in their respective countries in light of new scientific findings, as well as making determinations as to public funds should be distributed for worthy research.[306] In Bulgaria for example, the Bulgarian Central Ethics Commission reports directly to Bulgaria's Ministry of Health and as to ethical issues dealing with clinical trials and medical products and further advises on the ethics of research projects seeking public funds.[307] They are also charged with ensuring that Bulgaria's Health Act of August, 2004 and 28 on Assisted Reproduction Activities 2007 are complied with by researchers. Bulgaria even takes this step further and has created the Central Committee on Research Ethics at the Ministry of Education and Sciences who is charged with monitoring these potentially unethical projects to make sure they do not deviate from their proposed research.[308] Like other countries with permissive, but highly regulated laws, Bulgaria has included in Art. 135 of the Health Act of 2004 a provision which limits research and experimentation which would prohibit experimentation of embryos for the purpose of choosing its gender, unless this is being done for health reasons for the embryo.[309]

4.2.2.5 Switzerland

The Swiss have actually addressed embryonic stem cell research in a national referendum in 2004.[310] In 2003, this law was approved by the Swiss Parliament and was voted into law with a majority of 66%.[311] As a result, the Stem Cell Research Act came into effect on March 1, 2005.[312] This law cites the Reproductive

[303] *See generally* EuroStemCell (2008).

[304] *Id.*

[305] Ganeva and Small (2008).

[306] *Id.*

[307] *Id.*

[308] *Id.*

[309] *See* Aleksandrova-Yankulovska (2007); Busardò et al. (2014), pp. 1–15.

[310] Fleck (2004).

[311] *Id.*

[312] Koeferl Puorger et al. (2006), pp. 772–777.

Medicine Act of 18 December 1998 in keeping the ban on creating embryos for strictly research purposes; prohibits the creation of clones and chimeras; to use surplus embryos for research in areas other than the removal of embryonic stem cells; to place a surplus, unwanted embryo into a woman's womb to obtain more stem cells; and places a 7 day limit on when stem cells can be derived.[313] The laws in Switzerland also place a great emphasis on informed consent of the donors of surplus embryos, and on the need for ethical assessments to be made on each research project, prior to authorization to obtain and/or use embryonic stem cells.[314] There are even specific ethical and scientific requirements that must be met prior to authorization, which is used to determine if the proposed research is actually "designed to yield significant insights...;" other options are not available and/or viable; the project meets scientific requirements; and the project is deemed ethical.[315]

Even in the case of Switzerland, where the issue was approved through a national vote, there have still been many hurdles, including ones coming from the citizens themselves who are opposed to the use of embryonic stem cells.[316] Their chief argument was one which is similar in virtually all countries, which is that the opponents sought to protect the "life" of the embryo, while a minority also expressed fear that human cloning would not be far behind.[317] Fortunately for the supporters of the existing law, a counter-referendum that was introduced met with about 36% of votes, thus keeping the more progressive laws intact.[318]

4.2.2.6 Spain

One country that stands apart in regards to embryonic stem cell research from the rest of the European countries that allow this form of research is Spain.[319] Initially, Spain only allowed for research to be conducted on embryos for the purpose of aiding that particular embryo through diagnostics, and treatments.[320] In 2003, the previous law was amended to allow for people to choose to donate their embryos for research purposes.[321] Next, Spain further amended their laws to allow for surplus embryos, created for *IVF* purposes, to be donated for a specific research purpose, so long as informed consent is freely given.[322] Finally, Spain passed Spanish Law on Biomedical Research, 14/2007, which banned the creation of embryos for the purpose of

[313] Switzerland CH398 (2003).

[314] *Id.*

[315] *Id.*

[316] Ganeva and Small (2008); Fleck (2004).

[317] *Id.*

[318] *Id.*

[319] Murnaghan (2019).

[320] Law 35/1988, BOE-A-1988-27108.

[321] Law 45/2003, BOE-A-2003-21341.

[322] Law 22/2006, BOE-A-2006-9292.

research purposes, yet benefits researchers by allowing for any technique in creating new stem cell lines.[323] This legal change in Spain's research laws has made this country one of the most progressive in the entire European Union.[324] The same instrument provides for criminal punishments for those in violation of the law, and allows for imprisonment and the issuance of fines.[325]

Spain also has a counterpart to the NIH known as the Carlos III Health Institute, which created a new title of "Subdirectorate-General for Research in Cell Therapy and Regenerative Medicine" (the "Subdirectorate-General") who has the main job of overseeing oversees the Commission on Guarantees concerning the Donation and Use of Human Tissues and Cells.[326] This Commission is integral to Spain's success in this area because they are tasked with reviewing novel experiments on a case-by-case basis, which in this field of science, occurs very often.[327] Essentially, their review must begin when no other framework, or policy can be found in Spain's laws and regulations, and a special determination must be made on a unique issue.[328] One of the Subdirectorate-General's most important responsibilities is overseeing Spain's National Stem Cell Bank which was created to ensure that hESC lines were derived according to strict and safe guidelines, while also making these lines available to researchers in Spain.[329]

4.2.3 Asia

4.2.3.1 China, Singapore, South Korea, Japan, India

The countries above all support stem cell research to the same extent.[330] These countries are very permissive in terms of embryonic stem cell research which includes the granting of federal funds in order to better conduct the research.[331] The only real prohibition comes in the form of reproductive cloning, yet still allows for embryonic cloning for both research and therapeutic purposes.[332] When the abilities of stem cells were first being proven, many people from all over the world who were in need of some form of medical intervention went to China to have some sort of stem cell treatment, unfortunately, these include unapproved stem cell

---

[323] Raya and Izpisúa Belmonte (2009), pp. 483–486.

[324] *Id.*

[325] *See generally* Boggio et al. (2019).

[326] Raya and Izpisúa Belmonte (2009).

[327] *Id.* p. 485.

[328] *Id.*

[329] *Id.*

[330] *See generally* Svendsen and Ebert (2008), pp. 454–462.

[331] *Id.* p. 461.

[332] *Id.* pp. 461–462.

treatments as well.[333] Through China's Beijing Ministry of Health Medical Ethics Committee and the Southern Chinese Human Genome Research Centre Ethical, Legal, and Social Issues Committee, laws have been passed which allow for new hESC lines to be created, and the only real difference from the rest of the permissive Asian countries is that stem cell treatments need some form of government approval before being used on humans, but this approval is more likely to be given than not.[334]

4.2.3.1.1 China

China has had loose laws pertaining to embryonic stem cells for many years, and as a result, many rouge clinics sprang up which engaged in unethical, illegal, and irresponsible attempts at treatments.[335] As a result China enacted Ethical Guidelines for Human Embryonic Stem Cell Research in 2003 which still approves most forms of embryonic research, and even allows cloning for therapeutic purposes, but has banned all forms of reproductive cloning using the "Four Nos" proposed by the Ministry which state that: "Under no situation, under no circumstances, will human reproductive cloning experiments be 1) endorsed, 2) permitted, 3) supported, or 4) accepted."[336] Their loose laws are directly responsible for such investment in scientific research in China, but also tend to be responsible for the clinics being allowed to continue their illegal research.[337] There is a hard balance that needs to be found, but the rogue clinics tend to not care at all, and will continue to operate with impunity.[338] Critics argue that this is because the regulations in China provide no "bite" in that they do not even specify any form of punishment for violations.[339] On the other hand, research in China as a whole has seen many clinical successes when conducted and reported properly, to the point that clinics (both approved and rogue) have treated literally tens of thousands of people with various disorders to great success, prompting an overwhelming response by foreigners who are traveling to China for "stem cell tourism."[340] It also does not hurt that in 2006 President Hu Jintao gave an opening address to 4th National Congress for Science and Technology and explained that it was his belief that biotechnology should be a top priority for his country as it needs to catch up to much of their competition across the world,

---

[333] Cyranoski (2012), pp. 147–450.

[334] Liao et al. (2007), pp. 1107–1112.

[335] *Id.*

[336] *Id.*

[337] Cyranoski (2015).

[338] *Id.*

[339] *Id.*

[340] Cyranoski (2012), pp. 149–150.

and that they will use it to in conjunction with other practices to benefit the health of the people.[341]

Even with these loose laws, Chinese research organizations have still found a way to try to add protections for patients seeking stem cell therapies in their country.[342] The Chinese Society for Cell Biology, a stem cell and regenerative medicine institute under the Chinese Academy of Sciences (CAS), and China National Institute of Standardization have attempted to standardize research methods involving embryonic stem cells.[343] Although still in its infancy, the standards specify "…the basic qualities of human embryonic stem cells, the technical guidelines for quality control, and related requirements for the use and circulation of human embryonic stem cells."[344] This is a crucial step for embryonic stem cell research in China where there are more than 30 government approved embryonic stem cell research projects currently active, and many more which have not been approved.[345] Until now, most if not all of the current ongoing studies in China have utilized different methods for obtaining, storing, and preparing stem cells.[346] This problem is so pervasive in China, and gives the country a bad name in the foreign media in respect to stem cell tourism, which in part led to Sun Yanrong, the deputy director of the China National Center for Biotechnology Development to state that this standardization is "indispensable" and a "cornerstone" for China's development and future in the world of biotechnology.[347] This can be seen even in the laws that China's National Health and Family Planning Commission promulgated in 2015 pertaining to stem cell clinics, which only seem to require informed consent of the donors, and requiring that the stem cells used be deemed "clinical grade."[348] While this seems to some like a step forward, it the new rules failed to provide for any form of penalties for those who break such laws, and that the clinics will not even care or respect the new rules.[349]

4.2.3.1.2 Singapore

Much like China, Singapore has little to no law regarding embryonic stem cell research.[350] But in 2000, Singapore established two advisory committees to determine how to regulate embryonic stem cell research, the Biomedical Sciences

[341] Chen et al. (2007), p. 951.

[342] Xinhua (2019).

[343] *Id.*

[344] *Id.*

[345] *See* Rosemann and Sleeboom-Faulkner (2016), pp. 5–9.

[346] Xinhua (2019).

[347] *Id.*

[348] Cyranoski (2015).

[349] *Id.*

[350] Lee et al. (2017).

Initiatives, and the Bioethics Advisory Committee ("BAC").[351] These Committees sought input from its citizens (using Reaching Everyone for Active Citizenry at Home (REACH)) which comprise of many religious and national views, and after review, the BAC drafted the "Consultation Paper on Human Stem Cell Research" which it then sent to thirty nine religious and professional groups throughout Singapore.[352] After many discussions, focus groups and the reviewing of applicable comments, the BAC drafted its "Ethical, Legal and Social Issues in Human Stem Cell Research, Reproductive and Therapeutic Cloning."[353] In 2002, Deputy Prime Minister Tony Tan announced that the report would be accepted and put into force.[354] The report sought to limit experimentation on embryos that are more than 14 days old, but still allowed for the derivation of super numerous *IVF* embryos, and non-reproductive cloning, at the same time, the BAC allows for exceptions to be made for the creation of embryos for purely scientific reasons.[355]

With the loose laws in place, Singapore realized that there was great financial potential in stem cell research, especially when then U.S. president George W. Bush announced the banning of federal funds for embryonic stem cell research.[356] To capitalize on this information, Singapore decided to invest 500 million Singapore dollars into a research facility known as Biopolis.[357] The outpouring of international support for such a move by Singapore prompted many well renowned international scientists to move to Singapore and work in Biopolis.[358] There has been virtually no case law to come out of Singapore in relation to embryonic stem cell research, but the positive results and amount of research being conducted within the country speaks for itself.

4.2.3.1.3 South Korea

Unlike many other countries, especially those in the West, South Korea specifically been supporting stem cell research for many years.[359] The main law regulating research using embryos in South Korea is known as the Bioethics and Safety Act, which was drafted by merging two separate sets of regulations known as the Act on Bioethics and Safety, and reports made with input from South Korean citizens.[360] The resulting Act bans human cloning, human/animal hybridizations, and requires

---

[351] O'Brien (2014a, b).

[352] *Id.*

[353] *Id.*

[354] *Id.*

[355] *Id.*

[356] *Id.*

[357] *Id.*

[358] *See generally* Min (2014).

[359] Jung (2010), pp. 3140–344.

[360] Clay (2013).

those who collect donations to provide the donors with adequate information prior to receiving consent, and outlaws researchers from paying for eggs.[361] This country's laws seem to have found a balance that works for their nation because they were willing to listen to all interested parties prior to implementing the law, in fact, South Korea actually stated that in accepting the law, they gave great deference to the people's desires in relation to moral and ethical issues with the subject.[362]

4.2.3.1.4 Japan

Like many of the Asian countries discussed above, Japan also has loose laws relating to embryonic stem cell research.[363] This was not always the case though, until 2004, Japan had strict laws relating to this area of research even though they did not seem to be, the 2001 guidelines required an approval process for scientists to go through when seeking to experiment with embryonic stem cells.[364] The approval rule actually required a two-step approval process, where scientists first had to seek approval from a local institutional review board, and then by a science ministry committee.[365] In 2004 Japan's Council for Science and Technology Policy voted to permit scientists to engage in embryonic research when there was a therapeutic purpose behind it.[366] When Prime Minister Shinzo Abe was elected, he promised to invest an enormous sum of money into regenerative medicine and exclaimed that Japan was further along in their research than many other countries but was disappointed at the slow progress that had been made in the area.[367] Then in 2014, Japan passed the Act on the Safety of Regenerative Medicine which worked to remove research barriers for human studies testing for efficacy and safety.[368] This was seen as a move by the government in an attempt to attract more biotech companies to come to Japan and not have to worry about the costly testing that is required in many other countries, such as in the United States.[369] As time progressed, Japan's laws became looser and looser, and as of 2018, scientists and researchers in Japan can now use embryonic stem cells in regenerative medicines so long as the embryos are obtained through legal *IVF* embryos that would otherwise be destroyed.[370]

---

361 *Id.*

362 *Id.*

363 Hildreth (2018).

364 Cyranoski (2009).

365 *Id.*

366 *Accord* Ralston (2008).

367 *Id.*

368 Cyranoski (2019).

369 *See id.*

370 *See* Pacific Bridge Medical (2017).

4.2.3.1.5 India

India has until recently had only loose guidelines relating to embryonic stem cell research, created by the Indian Council for Medical Research, such as the banning of reproductive cloning, and cross hybridization of species, yet because these "rules" were only guidelines, there was no technical or real penalty for breaking them.[371] In 2017 the same council drafted an update to their guidelines, the "Indian National Guidelines for Stem Cell Research," which continue to prohibit the same as discussed above, but also ban "research on in-vitro culture of intact human embryos, beyond 14 days of fertilization."[372] The guidelines require committees to be set up and register labs that register with said committees; consent with proof recorded on video; mandatory testing of cells and donors; and further explained India's position on differing forms of cells and how they are collected.[373] As for the different forms of cells, India has separated stem cells into three categories, (1) "permissible" meaning cells that are typically induced pluripotent cells (2) "restrictive" which refers to surplus *IVF* embryos; and (3) "restrictive" which bans human reproductive cloning as well as germline gene therapy.[374] Unfortunately, this set of rules fails to adopt or even address any punitive measures that can be taken by the government against actors who fail to comply with said rules.[375] As a result of the lack of hard laws in India, their research sector has been soaring because there are no new laws that seem to be in sight that would change anything to make research more prohibitive.[376] Hinduism is the chief religion among the people of India with around 80% of its population belonging to this religion.[377] The reason by this being so interesting is that members who practice Hinduism typically believe that life begins at conception, and many have argued that: "Hindus reject the use of embryonic stem cells as Christians do."[378]

4.2.4 Israel

One of the most permissive countries in terms of embryonic stem cells laws is Israel.[379] Many of the beliefs surrounding the study of stem cells comes from the fact that Jewish law states that human life begins at some unidentified point after

---

[371] Mittal (2013), pp. 105–107.

[372] *Id.*

[373] Lahiry and Sinha (2017), pp. 149–150; Lahiry and Sinha (2019), pp. 153–154.

[374] Lahiry and Sinha (2019), p. 153.

[375] *See id.*

[376] Sharma (2009), pp. 133–140.

[377] *See* Kramer (2021); *See generally* Kurien (2015).

[378] Knowles (2009).

[379] Resmovits (2011).

conception, and if an embryo is harvested prior to reaching 40 days old, it is not technically a person yet.[380] Further, Jewish law puts an emphasis on the life already in being, over a life in development; Jewish law also recognizes that if anything is done to God's creation, meaning life, it is permissible, so long as it was done in a responsible and reasonable manner, and Jewish law also holds that procreation is a requirement for Jewish males.[381] Jewish law is relevant in this country and is almost directly responsible for shaping the nation of Israel's laws as Israel was created as a Jewish state to be a homeland for the Jewish people.[382] Israeli law further states that there is an "imperative" to help humans in need of medical treatment. Through the passing of the Use of Embryonic Stem Cells for Therapeutic Research bill in 2001, an Advisory Committee was established to determine the ethicalness of stem cell research projects.[383] This Committee has restricted only a few studies, most of which dealt with cloning of embryos for research purposes, but has at times, even allowed for this form of cloning when a compelling need was found to exist.[384] Cloning has been long considered illegal in Israel, thanks to Prohibition of Genetic Intervention Law which was passed in 1999 and along with its amendments mandates that Israel's Supreme Helsinki Committee (which was established under the abovementioned law) requires the Committee to make changes as scientific research and knowledge grows and changes.[385] Five years after the creation of this law, Israel made minor changes and reinstated the law, but ensured that the language used allows for future generations of engage in cloning once the techniques and methods are safe, and potential issues are understood and "solved."[386] Israel has been found to be one of the countries with the most amount of embryonic stem cell research conducted, and has created many new advancements in science through their permissive regulations. Israel has even faced some backlash for their extremely permissive stance on this subject based on the perception of the embryo's human dignity being harmed.[387]

Aside from the promise that stem cells potentially hold, various Israeli startup companies have sought to create new techniques to help benefit the lives of their citizens.[388] Israel actually currently leads the entire world in all forms of stem cell research per capita, which was recently confirmed by a German study conducted by the Central Library of the Research Center Julich.[389] The study found that from the

[380] *Id.*

[381] Prainsack (2006), pp. 180–184.

[382] *See generally* Elon (1969), pp. 80–140.

[383] *See generally* Knowles and Kaebnick (2007), pp. 31–37.

[384] *Id.*

[385] *See* State of Israel (1999).

[386] *Id.*

[387] Prainsack (2006), p. 174.

[388] Margulies (2006).

[389] *Id.*

years 2000–2004 Israel published 113 articles per every 1 million citizens, followed by Sweden.[390]

Citizens and researchers in Israel take great pride in the fact that they do not face the same restrictions as many other nations when dealing with embryonic stem cell research due to their cultural beliefs.[391] This can best be summed up in a quote by a top Israeli stem cell researcher named Dr. Daniel Offen who stated: "Medical research in Israel in general is very high," Offen told ISRAEL21c. "And we don't have the same ethical and religious problems here, especially with regards to embryonic research. Not only is stem cell research allowed Halachically [according to Jewish law], but it is considered a mitzvah because it saves lives."[392] This perhaps explains another interesting fact that seems to only pertain to Israel, and that is that aside from all of the private companies and universities studying stem cells for financial gain, they many of them are also working together to further this area of research with the hope that they can develop embryonic stem cells that will be approved for use by America's F.D.A.[393]

4.2.5 Islamic States

Although many Islamic states recognize embryos as living beings, most of the Muslim World League has stated that life does not technically begin at conception, but instead beings 120 days after at a time called "ensoulment."[394] Many strong Islamic states were also instrumental in causing an international convention, created by Costa Rica, and backed by the United States, to implement bans on human cloning for any purpose.[395] The Islamic nations took issue with the fact that a "Christian" notion as to when life is viewed as beginning was used, which stated that life begins during an embryo's earliest stages, which the Islamic nations disagreed with.[396] This has led to different laws coming from different countries who share similar faiths. While discussing these countries, it is necessary to discuss the leaders in the area such as Iran, Qatar, Turkey, Egypt and Pakistan.

---

390 *Id.*

391 *Id.*

392 *Id.*

393 *See* The Consortium Bereshit for Cell Therapy (2024).

394 Wadvalla (2012).

395 Corrigan et al. (2006), pp. 8–9.

396 *Id.* p. 8.

4.2.5.1 Iran

In Iran, embryonic stem cell research and therapies have long been approved by the government.[397] In fact, even Iran's Supreme Leader, Ayatollah Khamenei issued a *fatwa*, (A *fatwa* is a religious opinion issued as law about whether or not an action is permissible or what actions, if any need to be taken), which specifically addressed embryonic stem cell research in 2002.[398] In issuing this *fatwa*, Iran became the first Muslim country to produce, culture and freeze human embryonic stem cells.[399] Unlike many Western countries, Iran, like many other Muslim countries has actual laws bolstered by or joined with religious jurisprudence.[400] This removes the obligations that many other countries have to undertake to ensure the protection of embryos because Islamic Sharia already provides for their protection and further sets out goals that should be undertaken.[401] Islamic Sharia tends to encourage scientific advances, especially those that benefit humans by extending lives and finding ways to better lives such as finding cures.[402] This does not mean that debates have not occurred as to how and when an embryo should be considered alive, much like the arguments that occur in most Western countries.[403] The Holy Quran specifically states that embryos should enjoy protection and should be treated with the sanctity of human life, but distinguishes when an embryo is actually alive.[404] Clerics and religious scholars have decided that embryos should be considered technically alive at day 14, which when the differentiation of the blastocyst occurs in the embryo, and prior to day 14, the embryos are classified as "pre-embryos."[405] Pre-embryos are still deemed to be worthy of respect, yet lack the "sanctity" that those beyond the pre-embryo stage enjoy.[406]

As explained above, the Iranian laws in this field have primarily come from religious decrees, and must comply with the Qur'an and the Sunnah.[407] Yet, because the country of Iran is predominately made up of Shi'ite Muslims which are even more relaxed in their views as to potential therapeutic benefits.[408] Typically, this form of Islam allows for allows for a greater amount of access and abilities in creating new techniques which can benefit humankind. It is also of importance to

[397] Saniei (2013), pp. 345–350.
[398] Zeldovich (2016).
[399] Wadvalla (2012).
[400] Farajkhoda (2017), pp. 67–74.
[401] *Id.* p. 69.
[402] *Id.*
[403] *Id.*
[404] *Id.*
[405] Amaris and Sivaraman (2019), pp. 422–424.
[406] Farajkhoda (2017), p. 69.
[407] *See generally* Al-Hayani (2008), pp. 783–795.
[408] Serour (2013), pp. 949–952.

note that Iran has banned all therapeutic cloning for the sole purpose of research.[409] Only when *IVF* produces a surplus of embryos which would not be used for their initially intended purpose can be used for stem cell research, so long as they have not matured past day 14.[410]

4.2.5.2 Egypt

In contrast with Iran's majority of Shi'ite Muslims, Egypt's majority is made up of Sunni Muslims.[411] Egypt currently lacks laws pertaining to embryonic stem cells, but when former President of the Egyptian Medical Syndicate Hamdy al-Sayed approved adult stem cell trials, he disapproved embryonic stem cell trials.[412] In explaining his disapproval of embryonic stem cell research in 2005, Hamdy al-Sayed stated: "Destroying embryos for research is not ethically right, it's not morally right, and it does not conform to our Islamic religion as it stands now."[413] Yet in 2007, after Hamdy al-Sayed was replaced, the Egyptian Fatwa Council in Dar Al Ifta decided to allow embryonic stem cell research from surplus embryos for *IVF*, and legally aborted fetuses so long as parental consent is given.[414] In order to gain new understandings of stem cell, the Egyptian Ministry of Health decided to form the National Committee of stem cells to further the area of study so as to better understand how to formulate future laws.[415] It has been noted that Egypt is in a state of controversy over this topic, which can be seen from the differing views over the past 10 years, and how they change so regularly, but the main goal for Egypt to not pass legislative over stem cells has come from the desire to increase private sector businesses in Egypt to study this area.[416]

4.2.5.3 Qatar

Like Iran, Qatar has a majority of Shi'a Muslim residents.[417] Until recently, there was very little in terms of laws governing stem cell research.[418] Due to this issue, Qatar decided to issue regulations concerning this form of research. The government

---

[409] Saniei (2013), p. 349; Larijani and Zahedi (2007), pp. 399–408.

[410] *Id.*

[411] British Broadcasting Corporation News (2013).

[412] Viney (2012).

[413] *See* Sleem (2012).

[414] United Nations Educational, Scientific and Cultural Organization (2011), p. 19 (discussing Egyptian Fatwa Council: Research Issues on Human Organ Transplantation, May 2007).

[415] *Id.*

[416] *Id.*

[417] *See* Study Country (2024).

[418] Flynn and Matthews (2010), pp. 143–150.

of Qatar, through their Supreme Council of Health, has issued regulations concerning the use of embryonic stem cells.[419] The main goal of the Supreme Council of Health is to advance biomedical research.[420] As such, the laws pertaining to embryonic stem cells allow for donated embryos, so long as their use could lead to "responsible and scientifically worthy stem cell research."[421] Qatar also established the Qatar Biomedical Research Institute in 2007 with a hefty amount of money earmarked by Emir Hamad bin Khalifa Al Thani for scientific research.[422] In an interview by Jo Marchant with the foundation's Vice-President, Abdelali Haoudi, Mr. Haoudi explained that Qatar's foundation has joined with the James A. Baker III Institute for Public Policy at Rice University in Houston, Texas to help flesh out the moral and ethical considerations on the subject, and that an international workshop was held by Qatar which included scientists, ethicists, and religious experts from Islam, Judaism and Christianity.[423] In the end, it was found that "Islam is actually very permissive when it comes to using embryonic tissue." To investigate this finding further, Qatar held a similar session with just Islamic scholars to formulate official rules in the form of a fatwa.[424] The fatwa stated that tissues taken from embryos for up to 14 days after fertilization was acceptable, so long as parental permission was granted.[425] It also stated that embryos cannot be created specifically for this form of research, or other scientific research.[426] If all of the above conditions are met, "then destroying these fertilized eggs is nothing but the destruction of human cells that have not been enlivened with the human spirit, and using them is favourable to wasting them."[427] Mr. Haoudi was also asked what considerations were made when this fatwa was drafted and he explained that the Koran along with the Hadith and updated interpretations by Islamic scholars were all looked at.[428]

Qatar is trying to be at the forefront of this medical science in the Middle East, and with their favorable position seem to be poised to advance farther than many other countries.[429] In fact, Emir Hamad bin Khalifa Al Thani has donated a large sum of money from his county's GDP (approximately 2.8%) and has donated his own money to further the interests in this science.[430] The leaders of Qatar have stated that they would like to have an "open door" to researchers from around the world,

---

[419] Al-Tabba' et al. (2020), p. 2.

[420] Flynn and Matthews (2010), pp. 143–150; United Nations Educational, Scientific and Cultural Organization (2011).

[421] *Id.* 147.

[422] Chouchane et al. (2011), p. 2; Marchant (2011).

[423] Marchant (2011).

[424] *Id.*

[425] *Id.*

[426] *Id.*

[427] *Id.*

[428] *Id.*

[429] *See generally* Roumi (2012).

[430] Marchant (2011).

without exclusion, to attend their yearly seminars to discuss their views on embryonic stem cell research, and to join with their studies to further progress.[431] It is also worth noting that prior to the Emir's generous contributions, Qatar was not even listed as one of the top 12 countries conducting any form of stem cell research, but now is close to the forefront.[432]

4.2.5.4 Saudi Arabia

Saudi Arabia originally had a decree issued that sanctions the use of embryonic stem cells.[433] This was true for all forms of stem cell research including that which could be deemed therapeutic or medical in nature.[434] It was believed from very early on that the newer trends of the country that federal funding would end up being allowed within a very short time, and some have likened the Country's embryonic stem cell field as the their "new oil" in terms of it bringing the country a lot of money.[435] This belief came about due to the fact that Saudi Arabia had been expanding their biotech fields, which have included some other forms of stem cells that have not come from embryos, but only time will tell if anything is actually changed legally.[436] The Kingdom of Saudi Arabia is aware that their citizens want a longer life and a better quality of life, and it is believed that stem cells can lead to better treatments and techniques that will help to accomplish these lofty goals while respecting the ethical difficulties.[437] So, the 2003's Fatwa provided what the Kingdom called a "religious framework for stem cell research," which allows for the harvesting of stem cells so long as the they were obtained through one of the following sources: (1) adults, with consent (adult stem cells); (2) children, with parental consent (adult stem cells); (3) placenta or umbilical cord, with parental consent (adult stem cells); (4) embryos that have been miscarried or ones that were aborted for "therapeutic" reasons permitted by Sharia law, with "parental consent"; and (5) leftover fertilized embryos from *IVF* that are donated by the parents (embryonic stem cells).[438] The flipside to this law also provides that it is impermissible to obtain stem cells in other methods and states that the following is off limits: (1) fetuses aborted for non-medical reasons, permitted by Sharia; (2) willful fertilization for the purpose of obtaining stem cells; and (3) therapeutic cloning is banned.[439] In the creation of this law, Saudi Arabia claimed to have formulated it based on "four principles of human dignity,"

---

[431] *Id.*

[432] Flynn and Matthews (2010), pp. 143–150.

[433] Doherty (2012).

[434] *See generally* Isasi and Knoppers (2006); Mahmoud (2022).

[435] Waskey (2008), pp. 488–489.

[436] *Id.*

[437] *See generally* Alahmad et al. (2020), pp. 1–7.

[438] El-Awady (2008).

[439] *Id.*

and "five working principles."[440] The four guiding principles were described as "human dignity, being God's servants, [the idea that] God has placed the universe at our disposal and we are caretakers of the universe," and the five working principles are "protection of religion, preservation of life, protection of life, preservation of mental faculties, and preservation of wealth."[441]

4.2.5.5 Pakistan

Pakistan has a positive view on embryonic stem cells, yet has had very little in terms of research conducted, and little to no laws in place over the subject matter at all.[442] Punjab University organized a conference titled "Ethical Issues Related to Work on Stem Cells" where participants were tended to agree that this area of research should be allowed to be undertaken, and that there are no objectionable issues from their point of view in regards to Shariah.[443] It has been noted however, that due to the lack of regulations, Pakistan is in a similar position to many other countries where private investors are too fearful to start up research in Pakistan over fears of future negative laws.[444]

4.2.5.6 Turkey

Like many other European countries, Turkey is a state party to the Biomedicine Convention of Europe in 2003, yet the country also meets the criteria for being an Islamic state based on the country's chief religion.[445] As discussed above, the Convention allows for many forms of experimentation on embryos, so long as the embryos are not created for the sole purpose of being destroyed.[446] However, Turkey itself has failed to promulgate separate laws which many of the cases heard by the European Court of Human Rights usually defers to.[447] Turkey does have a penal code which has been used for illegal and unethical scientific experiments, such as Article 90 of the Turkish Penal Code which punishes human experimentation, but does not have the same or similar punishments for experiments conducted on embryos.[448] There is an exception to this law though which has been used to skirt around it, which under Article 13 of the Regulation of Assisted

---

[440] Wadvalla (2012); Zahra et al. (2015).

[441] *Id.*

[442] Zahra et al. (2015), pp. 1–8.

[443] Fadel (2012), p. 134.

[444] Al-Tabba' et al. (2020), p. 7.

[445] Vatanoğlu-Lutz (2012), pp. 191–195.

[446] *Id.*

[447] *Id.* p. 192.

[448] *Id.* p. 193.

Reproductive Treatment Centers Act ("RAPTCA") of 2009 can: "if it provides a direct benefit in terms of people treated and if there is written permission from the Ministry of Health, it is possible to conduct tests and trials."[449]

4.2.5.7 Jordan

In 2008 a Committee was established in Jordan, made up of religious scholars, physicians, scientists and others to discuss and adopt a framework which could be recommended for a future law regarding stem cells, including embryonic stem cells, and at the same time the Director of the Cell Therapy Center at the University of Jordan was working with other scientists and doctors to lobby the government to draft the same law.[450] There were two groups working on this issue due to the fact that prior to both groups coming together, a stem cell "regenerative" center opened in Jordan, and guidance was needed right away.[451] After consulting Jordan's Science and Technology Ethics Committee and the Ministry of Education, a draft of a statute was adopted and approved in 2014 in Jordan's General Iftaa' Department, which is made up of Muslim Scholars, which eventually became "Statute number 10 on Stem Cells was issued under Law number 47 on Public Health," and known as the "The Jordanian Stem Cell Statute."[452] In essence, the law stipulates that there must be substantive requirements met, prior to authorization of the study of an embryo, which are as follows: "(a) [I]t is used for a justified medical/therapeutic or research purpose and (b) is in line with Islamic and medical regulation in the country."[453] A list of procedural requirements was also included which allows for permissible sources of stem cells for research purposes and stipulates that the source must be from either an embryo created for *IVF* purposes, "from the date of fertilization and up to 5 days following the beginning of successive divisions;" adult stem cells sourced from umbilical cords, blood, teeth, other tissues, or "spontaneous legal abortion embryos and embryonic tissue inside the uterus or umbilical cord or placenta, and its tissue, amniotic fluid whether inside or outside the uterus;" induced pluripotent stem cells; somatic cell nuclear transfer stem cells; hybrid-human somatic cell nuclei transfer into a non-human animal egg, so long as it would never be implanted for "reproductive services."[454]

To ensure that the above laws and safety measures were implemented correctly, the Jordanian Stem Cell Law also vested authority in the Minister of Health who must approve the various studies conducted using stem cells.[455] In order to obtain

---

[449] *Id.*

[450] Zahra et al. (2015); Al-Tabba' et al. (2020).

[451] *Id.*

[452] *Id.*

[453] *Id.*

[454] *Id.*

[455] *Id.*

approval, those seeking it must present their hypothesis, proposed techniques, and even their employees' qualifications to ensure that the experiments are done in accordance with the law.[456]

4.2.6 South Africa

Although originally very restrictive, South Africa has since changed their views and laws regarding embryonic stem cells.[457] South Africa has very loose laws regarding embryonic stem cells now, and has passed legislation that only bans cloning for reproductive cloning, but allows for therapeutic cloning.[458] The National Health Act states that organic material may not be removed from living people that will cause them serious problem or death, but the current laws in South Africa state that an embryo is not a living being, but is a part of the mother.[459] This means that the mother would have the choice to donate her embryos for scientific purposes.[460] South Africa has tried to bring itself into the forefront of this area of research, and in 2004, South Africa became the first African nation to form a stem cell bank.[461] South Africa has also created legislation known as the Medicines Control Act which works in tandem with the National Health Act by creating clinical setting frameworks of the National Health Act.[462] One major issue facing South Africa is that there is no real implementation of their laws in regards to embryonic stem cells.[463] Another issue is that there are very few stem cell trials occurring in the country.[464]

4.2.7 Brazil

Brazil's stem cell laws are unlike that of most of South America.[465] In 2005 Brazil passed legislation that allows for most forms of embryonic stem cell research, even though there were many protests lodged primarily by Brazil's Catholic Church.[466] The Supreme Court of Brazil has also weighed in on the 2005 stem cell law by

---

[456] Al-Tabba' et al. (2020).

[457] Ralston (2008).

[458] Kassah (2011).

[459] Sithole (2011), p. 55.

[460] *Id.*

[461] *See* Pew Research Center (2008).

[462] Pepper (2015).

[463] Gaobotse (2018), pp. 1–3.

[464] *Id.*

[465] Ralston (2008).

[466] *Id.*

holding that stem cell law research does not violate the right to life.[467] Brazil is essentially experiencing a boom with biomedical research, especially in terms of embryonic stem cell with the government actually sponsoring more than 50 laboratories, and many institutions to study all aspects of stem cells.[468] Even more, Reinaldo Guimarães and Antonio Carlos Campos de Carvalho who are both former directors of Brazil's Department of Science and Technology at the Ministry of Health, with the help from Brazil's Funding Authority for Studies and Projects and the Brazilian Development Bank have created eight research centers strictly focusing on stem cells.[469] The freedoms afforded to researching embryonic stem cells is due in part to the Biosafety Act of 2005 which officially permits scientists to use surplus embryos created for in vitro fertilization for research purposes.[470] These permissive laws seem to be in stark contrast to Brazil's abortions laws which prohibit and criminalize abortions with imprisonment of up to 3 years, absent three exceptions.[471] These exceptions are not out of the ordinary and are: (1) the pregnancy was a result of rape; (2) the pregnancy will endanger the expectant mother; and (3) the baby would suffer from anencephaly.[472]

4.2.8 Cuba

One country that many would be surprised to learn has very advanced regenerative medical techniques is Cuba, despite their harsh economic problems.[473] Many news articles laud the successes of Cuba's regenerative medicine program, especially due to the fact that, in theory, all 11.3 million Cuban citizens are entitled to medical help from the country, including help using stem cells.[474] The major issue, or implication comes from the fact that there has been little to no cooperation between the United States and Cuba due to the embargo.[475] Cuba's Constitution (1976) requires the country to provide medical attention for those in need as well as while further requiring the country to make accessible health services.[476] Cuba's Public Health Law (1983) further states that "that the protection and improvement of the health of the population is a fundamental and permanent obligation of the State," and specifically, Article 4 stipulates that not only should the above be provided to the Cuban

---

[467] *Id.*

[468] Palma et al. (2015), pp. 785–797.

[469] *Id.* p. 792.

[470] de Sá Mde and de Oliveiro Naves (2008), pp. 177–191.

[471] Malta et al. (2019), p. E552.

[472] *Id.*

[473] Vickery (2010), pp. 10–14.

[474] *See generally* Healio (2013); Pineo (2019), pp. 33–37.

[475] *Id.*

[476] Evenson (2005).

people, but that scientific and medical advances must be used in helping their citizens.[477]

4.3 Countries with Restrictive Stem Cell Laws

4.3.1 France

As mentioned above, much of Europe has yet to fully agree on rules and regulations regarding the use of embryonic stem cells. However, in 1994, France passed a law which banned the creation of embryos for experimental purposes and banned the outright experimentation on embryos.[478] However, 10 years later, France passed a revised law through Parliament which upheld the band on embryonic and hESC research but created an exception to permit hESC research using surplus embryos created for in vitro fertilization, yet instituted an approval requirement which must be obtained by the Biomedicine Agency of France.[479] This revised law saw many challenges over the years, yet and as a result, the Senate attempted to make amendments which would make obtaining approvals by the Biomedicine Agency more difficult, but finally the issue was put to rest in 2011 when the National Assembly rejected the amendments and kept the law as it was created in and instituted in 2004 and kept the requirement of an exception to obtain approval for this form of research.[480] Most recently, President Hollande signed law 2013-715 which created a new and less stringent form of authorizations for embryonic stem cell research, which has now led to a rise in the amount of projects and research being conducted on embryos for new stem cell studies and treatments.[481] In doing so, President Hollande removed the need for researchers to obtain an exception and instead put into place a system which granted authorizations which would be far easier to obtain. For France to accomplish such a goal, France was required to amend their Public Health Code and did so by enacting Articles L2151-1 to L2151-8 which acts in a manner to direct and explain how and when exceptions can be given to allow for embryonic stem cell research.[482] In support of the new method of obtaining authorizations, the authors and proponents argued that it was necessary to change the old outdated laws due to them being "unduly restrictive," "hypocritical," and "absurd."[483] Further, Senator Jacques Mezard pointed out the necessity of allowing embryonic stem cell research so as to create new medical therapies for those in need,

[477] *Id.*

[478] Butler (2011), p. 277.

[479] Drabiak-Syed (2013), pp. 379–356.

[480] Weissman (2012), pp. 663–665.

[481] Hug (2006), pp. 28–36.

[482] *See* Bell et al. (2018), pp. 28–32.

[483] Drabiak-Syed (2013), p. 352.

and for France to remain competitive in global markets who do not seek to protect embryos.[484] However, it is very important to note that Senator Mezard, is very liberal with his beliefs in amending France's Public Health Code and believes that there should be practically no limitations to the types of research France allows.[485] To summarize, France allows for stem cell research to be conducted using embryos, but is more prohibitive than some other countries in Europe.[486] The French laws prohibit the use of embryonic stem cells unless certain conditions are met, which are very similar to those of the United States (to a lesser degree), but many are less restrictive in that they allow research to be conducted if there is a likelihood that medical advances will come about.[487] The French National Consultative Committee on Ethics, and the French Biomedicine Agency have broad oversight when dealing with these types of issues, and regularly issues opinions on what can and cannot be done based on bioethics.[488] So long as the experiments and research are in conformity with the laws of the country, France will usually allow for various experiments and federal funding.[489]

4.3.2 Italy

Italy has one of the most comprehensive set of laws dealing with embryonic stem cell research.[490] In 2009, a Ministerial Decree sought to prohibit the use of embryonic stem cells, and only allowed for adult stem cell research.[491] There have been scientists who have sought to challenge this Decree but have so far been unsuccessful.[492] Law 40 was passed in 2004 which is controlling as to this issue, and has also been challenged unsuccessfully.[493]

[484] *Id.* pp. 351–352.

[485] *Id.*

[486] Doherty (2012).

[487] *Id.* (translating the law of France, found in, Law on Bioethics, LOI n° 2011-814 du 7 juillet 2011 relative à la bioéthique).

[488] *Id.*

[489] *Id.*

[490] Al-Tabba' et al. (2020), p. 15.

[491] *Id.*

[492] *Id.* (describing how the ban of all forms of stem cell research have left many scientist to go to other countries who have "minimal ethical oversight" in comparison to Italy's harsh stem cell laws).

[493] *Id.* 15.

4.3.3 Germany

The laws regarding embryotic stem cell research were brought about through the passage of the Embryo Protection Act of 1991.[494] This act not only considers embryos to be humans who inherently have the right to life, but also criminalizes offenses where embryonic stem cells are used.[495] The only time embryonic stem cells can be used is if they are imported and derived before a certain date. Scientists seeking to import embryos for research must also show that doing so is "vital in developing new medical and scientific knowledge."[496] Germany is one of the rarest cases of having laws that are much more restrictive than that of the majority of the world. This may have a lot to do with Germany's constitution, also known as Basic Law (Grundgesetz), which under Article 1 recognizes basic human rights, and is further bolstered for this situation in Article 2(2) which states: "Every person shall have the right to life and physical integrity. Freedom of the person shall be inviolable. These rights may be interfered with only pursuant to a law."[497] Even further, Article 3 guarantees that all people are to be treated equal.[498] Unfortunately, the Basic Law does not define when the right to life attaches by not further explaining when life is said to actually begin.[499] In 1990, the reunification of Germany officially occurred but laws from both East and West were at times completely contradictory, so after years of debate, it was decided that abortions within the country would be deemed illegal, and punishable under section 218 of the Criminal Code.[500] This rule was created after Germany's Constitution vested in their Constitutional Court the power of control over all other governmental bodies within Germany, which has caused the Court to be also known as the "supreme guardian of the Constitution."[501] In doing so, the framers of the Constitution created the authority to protect each person's rights against any violation by public authority, and when an act of the government is challenged in these situations, the Court must balance the rights claimed by the individual and actual authority of the country, as well as take into account the legal issues at stake viewed in the context of the current "socio-political environment."[502] It is also important to note though, that Germany has provisions that allow for abortions that would otherwise violate the above mentioned law, so long as the mother would be in danger of great health risk, or death, the pregnancy was caused by rape, among other extreme social situations, but the abortion must

---

[494] Small (2011).

[495] *Id.*

[496] *Id.*

[497] *See* Grundgesetz für die Bundesrepublik Deutschland (Basic Law), Grundgesetz, GG.

[498] *Id.*

[499] *Id.*

[500] Hanschmidt et al. (2020), pp. 84–94.

[501] Kallman (1995), p. 529.

[502] *Id.* p. 56.

take place "before the end of the 12th week of pregnancy."[503] Those who violate this law can face imprisonment for 3 years, or 5 years if done so recklessly, as well as monetary fines.[504] This law was created after Germany's Constitutional Court heard a case during a time when all abortions were illegal, except for times when certain exceptions should apply, such as when the mother's life would be in jeopardy carrying the baby to birth, or if performed prior to the 13th day after conception.[505] Plaintiffs challenged the exceptions claiming that Germany's Constitution protected the unborn child's right to life as the Article 2(2) of the Constitution provides that all people have the right to life and "physical integrity" which is what the plaintiffs cite in an attempt to include the unborn.[506] The question the Court faced was whether the life of the unborn is to be protected as a life in the meaning contemplated by the Constitution.[507] After months of deliberation, the Court struck down the exceptions, choosing to protect the life of the unborn and allowing for only certain exceptions as follows: (1) abortions would be allowed for medical or eugenic reasons if done within the first 12 weeks of pregnancy; (2) pregnancies created due to a criminal assault, so long as the abortion is conducted within the first 12 weeks of pregnancy; and (3) in times of where a person can show an extreme hardship in keeping the baby, the Court would have the discretion to allow for an abortion so long as it was requested and could be done within the first 12 weeks of pregnancy.[508] In reaching their decision the justices stated that:

> human life is the highest value of the political order. Moreover, the Federal Constitutional Court was created so that this value would be observed by all organs of the state. Thus, in striking down the Abortion Reform Act, the Court is doing nothing less or more than upholding the Federal Republic's Constitution[509]

4.3.4 Norway

Unlike in Sweden, Norway started out with one of the most restrictive laws pertaining to human embryo research.[510] Norway's Act on Artificial Fertilisation of 1991 and banned the freezing of unfertilized eggs, donations the forms of eggs

---

[503] Hanschmidt et al. (2020).

[504] *Id.*

[505] Kallman (1995) (explaining that there was also an exception that was created prior to this decision that allowed for an abortion before the thirteenth day after conception); 39 BVerfGE 1 (1975); Cappelletti and Cohen (1979).

[506] Gorby (1976), pp. 557, 570.

[507] *Id.*

[508] *Id.* pp. 572, 590.

[509] Kallman (1995), p. 58.

[510] *Id.* p. 54.

and embryos, and all embryo research.[511] The topic of embryo research was again addressed in 1994 with Norway's Act Relating to the Application of Biotechnology in Human Medicine, but still kept the bans in place set forth in the 1991 Act.[512] The bans on this form of research was expected to be reversed when Norway's liberal Labour Party rose to power in 2000, yet the bans remained in place until January of 2007, when an amendment was made allowing for certain forms of embryonic research, the government had only lifted the bans on assisting with reproduction.[513] The 2007 amendment stipulated that the only embryos that were allowed to be studied were the surplus ones created for in vitro fertilization.[514] It also stated that the research conducted on the surplus embryos must have an aim relating to furthering in vitro fertilization procedures, developing new techniques of "preimplantation genetic diagnosis, or promote knowledge to combat serious diseases."[515] Further, the proposed experiments must also be approved by Norway's regional ethics committees.[516] One of the most interesting aspects of Norway's laws pertaining to embryonic studies is that informed consent from the parents is not all that is required, if there was a sperm donor involved, the donor must too consent.[517]

4.3.5 Austria

Until 2004, Austria had no laws regulating forms of embryonic research, this changed however with the introduction of Reproductive Medicine Act (Fortpflanzungsmedizingesetz) was passed which banned all forms of stem cell procurement from embryos.[518] The only exception to the total ban of embryo research is in the form research to help with reproduction.[519] This country is in an area of change regarding stem cell laws which can be seen from Austria's Bioethics Commission which was formed in 2001, but seems to have changed their opinions as to the ethics of embryonic research, and has actually advised that the laws should be amended to allow for generating embryonic cells for research purposes.[520] Yet, a smaller number in this advisory committee voted against the lessening of restrictions

---

[511] *Id.*
[512] *Id.*
[513] *Id.*
[514] *Id.*
[515] *Id.*
[516] *Id.*
[517] *Id.*
[518] Small (2012).
[519] Kostenzer et al. (2021), pp. 1–2.
[520] Al-Tabba' et al. (2020), p. 15.

and sought to impose further regulations on this area of research, dooming the chances for a consensus to officially recommend any legal changes.[521]

4.3.6 *Lithuania*

Lithuania has one of the most restrictive forms of law pertaining to embryonic research of all kinds.[522] In 2000 Lithuania passed the Law on Ethics of Biomedical Research (amended in 2007) which strictly prohibits all forms of embryonic research other than strict clinical investigations which must remain non-invasive.[523] The only time this law is somewhat flexible is when it can be shown that the medical risks to the embryo "...are not disproportionate to the potential benefits."[524] Further, all forms of cloning are strictly prohibited.[525]

4.3.7 *Ecuador*

Out of all of the South American countries, Ecuador has the strictest laws pertaining to embryonic stem cell research, and bans all forms of embryonic stem cell research, as well as cloning.[526] This ban also encompasses excess embryos created for *IVF* purposes.[527] Interestingly, a researcher investigated the reasoning for Ecuador's laws, and found a unique perspective on how the citizens of the country view an embryo and it was unlike virtually any other country.[528] Elizabeth F.S. Roberts found that both the country and religious institutions do not exert a significant amount of control or influence, and yet, these laws exist.[529] Ms. Roberts also noted a key difference in how the people of Ecuador view what others would consider fertilized embryos, as something akin to "pre-life," or more of a set of cells or cellular structures that have yet to begin to develop.[530] Even with these opinions, Ecuador's pertinent law specifically bans obtaining any human embryos for any experimental research.[531]

---

[521] *Id.*

[522] *See* Busardò et al. (2014), pp. 1–15.

[523] *Id.*

[524] *Id.*

[525] *Id.*

[526] Gardner and Watson (2005).

[527] *See* Roberts (2007), pp. 181–189.

[528] *Id.* p. 183.

[529] *Id.* p. 183–184.

[530] *Id.* p. 186.

[531] *See* Law no. 26, Health Law [Ley OrgAnica de Salud], 423 REG. OF. (Ecuador 2006), Art. 214.

4.3.8 Canada

Canada is a country with laws very similar to those that are in place in the United States.[532] In fact, many doctors have argued that federal regulations are causing major problems in advancing medicine.[533] The majority of legislation concerning stem cell research prohibits the use of embryos for virtually any reason.[534] However, discarded embryos that were left over from in vitro fertilizations can be used, so long as they were created for reproductive purposes, donated with free informed consent, and that they were not obtained through any financial transaction[535] When a clinical trial using embryonic stem cells is created, the researchers must first prove with "overwhelming evidence" of the trials "safety and efficacy."[536] Canada's Canadian Institutes of Health Research (the "CIHR") developed their Guidelines for Human Pluripotent Stem Cell Research, which became a part of the Tri-Council Policy Statement: Ethical Conduct for Research Involving Humans (TCPS 2), Chapter 12, section F, and created their Stem Cell Oversight Committee to review research proposals, and the actual research itself.[537] The rules stated above were issued as the CIHR's Stem Cell Guidelines in a similar manner to how the United States' NIH.[538] So long as these requirements are met, Canada will grant public funds to researchers.[539]

4.4 Countries Without Stem Cell Laws

4.4.1 Ireland

The laws of Ireland can be seen as a very different case in Europe. This is because they do not have any regulations regarding stem cells at all.[540] In fact, the only rules regarding any form of embryonic research comes from the "Guide to Professional Conduct and Ethics for Registered Medical Practitioners" which was first instituted on 2004 and amended in 2009.[541] There is virtually nothing specific in these guidelines pertaining to embryonic stem cell research, only assisted human

[532] Canadian Institute of Health and Research (2010); Canadian Institutes of Health Research (2022), pp. 241–248.

[533] Blackwell (2012).

[534] *Id.*

[535] Al-Tabba' et al. (2020), p. 14.

[536] Nelson (2010).

[537] Canadian Institutes of Health Research (2023).

[538] Johnston (2002), pp. 722–726.

[539] *Id.*

[540] Al-Tabba' et al. (2020), pp. 15–16.

[541] *See* Medical Council (2016).

reproduction treatments and gamete donation are discussed as somewhat regulated areas.[542] The only decision coming from Ireland was in the case of *M.R. v T.R.* where Ireland's high court ruled that embryos that are created from outside of the womb are not protected by the Irish Constitution, yet those who are created in the womb are protected as living beings.[543] There seems to be a belief that this will soon change due to the fact that there are proposed laws regarding stem cell experiments and research, but none of which have been signed into law.[544] In fact, Justice Brian McGovern stated that when the Constitutional Amendments on Article 40.3.3 were voted into law, it was not within the realm of beliefs of the voters that an unborn embryo meant anything other than an embryo within a womb.[545] It is forgivable for one to assume that this lack of laws would be beneficial for this area of study in Ireland, but that has in fact not been the case.[546] In an article written by Ciara Staunton, the chief legal officer of the Irish Stem Cell Foundation, Ms. Staunton has highlighted two major problems with Ireland's law: (1) Companies are unwilling to invest in an area of science which is under regulated for fear that regulations are on the horizon and they are still completely unknown; and (2) Due to the lack of regulations, rouge scientists have been able to more freely operate without fear of repercussions for their abusive, dangerous, and untested treatments and other scientists do not want to risk associating with them.[547]

Even though there are no hard laws in Ireland to cover this subject, Ireland still maintains two vital committees which are considered independent and autonomous of the government yet look to the philosophical and ethical issues pertaining to medical advances that some deem questionable or problematic.[548] The Irish Council of Bioethics was established in 2002, and has met almost annually to discuss the relevant stem cell laws and embryonic status in Ireland.[549] Both groups have joined together with another group as the Irish Stem Cell Foundation, and together they are trying to influence future laws, and studies by bringing together doctors, patients, scientists, advocates, students, teachers and others with an interest in the subject.[550]

---

542 *Id.*

543 Al-Tabba' et al. (2020), p. 15.

544 *Id.* p. 16.

545 Whelan (2007).

546 Staunton (2013).

547 *Id.*

548 *See* The Irish Council of Bioethics (2004).

549 *Id.*

550 *See* Molloy and Smith (2018), p. 75.

4.4.2 Croatia, Luxembourg, Malta, Poland and Slovakia

The above-named European countries have no laws pertaining to embryonic research.[551] Some however have indicated through talks with the European Union that if issues pertaining to embryonic research were to come up, they would not allow them to continue.[552]

4.4.3 Mexico

There are many problems surrounding the ability for legal stem cell research in Mexico. This is primarily due to the fact that there are no real laws regarding this field. As a result, there are many "unregulated, and questionable" forms of research being conducted using stem cells.[553] This lack of law, creates a whole separate issue.[554] This has also caused the usefulness of stem cells usage to be called into question. Many people hear and read about the problems that can come about using stem cells, but the real problem came from the fact that the experiments and treatments used were not approved, or likely even tried, prior to being used on a person who then suffered.[555] This is best illustrated by the fact that companies in Mexico claim to offer approved, or proven treatments, yet only a small handful of these claims has been actually approved and looked into by Mexico's Federal Commission for the Protection Against Sanitary Risk (Comisión Federal para la Protección contra Riesgos Sanitarios,) a branch of the Mexico's Ministry of Health who is charged with overseeing, inspecting, approving and authorizing activities related to human stem cells.[556] Mexico's Constitution promises access to health care, and as such has passed a law known as the General Health Law which contains the Biomedical Research Regulation which stipulates various requirements for medical experimentation for most samples derived from humans, yet somehow embryonic stem cells are not covered.[557] Further, the General Health Law prohibits false advertising pertaining to medical treatments that have not been supported by clinical trials, but unfortunately, this prohibition is not enforced.[558] Companies in Mexico have profited tremendously thanks to stem cell tourism, which includes the real

---

[551] Hovatta et al. (2010), pp. 1005–1007.

[552] Kallman (1995).

[553] de Jesus Medina Arellano (2012), pp. 63–78.

[554] *Id.* p. 73.

[555] Duffy (2002), pp. 1–6.

[556] *See generally* Board on Health Sciences Policy et al. (2014), pp. 1–9.

[557] *Id.* p. 6.

[558] *Id.*

treatments, as well as the many scams that are able to operate with complete impunity.[559]

Mexico's one seemingly applicable law related to the banning of the commercialization of human organs, tissues, and cells was amended in 2012. The original goal was to prevent organ and tissue trafficking, but in 2012 Mexico adopted a change to the law which now states "[T]hat the procurement, extraction, analysis, derivation, preservation, transportation, and supply of organs, tissues, and stem cells are not considered acts of commerce, making this area outside of Mexico's laws."[560] Many blame the powerful cartels and their abilities to pay off lawmakers or at the least lobby for their desires.[561] As a direct result of Mexico's lack of laws coupled with the tremendous amount of corruption, Mexico has ended up with a tremendous problem with human organ and tissue trafficking for a number of years.[562] To better illustrate this fact, one only needs to look to the well documented case of the cartel known as the Knights Templar who kidnapped a number of schoolchildren for the sole purpose of harvesting their organs, but luckily due to a mistake by the driver, was found prior to mutilating the children.[563] There are many other cases with much worse endings for the children, some of who have been found in shallow graves, with their organs missing.[564] To make matters worse, the politicians claim that this is not happening within their jurisdiction and actively work to cover the truth up.[565] As it turns out, many cartels have found it more profitable or at least easier to harvest human organs, tissue, or other human derived items than it is to trafficking drugs, especially in the areas that boarder the United States where there has been an increase in drug trafficking crackdowns.[566] With this in mind, it is not hard to imagine or see that this too applies to the harvesting of embryos to sell on the black market.[567] Mexico in particular is an extreme example but one which must be shown because in this instance, the lack of laws have egregious repercussions.[568]

---

[559] *See generally* Ryan et al. (2009), pp. 27–33.

[560] *Id.*

[561] *See generally* del Bosque (2017).

[562] *See generally* Agren (2017).

[563] Sedlak-Hevener (2019).

[564] *Id.*

[565] *Id.*

[566] *Id.*

[567] *Id.*

[568] *Id.*

4.4.4 Argentina

An interesting situation has occurred in Argentina where there are no specific laws pertaining to stem cells at all.[569] However, Argentina has attempted to regulate this field under their Transplant Act (1993) and the Medicines Act (1964).[570] The Transplant Act finds its regulatory authority in the Unique Central Institute for Ablation and Implantation ("INCUCAI"), and the INCUCAI has further promulgated resolutions which pertain to some aspects of stem cells, such as the proper means of storage, as well as who may be eligible to store them and at what times.[571] The Medicines Act, through the Administration of Drugs, Food and Medical Technology, Regulation 7075/2011, and 7729/2011 is in place in an attempt to regulate experimental procedures and research on the national level only, and not local.[572] As a result, there are situations where the law does not apply even on a national level such as the situation where there are pre-approved sites who derive stem cells strictly for use within their own location.[573] Aside from these two acts which only barely touches on issues relating to stem cells, there are no actual regulations that cover them.[574] This has led to many problems where researchers, and doctors make wild claims regarding their treatments which people believe and undergo even though the claims have no factual background and may have never actually been tested in the first place.[575] Because there are no laws pertaining to non-national research and treatments, those making these wild claims are effectively protected and are only bound to some form of unwritten ethical principles which are not enforceable.[576] Further issues have arisen due to the fact that Argentina, on a national level, and the provincial areas hold themselves out as "guarantors of health" for their The author of "Stem Cell Research and Therapies in Argentina: The Legal and Regulatory Approach" has expertly summed up the situation relating to embryonic stem cell laws in Argentina as such:

> [T]he existing regime establishes a poorly harmonized/coordinated regulatory patchwork that has proven to be insufficient to uniformly regulate regenerative medicine (and stem cell) research and its subsequent translation into clinical practice and treatments. Moreover, attempts by regulatory authorities to stop these practices, though valiant, have proven ineffective.[577]

---

[569] de Arzuaga (2013), pp. 40–43.

[570] *Id.*

[571] *Id.* pp. 42–43.

[572] *Id.*

[573] *Id.* p. 41.

[574] *Id.* p. 42.

[575] *Id.*

[576] *Id.* p. 43.

[577] *Id.*

4.5 Laws and Decisions from the United States

Prior to the case of *Dobbs v. Jackson Women's Health Organization,* many would have agreed that the case of *Roe v. Wade* from 1971 which held that as a right of privacy, women have the right to obtain legal abortions.[578] The general belief was that this issue is a fourteenth amendment one, and with this decision, one could see the applicability of this decision in embryonic stem cell research. Nearly four decades after *Roe v. Wade* was decided, this issue started to make headlines again with it being challenged in both the courts and in the States by creating new anti-abortion laws, or by new challenges coming to the courts.[579] The most dramatic shift came from a legal situation in Mississippi where a 2020 court decision essentially overturned *Roe v. Wade* by holding that: "The conclusion that abortion is a constitutional right has no basis in text, structure, history or tradition." (this case would continue until reaching the Supreme Court and eventually became *Dobbs v. Jackson Women's Health Organization.*)[580]

The *Dobbs* case initially began due to a Mississippi state law which banned most abortions after the first 15 weeks of pregnancy which was challenged by an abortion clinic (Jackson Women's Health Organization) and one of its doctors.[581] In attempting to argue against this law, various arguments were proposed by the abortion clinic, among the most pertinent was that the law in question violated the Due Process clause of the 14th Amendment, as it had been used in the past to apply rights that had not been expressly included in the Constitution, such as the right to an abortion.[582] This Clause was used because it has been held to provide protections for certain substantive rights, those being the enumerated ones in the first eight Amendments, and the others being rights deemed fundamental, but not mentioned specifically anywhere in the Constitution.[583] To make a determination of whether a potential "right" should be protected when not mentioned in the Constitution, the Court looks to the "deeply rooted in [our] history and tradition" and whether it is essential to our Nation's "scheme of ordered liberty."[584] The Court disagreed with this argument as "...abortion is fundamentally different, as both *Roe* and *Casey* acknowledged, because it destroys what those decisions called 'fetal life' and what the law now before us describes as an "unborn human being."[585] The Court went on to discuss how the term "liberty" should apply and looked to previous case law

[578] *Dobbs v. Jackson Women's Health Organization*, 2022 U.S. LEXIS 3057, No. 19-1392, 597 U.S. ___ (2022); *Roe v. Wade*, 410 US 113 (1973).

[579] Oxner and Borha (2021); Totenberg (2021).

[580] *Id.*

[581] *Dobbs*, 2022 U.S. LEXIS 3057.

[582] *Id.*

[583] *Id.* pp. 11–12.

[584] *Id.* p. 12.

[585] *Id.; Planned Parenthood of Southeastern Pa. v. Casey,* 505 U.S. 833 (1992) (discussing that the Court upheld *Roe v. Wade* due to the doctrine of *stare decisis* which requires courts to uphold

which to them, provided "little guidance" as the term can be very ambiguous.[586] For even further understanding, the Court looked to the legal history of abortions on a federal and state level and found that no law was found to protect the right to obtain an abortion, and no case law provided a similar right until a few years before *Roe v. Wade*, and was in fact, a crime in "every single state."[587] After fully analyzing *Roe v. Wade* and its subsequent case law, the Supreme Court held: "The Constitution makes no reference to abortion, and no such right is implicitly protected by any constitutional provision..." and with that the federal protection that provided the so-called "right to abortion" was gone.[588] The Court held that under substantive due process, a woman has the right to terminate her pregnancy under certain conditions such as the abortion must be done prior to viability of the fetus; the state has a right to prevent abortions post-viability, unless the woman's life or health would be at risk; and establishes that the state has a legitimate interest in protecting the health of the pregnant woman and the fetus.[589] The Court essentially pointed to a dividing line which would be the time of viability of the fetus.[590]

Up until *Dobbs*, *Roe v. Wade* could be used to show that cases where women want to donate their unused embryos, it may be found to be in compliance with the decision of this Case.[591] Now all of this has changed, and the Country is going to see a lot of challenges to this Holding, so things may change soon, as this decision was very polarizing, even on the Supreme Court.[592] Yet with *Dobbs* being decided in 2022, a lot has changed, it may be up to the states to determine when life begins, with obvious consequences for embryonic stem cell research. The *Dobbs* decision effectively abolished the federal right to terminate a pregnancy, but it not establish a right of the fetus to life.[593]

Under the laws of the United States, when dealing with constitutional issues such as the case here, Courts engage in a form of judicial review to determine if the limits set out by a challenged law are legal, or constitutional.[594] Typically, three levels of scrutiny exist when making a determination as to a law's constitutionality which range from the strongest level of scrutiny to the least, the strongest is known as "strict scrutiny," followed by "intermediate scrutiny," and the weakest, "rational basis."[595] The determining factor as to which level of scrutiny should be applied depends on

---

precedent, but in its holding, the Supreme Court Justices expressed reservations as to if the Constitution should truly be deemed to grant the right to an abortion.)

[586] *Dobbs*, p. 13.

[587] *Id.* p. 16.

[588] *Id.* p. 5.

[589] *Id.* pp. 846, 860.

[590] *Id.*

[591] *Roe v. Wade.*

[592] *See generally* Felix, et al. (2023).

[593] Alvarez Manninen (2023), pp. 357–362.

[594] *See generally* Chemerinsky (2016), p. 401.

[595] Ellis (2017), pp. 1325, 1333.

what rights the law in question restricts, when there is a law that allegedly discriminates against a person or group of people based on their race, religion, sexual preference, and national origin and requires the one who passed the law, i.e., the state to show that the law itself "further[s] a compelling interest and [be] narrowly tailored to achieve that interest," and prior to this showing, the law is presumed to be invalid.[596] Intermediate scrutiny typically applies when a challenged law allegedly discriminates against or adversely affects some sort of protected class, or works to create an improper classification of people, and can be seen in cases where a law discriminates based on gender or legitimacy, requiring a showing that the law in question furthers an important government interest through a means that is substantially related to the stated interest and that the intrusion that the law causes is used to significantly further said interest.[597] Finally, the weakest, rational basis is used when the law challenged law does not impact any rights that are considered fundamental, and when there are no suspect classifications used, and will be upheld if the law is "rationally related to any legitimate government purpose."[598] In dealing with the laws pertaining to embryonic stem cells, a right to life issue appears due to the fact that less restrictive laws would likely lead to the destruction of embryos, and on the other side, not allowing embryonic stem cell research impacts the lives of those in need, or who may one day been in need of treatments derived from embryonic stem cells and their research.[599] Laws that are argued to infringe on one's right to life have been held by the Supreme Court to require intermediate scrutiny.[600] For any challenged laws in the United States, this means that the plaintiff would have to show that their right to life is being infringed upon and either that the law was not adequately tailored in ensuring that law is in fact an important government interest, that the law does not further the government's stated interest or that the intrusion contemplated by the law does not work to further the stated interest.[601]

There are still federal laws that have applied still apply to embryonic stem cell research. They can be grouped into five distinct phases which correlate directly with the then current presidents.

4.5.1 Phase One: Laws Under President Bill Clinton

Even though stem cells were discovered well before President Clinton's time in office, there was little to no real clear, or established laws regarding what can and cannot be done with stem cells, or how and when the country would allow for federal

[596] *Id.* p. 1331; Winkler (2006), pp. 793, 815.

[597] *Id.* p. 1342; Stearns (2017), pp. 1043–1046.

[598] Chemerinsky (2016).

[599] *See generally* Meyer (2000), pp. 166–170.

[600] *See Thompson v. Hebdon*, 589 U.S. ___ (2019).

[601] *Id.* p. 14.

funding.[602] While President Clinton was in office, a number of laws were created regarding stem cell research.[603] Prior to President Clinton coming into office, President George H.W. Bush had allowed for a minimal amount of funding to be used for stem cell research.[604] At the time, stem cell research was still a new and its methods, and abilities were relatively unknown.[605] A year into President Clinton's time in office, the National Institute of Health Revitalization Act was passed thanks to the efforts of both President Clinton, and Congress.[606] This Act allowed for federal funding to be used in researching embryonic stem cell research, which was a first in America.[607] With this new Act in place, the National Institute of Health created a panel of scientists whose primary focus was to determine what kinds of research should take place, and the limits of it.[608] These recommendations stated that spare embryos that were left over from fertility clinics should be destroyed for use in this research.[609] President Clinton decided to not approve funding for the creation and destruction of new embryos for research, and President Clinton, and Congress approved the Dickey-Wagner Amendment which had a purpose of restricting federal funding for embryonic stem cell research.[610] This Amendment still left in place the laws that allowed embryos that had already been created to be used for experimentation, but prohibited new embryos from being created.[611] Interestingly, this Amendment actually requires it to be renewed each year, which is automatically as it is attached to the Department of Health and Human Services funding bill.[612] The creation of this Amendment came about due to the 1994 recommendations of the NIH Human Embryo Research Panel which brought together a number of researchers and specialists and were there to review the ethics of these forms of research, experts in policy, and even advocates for those who could potentially benefit from cures and treatments derived from embryonic stem cell research.[613] Their overall mission was to determine a time and situation where federal funds should be granted to certain embryonic research, and promulgated a report which stated that embryonic stem cell research should be allowed using surplus embryos and gametes from *in vitro* fertilization, so long as informed consent was given; and even recommended that embryos be created for research purposes.[614] Unfortunately,

---

602 *See generally* Spiegel (2013), pp. 94–110.

603 *Id.* pp. 104–105.

604 Dunn (2005).

605 *Id.*

606 Oh et al. (2015), pp. 1–9.

607 Hammond-Browning (2009), p. 251.

608 *Id.*

609 Dunn (2005).

610 Gardner (2010).

611 *Id.*

612 Kearl (2010a).

613 *Id.*

614 *See* National Institute of Health (1994).

President Clinton rejected this portion of the Panel's recommendation, and instead, the Dickey-Wagner Amendment became law.[615] Interestingly, the Dickey-Wagner Amendment became effective even without any formal House vote, and instead was approved by the House Appropriations Committee of the 104th Congressional session in 1995, and when a formal vote was actually administered, the Dickey-Wagner Amendment was already law.[616] Perhaps due to the fact that both the House of Representatives and Senate were controlled by a majority of conservatives, this Amendment passed easily, and President Clinton signed it into law on January 26, 1996, thus becoming Public Law 104-99.[617] In 2000, while President Clinton was in office, the NIH issued new guidelines based on their interpretation of the Dickey-Wagner Amendment, and stated that embryonic stem cells must be derived from the private sector, the embryos to be used were part of *IVF* in surplus of the donor's *IVF* needs, with informed consent.[618]

4.5.2 Phase Two: Laws Under President George W. Bush

On August 9, 2001 President Bush brought forth an executive order banning any federal funding for research on new stem cell lines but would allow federal funds to be granted for research using the already existing 21 embryonic stem cell lines.[619] While this might seem like a fair situation, most scientists actively objected to this decision, stating that the already existing embryonic stem cell lines were not "genetically or ethnically diverse" meaning that only certain diseases could be studied in relation to embryonic stem cell treatments.[620] Potentially, even more problematically, the stem cell lines in existence were created could only be attributed to certain ethnicities, meaning that the experiments would fail to take into account "different cellular processes in minority groups."[621] In 2005, during President Bush's presidency, Congress passed a bill called "Stem Cell Research Enhancement Act" with a vote 63-37 which had a goal of loosening federal stem cell research restrictions. The Bill's main goal was to allow for parents with unused embryos to be donated for the purpose of research, rather than just being destroyed.[622] This Bill was actually brought by a republican Sen. Arlen Specter and Majority Leader Bill Frist who argued that the laws at the time were way too restrictive.[623] This Bill,

[615] Kearl (2010b).

[616] *Id.*

[617] *Id.*

[618] Murugan (2009), pp. 101–102.

[619] *Id.*

[620] *Id.* p. 101.

[621] *Id.*

[622] Bush and Walsh (2006); Hopkins Tanne (2006), p. 216 (Jul. 2006).

[623] Check (2006), p. 391.

though passed, was vetoed by Bush in his first use of the presidential veto.[624] After vetoing the Bill, President Bush stated that he did so because "This bill would support the taking of innocent human life in the hope of finding medical benefits for others…"[625] Again, in 2007, President Bush vetoed a bill also known as the Stem Cell Research Enhancement Act and instead issued Executive Order 13455.[626] The Executive Order required the Secretary of Health and Human Services to provide information every year to the President about the advancements made in the field of embryonic stem cell research; it also stated that embryonic stem cell research could continue to receive federal funding, so long as the human life was not harmed or lost; including embryos, which this Order recognizes to be a "living being;" and bans the buying and selling of embryos.[627] President Bush's own bioethics advisory committee even stated that they would recommend using the method discussed above, where a single embryonic stem cell is removed from an embryo without the need to destroy the embryo.[628] Clearly, President Bush's veto was based on his view that life begins at conception, which at the time, and even to this day, has not been officially established by the United States for these purposes.[629] There were claims that a future bill would be put to vote which would take all considerations into account, but it never came to be.[630] Instead the law remained that only allowed for the 60 hESC lines to be studied, which though seems like a lot, actually amounts to very little.[631] Many, if not most were contaminated, and would never be deemed fit for human testing.[632] It has been estimated that in reality, only 20 or so hESC lines were made available to researchers.[633]

4.5.2.1 Armed Forces Institute of Regenerative Medicine

In 2008, while President Bush was in office, the United States Department of Defense established and funded the Armed Forces Institute of Regenerative Medicine ("AFIRM") with the stated "mission" of "The Armed Forces Institute of Regenerative Medicine is dedicated to repairing battlefield injuries through the use of regenerative medicine technology."[634] To accomplish such a difficult and complicated task, the group must be comprised of: "multi-institutional, interdisciplinary

---

[624] *Id.*

[625] Jackson (2007).

[626] *See* Exec. Order No. 13435, 3 C.F.R. 222 (2007).

[627] *Id.*

[628] *See* Kaplan (2012).

[629] *E.g.*, Khokhar (2010).

[630] *Id.*

[631] *Id.*

[632] Kaplan (2012).

[633] *See* Wadman (2013).

[634] U.S. Department of Defense (2008).

network of universities, military laboratories and investigators" all working closely together so that their knowledge could be put to good use quickly.[635] Unfortunately, AFIRM has stated that embryonic stem cells will not be included in this research, claiming that embryonic stem cells are not yet close enough to use in trials, so they should be precluded.[636] This was stated even though the first FDA approved human trial using embryonic stem cells was conducted barely a few short months after the previous mentioned statement was made.[637] It is important to note that even without the use of embryonic stem cells, AFIRM has been able to outlive its original duration of 5 years, has now received funding from a huge group of military entities, supported many clinical trials, and has led to the treatment of hundreds of patients.[638] As of 2013, AFIRM was renamed AFIRM II as it was newly funded, and the mission slightly refocused in terms of what treatments were most needed in 2013, as opposed to when AFIRM was first implemented in 2008.[639]

4.5.3 Phase Three: Laws Under President Barak Obama

President Barack Obama issued an Executive Order on March 9, 2009 which was titled "*Removing Barriers to Responsible Scientific Research Involving Human Stem Cells*" ("2009 Guidelines") which was used to override the laws put in place by President George W. Bush.[640] These guidelines greatly transformed the laws regarding embryonic stem cell research. This Executive Order was a great change in the law as it stood under President George W. Bush. Overall, the 2009 Guidelines discuss policies, and procedures under which the NIH may receive federal funding, and hopes to ensure the that the NIH acts in an ethical manner, for "scientifically worthy" causes, and in accordance with the law.[641] The Executive Order also retained the old rules which gave broad budget power to the Office of Management and Budget to reduce federal funding as they saw fit.[642] This Executive Order also kept a rule that has been in place since President Clinton first signed a federal law banning stem cell research in 1995, which prohibits the creation of new *hESC* cell lines from any viable embryo using federal funds. This also allowed for private investor backed scientist and researchers to submit new *hESC* lines for study, but doing so is very costly, and takes a very long time, making this option effectively

---

[635] Lafontaine (2014).

[636] Coombs (2008).

[637] *See*: Pollack (2009).

[638] Rauch (2013), pp. 10–12.

[639] *Id.*

[640] National Institute of Health (2009) (discussing Executive Order 13505, the changes made, the newly existing limits, and why not enough changed to actually allow for substantial benefits).

[641] *Id.*

[642] *Executive Order 13505* (2009).

prohibited. The Act also contained a set of ethical requirements that had to be met in order to receive any federal funding.[643] The requirements were that:

1. The embryo was discarded after in vitro fertilization;
2. Informed consent was obtained from the donors; and
3. The couple who is donating the embryo cannot receive any monetary, or medical compensation for their donation.[644]

The National Institute of Health than issued its own similar set of guidelines with the same restrictions in place called the "NIH Guidelines for Human Stem Cell Research" and made effective on July 7, 2009.[645]

During this time, the legality of the Dickey-Wagner Amendment was tested in relation to President Obama's Executive Order in the case of *Sherley v. Sebelius*.[646] In that case, the plaintiffs sought an injunction as to the Executive Order claiming that it violated the Dickey-Wagner Amendment and claimed that its guidelines when dealing with stem cell research was "contrary to law" in the manner that it allowed for federal funding.[647] The Plaintiffs were made up of Dr. Sherley, Dr. Deisher, Nightlight Christian Adoptions, the Christian Medical Association, among others, and sued defendant Health and Human Services Secretary, Kathleen Sebelius. Dr. Sherley conducts research using adult stem cells, and believed that federal funding of embryonic stem cells was contrary to the Dickey-Wagner Amendment.[648] Initially, the U.S. District Court hearing the case ruled in favor of the plaintiffs, and issued an injunction against the policy of using government funds for embryonic stem cell research, as at the time, it required the destruction of an embryo.[649] On appeal, the Court ruled that the Congress was within its authority to create the Amendment, and as it pertains to stem cells, the Court found that the Amendment was created to distinguish between creating embryos for research, and using already created embryos for research.[650] The initial plaintiffs sought a further appeal, so that their case could be heard in front of the Supreme Court, but the Supreme Court declined to hear the case, ending the 3 year legal battle over this issue.[651]

After President Obama's overturning of the previous, more restrictive laws, certain members of Congress sought to codify the new interpretation through the Stem Cell Research Advancement Act.[652]

---

[643] Bellayr and Li (2009), pp. 92–94.

[644] *Id.* p. 93.

[645] *Id.*; *see also* Kington (2009).

[646] *Sherley v. Sebelius*, 644 F.3d 388 (2011).

[647] *Id.* p. 393.

[648] *See generally* Cohen et al. (2015), pp. 2087–2088.

[649] Katsnelson (2010).

[650] *Id.*

[651] Wadman (2013).

[652] H.R. 2443 Stem Cell Research Advancement Act of 2013.

Further, in 2015, Rep. Suzanne Bonamici sponsored new legislation to come up with new treatments and cures for diseases and injuries that have had little to no success in treatments, using new techniques and approaches.[653] This Act was signed into law by President Obama in 2016, and became known as the 21st Century Cares Act.[654] Unfortunately, this Act does not specially provide for embryonic stem cell research to find cures, yet does specifically mention adult stem cells.[655] This Act created funding for three very important areas of biomedical research, which are Brain Research through Advancing Innovative Neurotechnologies (BRAIN), the Precision Medicine Initiative (PMI), and the Beau Biden Cancer Moonshot, all of which likely benefitted greatly from this legislation, but because the Act was created to use the most current forms of research, it is troubling and odd that embryonic stem cells were specially left out, even after President Obama's Executive Order on embryonic stem cells.[656]

President Obama's lenient legacy on embryonic stem cell research enabled researchers to have the ability to use approximately 200 hESC lines, which is a considerable amount more than what the previous president allowed for.[657]

4.5.4 Phase Four: Laws Under President Donald Trump

Former President Donald Trump did not immediately seek to restrict the study of fetal tissues including embryonic stem cells, but waited until July 26, 2019 to issue and enforce new rules (enforceable starting in September of 2019) and requirements promulgated through a notice by the NIH.[658] The NIH notice was titled "Changes to NIH Requirements Regarding Proposed Human Fetal Tissue Research" and effectively creates new requirements in order to obtain federal funding for the study of fetal tissue, as well as a more lengthy review process that looks to the ethics of those seeking grants.[659] Specifically, the new rules require applicants to describe how they obtained consent to use the fetal tissue from the donor, how they will dispose of the tissue, and provide a scientific justification for needing to use fetal tissue.[660]

Prior to this National Institute Of Health Notice, the Department of Health and Human Services (the "HHS") announced in September of 2018 that they would be conducting a review of federally funded research using human tissue.[661] The HHS

---

[653] Hudson and Collins (2017), pp. 111–113.

[654] H.R.34—21st Century Cures Act.

[655] *See* H.R.34—21st Century Cures Act §1001(4)(D), 130.

[656] *See* National Institute of Health (2016).

[657] Wadman (2013).

[658] National Institute of Health (2019).

[659] Kaiser and Wadman (2019).

[660] *Id.*

[661] *Id.*

finished their review and announced their finding on June 5, 2019 which banned all studied being conducted by National Institute of Health scientists from using any fetal tissue, and required that non-NIH workers would be subjected to a review by an HHS picked Ethics Advisory Board of 14–20 members.[662] Further, the HHS stated that these board members "must" include "a theologian, an ethicist, and an attorney," and restricted the board by not allowing more than half of its members to be scientists.[663]

In looking at these phases, and the times in which they occurred, it is easy to see why the laws were put in place. When President Clinton was in charge, stem cell laws were generally looser and less restrictive law. One of the primary reasons for this was due to the fact that the President, and Congress were made up of a larger liberal group. In the President Bush years, the exact opposite can be seen as far as the President. During that time, Congress agreed to loosen stem cell laws, yet the President vetoed that law. During President Obama's time as the President, and being that he is politically liberal, it can be easily understood that the laws were changed to reflect the liberal position. The same can be said for President Donald Trump's stance on embryonic stem cell research as being politically motivated as he is a conservative, and as explained above, conservatives have typically worked against these forms of research.

One major benefit conferred to the American public by President Trump is the Trickett Wendler, Frank Mongiello, Jordan McLinn, and Matthew Bellina Right to Try Act/Right To Try Act which allows for terminally ill patients to obtain access to still experimental drugs, treatments and medical devices, so long as the novel treatments have passed the FDA's Phase I Trials.[664] Prior to the passage of this Act, the FDA would have to evaluate individual applications for terminally ill patients which usually proved to be fruitful, but due to the time delay, the patients were left with a higher risk of passing before obtaining the proper authorizations.[665] The Act works by allowing biologic, or pharmaceutical company to submit an Investigational New Drug application which is reviewed by the FDA as to the safety of the drug or treatment before its ever used on a patient.[666] The process actually only requires the acceptance/permission from a patient's doctor, and the approval of the biologic company.[667] However, if the company working on the treatment or drug denies a patient's request, than unfortunately, the FDA has no ability to challenge this decision.[668] It must also be noted that the biologic company is not under any requirement to pay for the costs of the treatment or drugs.[669]

---

[662] *Id.*

[663] *Id.*

[664] *See* U.S. Food and Drug Administration (2020a).

[665] *See generally* Kiely (2019).

[666] *See* U.S. Food & Drug Administration (2020b).

[667] *Id.*

[668] *Id.*

[669] *Id.*

A completely new law came into effect in 2016 through an appropriations bill which indirectly bans research "in which a human embryo is intentionally created or modified to include a heritable genetic modification."[670] The law further states that if such a submission appears before the FDA seeking an exemption to this rule, they are not only to deny its use, but also cannot give any acknowledgement of receipt of such an application.[671] "Any such submission shall be deemed to have not been received by the Secretary, and the exemption may not go into effect."[672]

In July of 2019 a Human Fetal Tissue Ethics Advisory Board was created under the direction of the Health and Human Services (HHS) Secretary Alex Azar, which was set up to further review the applications of those seeking to use fetal tissue.[673] After meeting approximately a year after its creation in 2020, the Human Fetal Tissue Ethics Advisory Board issued their decision to withhold funding from 13 of the 14 grantees, that had already received NIH approval.[674] Many have accused this Board of an actual explicit bias, as no less than 10 of the 15 members have publicly argued against all forms of fetal tissue research including the research of embryonic stem cells.[675]

It should be interesting to note though, that although President Trump worked to restrict embryonic stem cell research along with other research into fetal tissues, he did implicitly authorize and even benefit from the study of fetal tissue.[676] President Trump was given a treatment for Covid-19 in October of 2020 which itself was created using fetal tissue.[677] In fact, many of the vaccines that were funded thanks to "Operation Warp Speed" were similarly working with cells created from fetal tissue.[678] Operation Warp Speed was an initiated that the government describes on their website as "a partnership between the Departments of Health and Human Services (HHS) and Defense (DOD)—aimed to help accelerate the development of a Covid-19 vaccine."[679]

A direct consequence of President Trump's winning of the election in 2016 was that other countries sought to take advantage of the fact that it was easily foreseeable based on President Trump's politics that government funding of embryonic stem cells would likely be further restricted, and that they could take advantage.[680] Canada realized that they could attract frustrated embryonic stem cell researchers,

---

[670] *See* Consolidated Appropriation Act, H.R 133, 116th Cong. (2020).

[671] *Id.*

[672] *Id.*

[673] Santoro et al. (2021), pp. 933–935.

[674] *Id.*

[675] Wadman (2020).

[676] Mandavilli (2017).

[677] *Id.*

[678] *Id.*

[679] *See:* U.S. Government Accountability Office (2011).

[680] Semeniuk (2016).

which would help their economy, and keep Canada competitive on a global level in new medical developments.[681]

4.5.5 *Phase Five: Laws Under President Joe Biden*

With the 2020 election over, many believe that now President Joe Biden will help to further expand embryonic stem cell laws, based on his previous statements, actions, accolades and even party affiliation.[682] President Biden has also issued an Executive Order 14081 titled "Advancing Biotechnology and Biomanufacturing Innovation for a Sustainable, Safe, and Secure Bioeconomy" on September 12, 2022.[683] This Executive Order seeks to use existing agencies to determine where, if any uncertainties in the current laws impacting "biotechnology products" by seeking advice and comments from contributors in the industry.[684] At the same time, these agencies will continuously monitor publications for new and beneficial biotechnology products, and with all of this information, the agencies will create a framework and determine which groups should answer to which oversight committees, as prior to this, everything was more convoluted.[685] Executive Order 14081 also seeks to update, streamline and provide further guidance on already enacted legislation that impacts these scientific areas and will also set up processes to foster cooperation among more than just the national interested groups, but also with the international community as a whole.[686] The end goal of this Order is to lead to an advancement that would allow scientists to

> . . .develop genetic engineering technologies and techniques to be able to write circuitry for cells and predictably program biology in the same way in which we write software and program computers; unlock the power of biological data, including through computing tools and artificial intelligence; and advance the science of scale-up production while reducing the obstacles for commercialization so that innovative technologies and products can reach markets faster. . . Simultaneously, we must take concrete steps to reduce biological risks associated with advances in biotechnology. We need to invest in and promote biosafety and biosecurity to ensure that biotechnology is developed and deployed in ways that align with United States principles and values and international best practices, and not in ways that lead to accidental or deliberate harm to people, animals, or the environment.[687]

The Order also seeks to ensure that the studies in question remain ethical, and that the use of and benefits derived from biotechnology should be provided in an

---

[681] *Id.*

[682] *See* Vote Smart (2021a).

[683] Executive Order 14081, 88 FR 25711 (April 27, 2023).

[684] *Id.*

[685] *See Id.*

[686] *Id.*

[687] The White House (2022a).

equitable manner so that all have the same access regardless of any differences.[688] This points to the fact that the current administration realizes the potential for advancements in biotechnology that not only provides medical benefits but that, one must also understand and protect ourselves from potential risks that can arise from the use of biotechnology.[689] Another consideration that was brought up in the issuing of this Order is what is called the "bioeconomy" as there are many potential monetary and investment benefits as a result of successful research, and that there are foreign interests that need to be prevented from stealing the research results for their own financial gain.[690] In 2021, the HHS removed the requirement that HHS review requirement on proposals for NIH funding when using human fetal tissue.[691] This did not remove the other constraints mandating that there must be a justification for the use of human fetal tissue, and maintained the rule that researchers must first obtain informed consent from the donors of the human fetal tissue.[692]

Another effort that can be very telling as to President Biden's stance on embryonic stem cell research is the President and First Lady's Cancer Moonshot.[693] The stated goal of this program is to work with all areas involved in fighting cancer, such as "federal agencies and departments, private companies, healthcare providers, patient groups, philanthropies, and all Americans."[694] There are five "priority actions" highlighted as what is needed to be done in order for this lofty goal of ending cancer to be achieved.[695] These priorities require the federal government to lead and create a drive for "innovative action and life-saving progress" to benefit all Americans; for scientists working on any research in this area to think "boldly;" to help patients and their families as well as survivors with having the best possible experiences during such a terrible time; for the private sector to create and test new potential therapies while actively collaborating with others by sharing "more data and knowledge;" and for those impacted by cancer in any way to share their "perspectives and experiences".[696] As the title of the program even suggests, curing cancer is a longshot, but it is one which requires more than just guidelines and goals, so a "Cancer Cabinet" was created to bring the many different agencies, departments

[688] *Id.*

[689] *Id.*

[690] *Id.*; *see* Congressional Research Services (2022) (defining bioeconomy as "the share of the economy based on products, services, and processes derived from biological resources [e.g., plants and microorganisms]. The bioeconomy is crosscutting, encompassing multiple sectors, in whole or in part [e.g., agriculture, textiles, chemicals, and energy].")

[691] National Institute of Allergy and Infectious Diseases (2021).

[692] *Id.*; Jorgenson (2016).

[693] The White House (2022b).

[694] *Id.*

[695] The White House (2022a).

[696] *Id.*

and others working in areas that impact this goal together so that the government processes run more efficiently in making this goal into a reality.[697]

One can also look to President Biden's long history in politics which provide more elucidation on his stance towards this area of science.[698] First, President Biden has actually voted in favor of previous bills seeking to expand embryonic stem cell laws, which he did for the Stem Cell Research Bill of 2005, Stem Cell Research Hope Act of 2007, and the Stem Cell Research Act of 2007, and actually made public his letters written to then President Bush explaining the need to expand such laws and asking for his help in the passage of bills seeking to do so.[699] Further, President Biden has removed a large burden placed under the Trump administration, which (as discussed above,) required an ethical review process for NIH grants for experiments using fetal tissue from abortions on federally funded medical experiments.[700] While discussing this new rule and in its defense, Secretary of Health and Human Services Xavier Becerra stated: "We believe that we have to do the research it takes to make sure that we are incorporating innovation and getting all of those types of treatments and therapies out there to the American people..."[701] One can also look to President Biden's political rivals critiques and see that President Biden has a long history of supporting further research into this area of research, which as a result has led people to accuse him of seeking to vote against certain religious values.[702] It is also important to point out that President Biden was the Vice President under former President Obama who, as discussed above, passed an Executive Order expanding embryonic stem cell laws, and he, and his wife have also been honored by the World Stem Cell Summit by receiving the Advocacy Award at the World Stem Cell Summit Gala in 2018.[703] However, it is of the utmost importance to point out that even though President Biden has expressed unwavering support for the expansion of laws pertaining to embryonic stem cell research, he has expressly stated that there are limits to his support and through his "nay" vote cast on the S 1601—Human Cloning Research Bill, affirmed such beliefs.[704] One can also look to President Biden's rhetoric in the past in regards to this area of science, such as the time when he was a Senator, and stated that it was his belief that those who oppose this area of science are proving their lack of care, and dedication to those who are in need.[705] While running for Vice President with President Obama, President Biden made a statement in contrast with his then vice presidential rival's stance on the issue of special needs children: "Well, guess what folks? If you care about it, why

---

[697] The White House (2022c).

[698] *See* Biden (2004); Biden (2006).

[699] *See Id.*

[700] Ollstein (2021).

[701] *Id.*

[702] McClusky (2020).

[703] Barrios (2018).

[704] *See* Vote Smart (2021b).

[705] Holmes (2018).

don't you support stem-cell research?"[706] While this may seem like Biden was referring to any form of stem cells, the context in which it was said, clearly shows his desire to expand the laws regulating embryonic stem cells because at the time, then Republican nominee John McCain stated his support for embryonic stem cells research, but with a caveat that banned experimentation on embryos, making adult stem cells the only true contender for the then candidate.[707] Because it is so early into President Biden's first term, one can only use the information at hand to make any kind of prediction as to which direction President Biden will go to, but with such a clear history and track record, one can easily make an educated guess that he would support the further expanding of embryonic stem cell laws and funding. This can be further demonstrated by President Biden's choice to advise his administration on science and technological policy.[708] Specifically, President Biden chose Dr. Arati Prabhakar to:

> serve as Director of the Office of Science and Technology Policy, (OSTP), and once confirmed to this position, also as Assistant to the President for Science and Technology. In this capacity, Dr. Prabhakar will be the President's Chief Advisor for Science and Technology, a co-chair of the President's Council of Advisors on Science and Technology, and a member of the President's Cabinet.[709]

During Dr. Prabhakar's Nomination Hearing, she made it very clear that she supports embryonic stem cell research with certain limits.[710] This Hearing took place on July 20, 2022, Senator Mike Lee specifically asked Dr. Prabhakar "Do you support the use of aborted fetal tissue for research purposes?"[711] To which Dr. Prabhakar positively replied and stated:

> This is a specific research approach, for which there are limited or no alternatives, that can help find therapies for otherwise incurable diseases and conditions. I support the Biden Administration's decision to lift an earlier ban on this research because of its potential to unlock solutions for cancers, Alzheimer's, Parkinson's, and other diseases and conditions that afflict so many Americans.[712]

Senator Lee's questions continued, with the next one directly speaking to embryonic stem cells, when he asked "Do you support the use of embryonic stem cell research?"[713] Again, Dr. Prabhakar responded positively stating: "I support the Administration's reaffirmation of the long-standing support of responsible, scientifically worthy, human embryonic stem cell research allowable under the law, subject to the NIH Guidelines on Human Stem Cell Research."[714] The next very relevant

---

[706] *Id.*

[707] Raoux (2008).

[708] The White House (2022d).

[709] *Id.*

[710] U.S. Senate Committee on Commerce, Science, and Transportation Nominations (2022).

[711] *Id.* p. 11.

[712] *Id.*

[713] *Id.*

[714] *Id.*

question the Senator asked Dr. Prabhakar: "Do you support the use of germline genetic engineering to create genetically modified embryos?"[715] The Doctor's answer was very explicit on her beliefs when she explained:

> While this has been proposed as an approach for treating extremely serious genetic conditions, there are significant safety and ethical issues raised by this application of gene editing technology, and I do not support it. In addition, Congressional prohibitions prevent funding of this research or regulatory approval of its clinical use.[716]

Senator Lee asked other very pertinent and specific questions which will be discussed in order.[717] Senator Lee next asked a series of questions to which Dr. Prabhakar gave similar replies essentially stating that she understands that there are currently laws in place that prohibit such conduct.[718] The questions were: "Do you support the use of live embryos in federally funded research?; Do you support the creation of human-animal hybrid embryos?; and Do you support allowing an embryo to gestate in an artificial womb environment for research purposes for less than 14 days? Do you support allowing such an embryo to gestate in an artificial womb environment for longer than 14 days?"[719]

Perhaps as a result of hiring Dr. Prabhakar in this role, President Biden's administration has reversed former President Trump's fetal tissue research restrictions, and instead allows research to be conducted by the National Institute of Health, and removed the ethical review process per applicant.[720]

Clearly, President Biden's choice to advise him in scientific matters has a very clear understanding and direction for where the administration should go in regards to embryonic stem cell research which allows for room for extended research rules, while also remaining ethical and within the bounds of the current laws.[721]

4.6 Health/Research Organizations in the United States

The United States has two primary agencies as well as other organizations that have been tasked with the responsibility over certain aspects pertaining to embryonic stem cells.[722] The Food and Drug administration typically works to ensure "the safety and effectiveness of" medical products, as well as other products, where the National Institute of Health "supports and conducts biomedical and behavioral research."[723]

---

[715] *Id.*

[716] *Id.*

[717] See *Id.*

[718] *Id.*

[719] *Id.*

[720] National Institute of Health (2021a, b); Sevrick (2021).

[721] *Id.*

[722] *See generally* National Institute of Health (2021a, b).

[723] *See* U.S. Food & Drug Administration (2018).

The two agencies are meant to act in a collaborative manner where their unique functions are meant to be "complementary in their roles".[724]

4.6.1 The Food and Drug Administration

The Food and Drug Administration also has a regulatory role which, grants the agency "the authority to regulate all stem cell products in the United States."[725] This means that the product must go through FDA testing to prove that it's safe to use, untainted, and is of the correct strength due to the known intended scientific uses.[726] These "products" could be anything from tools developed to work with stem cells, to the actual created stem cell lines which could very well be implanted into a patient, so stringent tests are crucial.[727] A case was brought that put stem cell treatments and other cellular based therapies into the FDA's jurisdiction through the Federal Food, Drug, and Cosmetic Act (the "FFDCA").[728] In *United States of America v. Regenerative Sciences*, the Court heard arguments from the appellants, Regenerative Science, who created procedures known as "Regenexx Procedures," which used adult stem cells taken from the patient and mixed with antibiotics and other biologics "biological product", but explicitly brings into the FDA's purview, all forms of stem cells.[729] The FDA claimed that this mixture was itself "unsafe" for use in humans, but the appellant's argument was that their treatments should not require any interference or require any forms of permission from the FDA because their treatments were cell based, and not drug or medicine based, making it a medical treatment, and not something under which the FDA has authority to regulate.[730] The FDA argued that the treatment created a "biologic" which typically is something that is controlled by the FDA under the FFDCA, which defines biologics as,

> Biological product means a virus, therapeutic serum, toxin, antitoxin, vaccine, blood, blood component or derivative, allergenic product, protein, or analogous product, or arsphenamine or derivative of arsphenamine (or any other trivalent organic arsenic compound), applicable to the prevention, treatment, or cure of a disease or condition of human beings... A therapeutic serum is a product obtained from blood by removing the clot or clot components and the blood cells.[731]

---

[724] *Id.*

[725] Fink Jr (2009), pp. 1662–1663.

[726] *See* U.S. Food & Drug Administration (2019).

[727] *Id.*

[728] *See United States of America v. Regenerative Sciences, LLC*, 741 F.3d 1314 (D.C. Cir. 2014).

[729] *See generally* Cohen (2015).

[730] Daniels (2010).

[731] 21 C.F.R. 600.3 (2010).

More so, appellants argued that the FDA's focus on the treatment itself was at issue, and not the mixture of cells, which the Court found to be untrue.[732] Accordingly, the Court held that the FDA was not even seeking to regulate the treatment using the mixture itself as unsafe, but was claiming that the injectable mixture was itself unsafe.[733] The Court further held in favor of the FDA stating that t the mixture of stem cells in the solution to be injected into the patient is in fact a biologic, which directly falls under the FDA's jurisdiction, and to add to this interpretation, the Court looked to the Commerce Clause of the United States Constitution due to the aggregate effects, in this case, the effects in this case implicate, "the public health and safety… it seeks to keep interstate channels free from deleterious, adulterated and misbranded articles of the specified types. of such treatments on interstate commerce."[734]

The FDA's agency known as Center for Biologics Evaluation and Research (the "CBER") is also tied to embryonic stem cell research.[735] The CBER promulgated their own rules pertaining to "human cells and cellular-based products…intended for tissue repair or replacement."[736] Two rules were created by this agency, the first being found in the Title 21 of the Code of Federal Regulations ("CFR") 1270, and the second is CFR 1271. CFR 1270 was issued by the FDA to revoke the previous rules that regulated "human tissue intended for transplantation and human dura mater recovered prior to May 25, 2005," and specifically states "The regulations in this part apply to human tissue and to establishments or persons engaged in the recovery, screening, testing, processing, storage, or distribution of human tissue."[737] CFR 1271 seeks to create a federal database and listing system those that manufacture human cells and tissues as well as establishing donor-eligibility, and has the lofty goal of creating procedures to be implemented when dealing with these forms of human tissue to "prevent the introduction, transmission, and spread of communicable diseases…"[738] These rules also apply to products derived from human tissues and cells, and this would include any and all treatments created using embryonic stem cells if used for implantation into a recipient, such as in the potential case when cancer vaccines.[739] As time goes on and new studies involving embryonic stem cells prove successful, there can be a large number of products created to vaccinate or treat many different diseases and viruses which will necessitate the companies to abide by this agency's rules.

As the FDA is charged with monitoring, prosecuting, and shutting down unscrupulous actors claiming to have novel treatments which remain currently unproven, it

[732] *Id.*

[733] *Id.*

[734] *United States of America v. Regenerative Sciences, LLC,* p. 1316.

[735] Crook, et al. (2007), pp. 490–494.

[736] *Id.* p. 490.

[737] *See* Title 21, Volume 8 CFR 1270.

[738] *See* Title 21, Volume 8 CFR 1271.

[739] Bross et al. (2015), pp. 133–136.

is important to show that they have been actively doing so.[740] FDA Commissioner Scott Gottlieb stated that:

> There are a small number of unscrupulous actors who have seized on the clinical promise of regenerative medicine, while exploiting the uncertainty, in order to make deceptive, and sometimes corrupt assurances to patients based on unproven and, in some cases, dangerously dubious products. . .[741]

It is crucial to understand that the clinics that the FDA shutdown did not engage in embryonic stem cell treatments, but did use adult stem cells in a dangerous way, falsely claiming that there were treatments for diseases such as Parkinson's, amyotrophic lateral sclerosis, different forms of blindness, among others, thereby not only getting the hopes of the patients' and their families' but also implanting stem cells into people in a manner that could very well prove dangerous.[742] Some of the clinics that were subjected to the FDA's harshest punishments were not even collecting money for their treatments, in fact, cancer patients went to a particular stem cell clinic in California, where they were only attempting to help people and to assist in gaining more knowledge in the use of stem cells.[743] One of the best takeaways from the FDA actively shutting down multiple stem cell clinics in multiple states, is that if, and when embryonic stem cells are studied in a similar manner, offering novel treatments, the FDA is ready, willing and able to protect the citizens of the United States from dangerous, unproven "cures."

4.6.2 National Institute of Health

The National Institute of Health is an agency that operates under the Department of Health and Human Services in the United States, and is considered to be "the steward of medical and behavioral research for the Nation. Its mission is to seek fundamental knowledge about the nature and behavior of living systems and the application of that knowledge to enhance health, lengthen life, and reduce illness and disability."[744] The NIH issued their formal guidelines pertaining to embryonic stem cells shortly after President Obama's Executive Order 13505 which creates a means to seek grants for this type of research, requires informed consent, and highlights an enforcement tool.[745] An interesting area covered by this Order is that no federal grant money would be given if It is also imperative to note that because this agency operates under the Department of Health and Human Services, there are methods of

[740] Stein (2017).

[741] *See* U.S. Food & Drug Administration (2017).

[742] Stein (2017).

[743] *Id.*

[744] *See* National Institute of Health (2015).

[745] *Id.*

punishing those who break these rules.[746] Executive Order 13505 also firmly established a mechanism that allowed for applicants to add embryonic stem cell lines into the registry so that they could be eligible for federal funding.[747] The process going forward from the time of this Executive Order requires applicants to submit an "assurance of compliance" with the NIH guidelines, and that the cells themselves must be derived from surplus embryos created for *IVF* purposes; the cells came from an embryo donated through voluntary, written consent that the embryo would be used for research purposes; the donors must not have been paid to make their donation; the embryos created were only created for reproductive purposes, and that there was no chance that donor created the embryo for donating purposes only; and it must be shown that the donors were made aware of their ability to withdraw consent.[748] Donors must also be made aware of what will actually happen to the embryo post consented to donation, such as what the embryo would be used for; what happens to the cells derived from the embryo; the research obtained through use of the embryo would not be directly medically beneficial to the donors; and the donors must be told if their personal information would be given to researchers.[749]

This Executive Order also set up a method for an approval process for new embryonic stem cell lines that were derived from embryos prior to this Order, which calls for the applicant to submit their compliance with the above mentioned requirements, as well as to submit all research materials pertinent to the applicant's request for approval to the Working Group of the Advisory Committee to the Director (the "ACD").[750] The ACD would then review all of the information presented and make a final recommendation to the NIH Director that will be charged with making the ultimate decision on eligibility.[751] Before getting to the NIH Director, the ACD will have to have documents from the applicant proving that: "1) that were created using in vitro fertilization for reproductive purposes and were no longer needed for this purpose; and 2) that were donated by donor(s) who gave voluntary written consent for the human embryos to be used for research purposes."[752] There is also a similar method for embryos donated outside of the country which is very similar to those who submitted prior requests for embryos donated prior to this Order, yet requires proof that "the alternative procedural standards of the foreign country where the embryo was donated provide protections at least equivalent" to the requirements set out in this Order.[753]

There were also new bans set in place by this Executive Order which now denies funding for research using human embryonic stem cells that would be transferred

[746] *Id.*

[747] *Id.* at paragraph (II)(A).

[748] *Id.*

[749] *Id.*

[750] *Id.* 648 at paragraph (II)(B)(2).

[751] *Id.*

[752] *Id.*

[753] *Id.* at paragraph (II)(C).

into "non-human primate blastocysts;" as well as research using human embryonic stem cells to be introduced into animals bred for research purposes; human embryonic stem cells "derived from other sources, including somatic cell nuclear transfer, parthenogenesis, and/or IVF embryos created for research purposes;" and those that would still otherwise violate the Dickey-Wagner Amendment.[754] All of these ineligibilities still control, even if the other criteria set forth in this Order were met.[755]

This Executive Order was updated and revised in December of 2022 to allow for this Executive Order to apply for all NIH grant and cooperative agreements between the NIH and other private or public entities to accommodate newer applications.[756]

Embryonic stem cells have been found to be in the purview of the NIH, which as discussed, is an agency operating under the Department of Health and Human Services, which means that the Code of Federal Regulations can apply, and can be used as a means to enforce the NIH's guidelines.[757] Specifically, 45 CFR 74.14 allows the NIH to "impose special conditions on an award, including but not limited to increased oversight/monitoring/reporting requirements for an institution, project, or investigator."[758] Further 45 CFR 74.62 allows for the government to be able to withhold grant funds for either a short term which can discontinue once the grantee provides evidence of compliance, withholding funds for certain expenses, or even outright terminating the grantor/grantee relationship.[759] Finally, 2 CFR Part 376 and 48 allows for the total barring of eligibility to receive federal grant money to individuals and even institutions.[760]

4.6.3 The National Research Council and Institute of Medicine

The National Research Council and Institute of Medicine has created the Human Embryonic Stem Cell Research Advisory Committee who has also established their own guidelines in 2010.[761] This Committee established that there are certain experiments and procedures that are permissible, permissible pending further information, or banned which is to be decided by the Embryonic Stem Cell Research Oversight Committee (also established by these guidelines).[762] These guidelines are similar to

754 *Id.* at paragraphs (IV)(A), (IV)(B), (V)(A), (V)(B).

755 *Id.*

756 *See:* Executive Order 13505, Fed. Reg. Vol. 74, 46, pp. 10667–10668.

757 Kearl (2010a).

758 *See* United States Congress 45 CFR § 74.14.

759 *See* United States Congress. 45 CFR § 74.62.

760 *See* United States Congress. 2 CFR Part 376.

761 *See generally* Board on Health Sciences Policy Institute of Medicine (2010).

762 *Id.*

that of the ISSCR in that they too have a total ban on reproductive cloning due to the dangers of failure.[763]

4.6.4 National Bioethics Advisory Commission, the President's Council of Bioethics, and the Presidential Commission for the Study of Bioethical Issues

From the years 1996–2001, the National Bioethics Advisory Commission (the "NBAC") was used to advise on emerging biotechnologies and treatments, and their ethical implications.[764] In 1999 the NBAC met to discuss the legal and ethical implications of embryonic stem cell research, as well as their potential to create new treatments and cures for those in need.[765] To accomplish this goal, the NBAC issued recommendations which stated that embryonic stem cell research should be funded by the federal government, so long as the embryos in question were derived from *IVF* treatments, and were considered surplus, and not fund research where embryos are created for the sole purpose of embryonic stem cell research.[766] Finally, the NBAC also advised that operations involving embryonic stem cells in the private sector should have to follow strict, and consistent ethical standards, while requiring the researchers to obtain informed consent from the donors, and banning the purchase and sale of embryos.[767] Unfortunately, President Clinton decided to reject these thought out and thoughtful recommendations made by a group of people from varying backgrounds.[768]

The President's Council of Bioethics was created by President George W. Bush in 2001 in order to replace the NBAC through Executive Order 13237.[769] The Executive Order specifically states that: "[t]he Council shall advise the President on bioethical issues that may emerge as a consequence of advances in biomedical science and technology."[770] President Bush charged the Council to review the moral and ethical significance in new research and technologies, and the Council would be made up of 18 members with varying backgrounds in law, government, science, philosophy and theology and all would be chosen by President Bush.[771]

---

[763] Board on Health Sciences Policy Institute of Medicine (2010), p. 21.

[764] *See generally* O'Brien (1999).

[765] *Id.*

[766] *See* National Bioethics Advisory Commission [NBAC] (1999), p. 11.

[767] *Id.*

[768] O'Brien (1999).

[769] United States Congress 3 CFR 13237.

[770] *Id.*

[771] *Id.*

After President Obama was elected, he issued Executive Order 13521, to supersede President Bush's previous Executive Order 13237 and removed the President's Council of Bioethics and replaced it with the Presidential Commission for the Study of Bioethical Issues.[772] This Commission would be used to advise the President of emerging scientific advances in relation to ethical issues as well as work to ethical scientific innovation.[773] Stem Cells were specifically mentioned as an area that the Commission should directly "examine."[774] The members of the Commission were required to be made up of at least 13 individuals from similar backgrounds required by President George W. Bush's Council, and would also be picked by the President.[775]

Interestingly, President Trump has not issued an executive order to replace President Obama's Commission, but has instead utilized the NIH's NIH Human Fetal Tissue Research Ethics Advisory Board to be his advisory committee, even though it is made up of people "known to oppose use of tissue from induced abortions, regardless of the scientific necessity and regardless of the fact that using such tissue does not in any way affect whether an abortion will take place."[776] However, the President does not require a unanimous decision in relation to the group's recommendations, it is possible that some studies will have a chance to be approved.[777]

4.7 Applicable State Laws

As with many other hotly debated topics, embryonic stem cell laws have played out in very different ways throughout the states which can be seen through their various laws. Looking to the state laws, one can find that there are two categories as to the states' laws in studying embryonic stem cells. First, there are those states that explicitly, and broadly support most, if not all forms of embryonic, and fetal research and issue grants, at times, there are those states that have somewhat permissible laws with certain restrictions, even though they there is nothing explicitly stated in their respective laws, and finally, there are those states with strict laws that ban virtually all forms of state funding, and experimentation using embryonic and fetal cells, some even going as far as completely forbidding all forms of similar research.[778]

---

[772] *See* Executive Order 13521 (2009).

[773] *Id.*

[774] *Id.*

[775] *Id.*

[776] Wadman and Kaiser (2020).

[777] *Id.*

[778] *See generally* Acosta and Golub (2018), pp. 419–436.

4.7.1 Federation of State Medical Boards

The Federation of State Medical Boards (the "FSMB") is an independent body that represents the 71 state medical boards, with a mission stating that: "[i]t supports its member boards as they fulfill their mandate of protecting the public's health, safety and welfare through the proper licensing, disciplining, and regulation of physicians and, in most jurisdictions, other health care professionals."[779] In May of 2017, U.S. Senator, and Chairman of the U.S. Senate Health, Education, Labor, and Pensions, Senator Lamar Alexander (R-TN) requested a framework for the best practices for the various state medical boards "in regulating the promotion, communication, and practices of treatments received at stem cell clinics in the United States."[780]

The FSMB did not directly discuss embryonic stem cells, but did offer recommendations for how stem cell clinics should operate and determine if a procedure should take place.[781] First, the FSMB stated that those engaged in regenerative medicine involving stem cells should balance the need of the patient with the state's requirement to keep their citizens safe from unproven treatments, then recommended that in the absence of proof of any stem cell benefits for a particular medical issue, "[n]ovel, experimental, and unproven interventions should only be proposed when traditional or accepted proven treatment modalities have been exhausted. In such instances, there must still be a basis in theory or peer-acknowledged practice."[782] Meaning that prior to any experimental treatment, specifically using stem cells, one must first attempt all known potential treatments and cures.[783] Next, the FSMB recommended that the state medical boards should assist in raising awareness of new developments in treatments using stem cells, and this information should be provided to those with medical licenses, as well as the public at large.[784] The FSMN then recommended expanding the informed consent guidelines to include more information, and a presentation of various options that the patient may have that could be used in the place of a stem cell treatment.[785] The state medical boards also received a recommendation stating that they should be responsible for viewing and monitoring the materials that stem cell clinics distribute regarding their work and treatments; should monitor and when the state medical board agrees, should investigate within the State's "laws, policies and procedures.[786] The report further makes

[779] *See* Federation of State Medical Boards (2018).

[780] Federation of State Medical Boards' Workgroup to Study Regenerative and Stem Cell Therapy Practices (2019).

[781] *See generally id.*

[782] *Id.;* White and Gelinas (2016), pp. 191–211.

[783] Federation of State Medical Boards' Workgroup to Study Regenerative and Stem Cell Therapy Practices (2019).

[784] *Id.* p. 10.

[785] *Id.*

[786] *Id.* pp. 10–11.

recommendations to doctors and physicians and states that they should only treat patients where they have a true "bona fide" doctor/patient relationship; physicians were also told that they should not engage in deceptive practices in an attempt to lure patients to a treatment that will likely not work; should not use "gag orders (rulings that a case must not be discussed publicly) or disclaimers as a way to circumvent liability;" physicians should seek to educate the public while providing resources regarding the proposed treatment, and other relevant information; and that the doctors should not charge excessive fees among other recommendations that are automatically implicated by the previous recommendations.[787]

4.7.2 *State Laws Explicitly Supporting Embryonic Stem Cell Research*

Even though there are states that protect the study and research of fetal and embryonic cells, there are still a number of categories that can be denied to researchers, depending on their state laws.[788] It is therefore important to understand the main differences between the state laws in order to further understand the legislative intent and reasoning, and the scope of the applicable laws. The vast majority of states that treat embryonic and fetal research as permissible, still have a few caveats and exceptions.[789] If a state were to allow the most broad and full spectrum of study on these forms of tissue, then nothing would be considered illegal, yet even some states where the laws can be described as favoring embryonic and fetal tissue studies, have placed some bans on certain forms of tissue acquisition, consent provisions, and even the outright ban of purchasing or selling these forms of tissue.[790] It is also crucial to understand that the states have a truly vested interest in novel research approaches. New forms of research can bring in a lot of outside investments, and can increase tax revenues exponentially. Further, when new research facilities are set up, new jobs are created for the local market from the very beginning stages of building the facility to it being fully occupied.

One of the most difficult issues when dealing with these laws comes from the fact that the federal government has offered very little guidance and has enacted stringent hurdles in an attempt to reduce or even prohibit stem cell based experiments.[791] This is true even though the federal government offers little in terms of actual laws and even contradicts the idea of the Supremacy Clause of the United States Constitution which places the states in a subordinate position to the federal government but allows for states to act in areas where the federal government has left open for states to do

[787] *Id.* p. 11.

[788] *See generally* Verginer and Riccaboni (2021), pp. 163–189.

[789] *Id.*

[790] *Id.*

[791] Acosta and Golub (2018).

so.[792] Further, some states have actually rose to the occasion and have expanded upon the federal stem cell research laws and encourage these experiments for the general purpose of receiving the benefits that come with new businesses in their states.[793] This is due to the fact that our country operates on a federal system that allows states to create laws that are not prohibited by the federal government, as the state is better able to determine the needs and wants of their citizens then the federal government.[794] Justice Brandeis's dissent in the case of *New State Ice Co. v Liebmann*, 285 U.S. 262 (1932), summed up the federal-state relationship which fits this topic perfectly.[795] Justice Brandeis's wrote: "It is one of the happy incidents of the federal system that a single courageous state may, if its citizens choose, serve as a laboratory; and try novel social and economic experiments without risk to the rest of the country."[796] Currently there are a handful of states that fund embryonic stem cell studies such as California, Connecticut, Maryland and New York who have produced thousands of studies on the potential for embryonic stem cells which have generated enough revenue for certain states such as California that the state voted in 2004 to fund 3 billion dollars to this field.[797] In expending funds to a cause such as embryonic stem cells research, states have effectively made themselves much more attractive to new laboratories and businesses which helps to generate revenue for the state.[798] Aside from the obvious tax benefits that accompany having new companies form and succeed in one's state, there are also various levels of prestige that come with the conducted research.[799] There have been many research papers and articles that have come from these states which have the potential of attracting new students to the state schools who receives funding for embryonic stem cell research.[800] Some states like California have taken a larger interest in not only increasing the ability of embryonic stem cell research, but have also cracked down on those who abuse the laws and practice without a license or who use stem cell treatments improperly.[801] The following states have been highlighted for the uniqueness of their respective laws, and for their policies in promulgating such laws.

---

[792] U.S. Const. Art. VI.

[793] Acosta and Golub (2018).

[794] *See* US. Cons. Art. I, Section 8.

[795] Acosta and Golub (2018); *New State Ice Co. v Liebmann*, 285 U.S. 262 (1932).

[796] *Id.*

[797] Levine (2015).

[798] *Id.*

[799] *Id.*

[800] *Id.*

[801] Hiltzik (2017).

4.7.2.1 California

Due to the fact that the federal government is so slow when authorizing funds, many American states have had to take over and provide funding for this form of research. California has spent at least $300 million dollars in funding for stem cell research, most of which is for embryonic stem cells.[802] California has issued what has become known as "Prop 71" which stands as one of the best examples of a state law benefitting the stem cell sciences.[803] Voted on and passed in California's state election of 2004, the Act which is formally known as "California Stem Cell Research and Cures Act" amends California's state constitution to include stem cell research as a right, created California Institute Of Regenerative Medicine, ("CIRM"); authorized California to issue $3 billion dollars in General Obligation Bonds to fund the grants for research for cures using stem cells of all kinds.[804] Under this Amendment, California recognizes and specifically allows for research to be conducted both on and using embryonic and fetal tissues, yet at the same time, still requires donors' consent, and prohibits the sale of embryos while setting a limit on reimbursements to research subjects.[805] Prop 71 has caused many research facilities to leave their more restrictive home states for California, where they can obtain grants from the state, and are not restricted by federal laws.[806] In fact, between the years of 2004 and 2012, CIRM spent over $1 billion dollars in recruiting out of state scientists, building new research facilities and issuing grants.[807] Many scientists, and researchers have agreed that Prop 71 should act as a guiding light for other states to protect experiments that are publicly funded, especially when the federal government bans such funding.[808] Right now CIRM just approved a third clinical trial for Covid-19 which would use blood stem cells obtained from placenta.[809]

Even with all of the great help and hope that Prop California gave through Prop 71, it has not been without its fair share of problems.[810] One key issue is funding, CIRM has been creating and expanding much scientific research infrastructure, and in 2017 CIRM's funding ran out, so in early 2019 CIRM announced that they were practically bankrupt and would need to create a ballot initiative to try to raise an additional $5 billion dollars in the 2020 election, yet the initiative alone would cost around $10 million dollars.[811]

---

[802] Palca (2007).

[803] California's Institute of Regenerative Medicine (CIRM) (2004).

[804] Russo (2005), pp. 1167–1171.

[805] *See* California Proposition 71 §§ 123440 (2004); California Proposition 71 §§ 24185 (2004); California Proposition 71 §§ 125300-320 (2004).

[806] O'Brien (2004).

[807] *Id.*

[808] Goldstein (2017), pp. 3943–3944.

[809] Villa (2020).

[810] Shanks (2019).

[811] *Id.*

4.7.2.2 Connecticut

Connecticut's General Assembly passed a statute in June of 2005 which worked to set up funding for embryonic and adult stem cell research which was set to last for a minimum of 10 years.[812] Connecticut's laws allow for most forms of embryonic research, so long as the embryo has yet to mature to a certain point known as "gastrulation" which occurs when the embryo is converted from a "blastula;" found as an embryo with a single layer of cells, to a "gastrula"; which is when the blastula folds in on itself, or divides to form an additional layer of cells.[813] This ban is extremely important to understand, because the timing of the gastrulation's effect of adding a layer of cells means that the embryo is in a state of development that will soon begin to form specialized tissues that will inevitably become organs, skin, the nervous system, or other highly specialized, and functional tissues.[814] The process involves what is known as "germ layers" which is what is actually created by the folding or dividing of the blastula, and in practice, these germ layers will take the form of one of three categories.[815] The first germ layer is known as the "ectoderm" and is the outer most layer which will eventually form into skin and a nervous system; the middle layer is known as the mesoderm which in most organisms will eventually turn to the connective tissues and their supporting musculature systems, and the bottom most layer is known as the endoderm which is charged with creating organs.[816] It is very important to understand that at the point of gastrulation, an embryo starts to develop into something more than just a single layer of cells and many believe gastrulation is the first true developmental milestone after the initial milestone of fertilization.[817] The state of Connecticut also sets up provisions that require consent from the donor, and ban certain compensation for embryonic tissues and their sales.[818]

Connecticut's applicable statutes are very interesting, as they are the only state that has an exception created by gastrulation and thereby creates a specific cut-off for when an embryo is no longer a candidate for study due to actual developmental stages.

The Connecticut statutes also provide for state funding and grants of embryonic and adult stem cell studies, by creating a "non-lapsing, general fund" account where the state initially placed $20 million dollars into, and where other private and public

---

[812] *See generally* Connecticut General Statutes §19a-32d (Supp. 2007); Connecticut General Statutes §19a-32e (Supp. 2007); Connecticut General Statutes §19a-32f (Supp. 2007); Connecticut General Statutes §19a-32g (Supp. 2007).

[813] Abas (2022), pp. 6–14.

[814] *Id.* pp. 6–7.

[815] Williams and Solnica-Krezel (2018), pp. 33–39.

[816] Kiecker et al. (2016), pp. 923–925.

[817] *See:* Zhai et al. (2022), p. 19.

[818] *See* CT Gen Stat § 32-41jj.

donations can also be placed.[819] Interestingly, the State's legislative branch decided to fund their share in part from the Tobacco Settlement Fund which was created in 1998 through a settlement between the four largest tobacco companies and 46 beneficiaries.[820] In total, Connecticut pledged approximately $100 million dollars over 10 years to give as grants for all forms of stem cell research.[821]

4.7.2.3 Illinois

Illinois passed Public Act 095-0519, also known as the Stem Cell Research and Human Cloning Prohibition Act allows for the study of embryonic stem cells and also allows for grants using public funds to be provided for embryonic stem cell research.[822] Section 3 of the Act explains that Illinois believes that all forms of stem cell research are important and valuable to one's health, and the state should not act to hinder "research materials" among qualified researchers.[823]

The Act established the Illinois Regenerative Medicine Institute Oversight Committee, which would be charged with determining the validity and appropriateness of awarding grants to qualified applicants and sets up that there must be seven members appointed to the Committee by the Governor, and that they must act with the advice and consent by Illinois Senate.[824] The members of this Committee are also required to be made up of members from professional medical organizations, volunteer medical organizations and private research institutions, but stipulates that at least one of each person out of these categories be members at all times.[825] Even with such permissible and favorable legislation, the Act still places certain bans and restrictions, such as the ban on human cloning, and the purchase or sale of embryonic or fetal tissue.[826]

4.7.2.4 Massachusetts

Massachusetts is another state with a unique caveat in a similar vein to that of Connecticut's, and allows for the study and research of embryonic stem cells, so

[819] sSB 934 (2005).

[820] Tragakiss (2009).

[821] *See* Connecticut State Department of Public Health (2024).

[822] *See Illinois* Stem Cell Research and Human Cloning Prohibition Act §§ 1–3.

[823] *See Illinois* Stem Cell Research and Human Cloning Prohibition Act § 3.

[824] *See Illinois* Stem Cell Research and Human Cloning Prohibition Act § 20(a).

[825] *See also* Illinois Stem Cell Research and Human Cloning Prohibition Act § 20(b)–§ 20(c).

[826] Illinois Stem Cell Research and Human Cloning Prohibition Act §§ 40; *Illinois* Stem Cell Research and Human Cloning Prohibition Act §§ 45.

long as the embryos have not reached more than 14 days of development.[827] The Act was passed because it recognized the potential in embryonic stem cell research in benefitting all people with new knowledge to create novel therapies and treatments for patients "suffering from disease or injury…"[828] Further, it recognized the potential benefits that embryonic stem cells can provide that other forms of treatment would not be able to.[829] It is important to note that this state's 14-day rule was adopted due to what became a hallmark of embryonic research and was derived from two studies, the first being 1979s Ethics Advisory Board to the Department of Health, Education and Welfare (HEW), and the second is the United Kingdom's Warnock Committee of Inquiry into Human Fertilisation and Embryology.[830] In essence, the 14 day rule was derived due to the above mentioned studies which found that "sentience," or the ability to feel pain, starts to develop around day 17, and to be safe as to not issues, the studies felt that giving a window of a few days subtracted from the 17 would be safest for the embryo.[831] This rule has more recently been shown to be somewhat problematic, at the time when the rule was first explained and adopted, embryonic science was not as well developed as it is now, and embryos could not be kept viable for more than a few days, and now with our culture techniques, it is possible to keep an embryo viable past the 14 day mark.[832] It is also important to note that the discovery of human pluripotent cells, can under certain laboratory conditions, form into something very similar to an embryo, and yet the 14 day rule still may apply, even though no embryo is actually involved.[833]

It is also important to understand that there is very little research of what occurs to a fetus from the 14th to the 28th day, and for this reason, has been known as the "'black box' of human development…"[834] As a result of this lack of knowledge, among other issues, there are many disagreements in science as to when a true "cut-off" should be recognized.[835] Some have even attacked the process of how the 14-day rule came into existence in the first place.[836] Some of the claims stem from the fact that the 14-day time period was created as a moral compromise without taking more specific facts into account, such as the fact that there is an actual impossibility to "draw a morally and legally significant distinction" between

---

827 Commonwealth of Massachusetts An Act Enhancing Regenerative Medicine In The Commonwealth Chapter 111L § 2.

828 Commonwealth of Massachusetts An Act Enhancing Regenerative Medicine In The Commonwealth Chapter 111L § 1(a).

829 *Id.*

830 Chan (2018), pp. 228–234.

831 Appleby and Bredenoord (2018), pp. 1–4.

832 Chan (2018).

833 Pera et al. (2015), pp. 917–919.

834 Appleby and Bredenoord (2018).

835 Cavaliere (2017), pp. 1–12.

836 *Id.* p. 3.

embryos from the 14-day period, and those who are a little newer and those that are a little older.[837] Other critics essentially parrot the conflicting claims discussed above.[838]

States including California, New Jersey, Missouri, and others allow for all aspects of embryonic stem cells to be publicly funded, including the use of therapeutic cloning. There are other states who allow for public funding yet prohibit therapeutic cloning also exist and are made up of a number of states. Some states allow for public funding but restrict all forms of cloning, and some states such as Arkansas, North, and South Dakota, Virginia, and others completely prohibit any state funds to be used for stem cell research. This was an attempt to prevent future legislation that can restrict stem cell laws. This bill has not gone further yet, but a legal challenge to President Obama's Executive Order was attempted to be brought to the Supreme Court, but the Supreme Court declined to hear the case.[839] This action by the Supreme Court can be seen as a positive affirmation that this Executive Order cannot be challenged in the courts.[840] However, it must be noted that there are a few methods to overturn or invalidate an executive order.[841] First, any incumbent president may revoke an executive order issued by a previous president, second Congress can effectively revoke an executive order by passing newer legislation that invalidates it.[842] One must note though that this new legislation passed by Congress invalidating the executive order can itself be vetoed by the President, which would than require Congress to override the President's veto with a "two-thirds majority," and finally the Courts can find an executive order to be unconstitutional when it exceeds the President's "constitutional authority."[843]

With these facts and the rules as to how to revoke or override an executive order, it would seem that the Court did not think it was beyond the President's scope of authority to pass such an order, and that in order for it to be changed, it would be either up to a future president, or for Congress to take action.[844]

4.7.2.5 New York

New York is another state that has passed its own laws that permit embryonic stem cell research which came about in 2007.[845] The New York Senate passed Public Health Law § 265 which created what is known as the Empire State Stem Cell Board

[837] *Id.*

[838] *Id.* p. 4.

[839] Savage (2013).

[840] *Id.*

[841] *See* American Bar Association (2021).

[842] *Id.*

[843] *Id.*

[844] *Id.*

[845] Acosta and Golub (2018).

(the "ESSCB") to be charged with administering the Empire State Stem Cell Trust Fund, that is in place to advise the Commissioner of Health regarding findings, and issues with stem cell sciences.[846] This Board is itself made up of two separate committees, the "funding committee" and the "ethics committee."[847] Already, through the use of this law, the State of New York has awarded close to $400 million dollars to almost 40 New York based institutions, and also creating close to 1000 new jobs in this field.[848] Unfortunately there are no hard numbers for how much money went to directly benefit embryonic stem cell research in relation to adult stem cell research, but the fields that have been benefitted include but are not limited to the following: Eye diseases; liver diseases; psychiatric disorders; diabetes; cancer; cardiovascular diseases; musculoskeletal diseases; hematopoiesis and related diseases; neurological disorders; and other forms of research.[849]

New York has also had interesting and unique challenges to their liberal practices regarding embryos and fetal tissue, and even upheld certain compensations for donations of embryos which could be used for embryonic stem cell research.[850] In the case of *Matter of Feminists Choosing Life of N.Y., Inc. v Empire State Stem Cell Bd.*, an aspect of Public Health Law § 265 was challenged, as to the issue of providing donors with compensation for donating ESSCB violated Public Health Law § 265 (a) (2) which bans the granting of ESSCB funds to research that "directly or indirectly" is to be utilized for various forms of human cloning research, including reproductive cloning.[851] Even though this law specifically mentions a ban on grants to research involving "human reproductive cloning," when using the somatic nuclear transfer method, the state legislature did not include a definition of what "human reproductive cloning" means in relation to the law, leading petitioner to cite case law that bars the Court from giving deference to an agency in determining their own scope.[852] In this situation, the Court found that the meaning of the term "reproductive cloning" is not something that is easily understood, so that the interpretation of *Kennedy v. Novello*, 299 AD2d 605, 607 (2002); and *Matter of New York State Assn. of Life Underwriters v. New York State Banking Dept.*, 83 NY2d 353, 359–360 (1994) was more on point in that they stand for the proposition that: "[W]here the statutory language is special or technical and does not consist of common words of clear import, courts will generally defer to the agency's interpretative expertise unless that interpretation is unreasonable, irrational or contrary to the clear wording of the statute."[853] The ESSCB argued that their interpretation was not meant to

[846] *See* New York State (2020a).

[847] *Id.*

[848] New York State (2020b).

[849] *Id.*

[850] *Matter of Feminists Choosing Life of N.Y., Inc. v Empire State Stem Cell Bd.*, 2011 NY Slip Op 05160 [87 AD3d 47] (2011).

[851] *Id.*

[852] *See id.* (citing *Kurcsics v Merchants Mut. Ins. Co.*, 49 NY2d 451, 459 (1980)).

[853] *Id.*

include any form of human reproductive cloning as alleged by petitioner, in violation of the ESSCB bylaws, which led to the Court upholding the ability to compensate donors of oocytes.[854]

New York was so pro-stem cell research, that when former President George W. Bush enacted his restrictive policies, the State of New York sought to find a way to allow for further stem cell research while also helping to fund it.[855] The answer to this issue was to create the New York State Stem Cell Science program under the direction of the Department of Health in 2007.[856] While in operation, NYSTEM awarded approximately $400 million dollars in grants, as well as:

> more than $481 million in additional support from other sources and supported over 750 jobs across the state. NYSTEM's successful efforts to increase New York's capacity for stem cell research have enabled our funded institutions to bring in over $142 million in philanthropic donations.[857]

Unfortunately, as of 2022, NYSTEM was longer funded by being removed from that year's budget.[858] While this may seem like a set-back, it in fact was unneeded as President Biden won his election after promising to ease fetal tissue research restriction, and the money was needed elsewhere while dealing with Covid-19.[859]

4.7.2.6 Iowa

Yet another state to pass its own laws relating to embryonic stem cell research is Iowa, who passed Chapter 707C, or "Iowa Stem Cell Research and Cures Initiative."[860] The stated purpose of the law is found directly in the law in § 707C.2 which explains that there is a noble purpose behind the passage of this law which is to grant access to patients using stem cell therapies, but also has a complete bar on human reproductive cloning using somatic nuclear transfer method.[861]

4.7.2.7 Maryland

In 2006 Maryland too enacted a pro-embryonic stem cell research law which requires any and all stem cell research to be conducted in an appropriate manner which remains ethical and prohibits human cloning.[862] In addition to the

---

[854] *Id.*

[855] *See*: New York State (2020c).

[856] *Id.*

[857] *Id.*

[858] Hamilton (2021).

[859] *See generally id.*

[860] *See* Iowa Code 2017, Chapter 707C.

[861] *See* Iowa Code 2017, Chapter 707C.2.

[862] Maryland Stem Cell Research Act of 2006, S.B. 144, (Md. 2006).

aforementioned ban on cloning, the law also mandates that an independent body of 15 members within the Maryland Technology Development Commission be created and made up of scientists, ethicists, and patient advocates, and has an interesting provision that is fairly unique among other similar laws that requires certain members to represent certain institutions within the state such as Johns Hopkins University, and further that another "[t]wo individuals with expertise in the field of biomedical ethics... as it relates to religion..."[863] There has been some debate as to the religious aspect of this law as it may very well violate the Establishment Clause of the United States Constitution in that it seems to favor a particular religion and that it seeks guidance in the law while in relation to religion.[864]

4.7.2.8 New Jersey

New Jersey is widely considered one of the most lenient states when dealing with granting public funds for embryonic stem cell research.[865] New Jersey passed and approved an Act which took effect in 2004 and effectively permitted embryonic stem cell research with a few caveats.[866] First, embryonic stem cell research, and research conducted on other forms of fetal tissue are permitted, but must follow the federal regulations, and the research must give deference to the moral and ethical considerations that working with such forms of tissue require.[867] Next, the law requires informed consent of the perspective donor, and the donor cannot receive "valuable consideration" for such donation.[868] This Statute even prescribes a penalty for those who break this law, which states that the person in violation can be charged up to $50,000 per each violation.[869] This Statute further bans all forms of human cloning.[870] Interestingly, the legislature provided the public policy behind the creation of this Act, which found that there are approximately 128 million American citizens that are currently suffering from a disease; that the costs associated with treating and/or curing such diseases costs "hundreds of billions of dollars" yearly; that embryonic stem cells seem to offer promise, which should be further investigated; that biomedical research should be fostered within the state as it is considered a "growing component of New Jersey's economy;" and that due to the novel, moral and ethical issues associated with embryonic stem cell research, the state must promote a public policy that works to balance the

[863] Maryland Code, Economic Development § 10-431 (2022).

[864] Acosta and Golub (2018).

[865] Garfinkel (2004).

[866] P.L. 2003 Chapter 203, An Act Concerning Public Funding of Stem Cell Research (2003).

[867] *Id.*

[868] *Id.*

[869] P.L. 2003 Chapter 203(C)(2).

[870] Codey (2003).

> ethical and medical considerations, based upon both an understanding of the science associated with stem cell research and a thorough consideration of the ethical concerns regarding this research; and be carefully crafted to ensure that researchers have the tools necessary to fulfill the promise of this research.[871]

New Jersey is one of the rare states that specifically outlines their reasoning for such a permissive law. In fact, as described above, the authorizing Statute even explains that there is a benefit to the state, in the form of attracting biomedical companies to move to New Jersey.[872] Further, the citizens of New Jersey have argued that this Act does not go against the right to life, but it actually is about the right to life itself, in that it works to better the lives of those in need, in fact, at the time of this bill's passage, then Democratic Senate President, Richard J. Codey wrote:

> This bill is not about cloning. This bill is about life. It is about advancing a science full of potential for sustaining human life which will be stifled in our state if this legislation fails. Other groups with alternative agendas have tried to confuse you with misinformation about this legislation. I have no alternative agenda. I urge you to put all of the misinformation aside and join me in supporting S-1909, simply because it is the right thing to do as a human being.[873]

Soon after the passage of this Act, Richard J. Codey became New Jersey's governor which shows that he had much support in the passage and advocation of this law.[874]

4.7.3 States with Restrictive Embryonic Stem Cell Laws

In recent times, states have decided to either fund embryonic stem cell research programs, allow them to exist in their state, or outlaw their research, or just outlaw the state funding of embryonic stem cell research. For example, Louisiana has banned all research pertaining to unused *IVF* embryos; Michigan, Iowa, Arkansas, North Dakota and South Dakota have banned research of cloned embryos; and with the most restrictive of laws, South Dakota has completely prohibited any embryonic stem cell research regardless of their initial source.[875] Although some of the laws have changed, it is important to note that most of these states with their bans are typically considered "red" states, meaning that they are usually politically conservative.[876] It is important to note that since the above discussed decision in the *Dobbs* case, states have started implementing new "personhood" laws.[877] The concept of "personhood" is a unique one that has its roots in the *Roe v. Wade* case which

---

[871] *Id.*

[872] *Id.*

[873] *Id.*

[874] *Id.*; New Jersey Legislature (2018).

[875] Goody (2006).

[876] Jones (2016).

[877] *See generally* Amy (2022).

essentially held that the unborn do not amount to a person, as a "person" described in the 14th Amendment.[878] Those who are against abortion, have argued that this should not be the case, and that the unborn should be considered a person according to law.[879] For certain states which will be discussed below, "personhood" laws came about by States that seek to further restrict access to abortions, in almost any and all cases, including situations where the unborn was created due to rape or incest.[880] Interestingly, these states have chosen to follow their voters' conservative and religious beliefs over the potential benefits of new businesses forming for the purpose of studying stem cells in the states.

4.7.3.1 Louisiana

One of the strictest states in regards to the laws contemplating embryonic stem cells is Louisiana who has an almost total ban on all forms of research conducted using embryos, with a narrow exception which allows for *IVF* related embryonic experiments, so long as it's for the actual purpose of future implantation to lead to pregnancy.[881] Louisiana's law further specifically states:

> The use of a human ovum fertilized in vitro is solely for the support and contribution of the complete development of human in utero implantation. No in vitro fertilized human ovum will be farmed or cultured solely for research purposes or any other purposes. The sale of a human ovum, fertilized human ovum, or human embryo is expressly prohibited.[882]

This is one of the most strongly worded of state laws pertaining to embryonic experimentation, and in the next sections of the statute state that an *in-vitro* embryo is to be considered a juridical person, meaning that it is known to not be a single, living person, but should be treated in a manner similar to a corporation, where the corporation is still considered an entity but it is recognized as a human being, with duties and rights.[883] The statutes also go on to state that an embryo created for *IVF* purposes, is not owned by the biological parents, unless they state their identities as the biological parents, otherwise, it is to be deemed to be in the care of the physician, as its temporary guardian, but the physician cannot claim ownership.[884] Section 9 §129 of the Louisiana Statutes goes on to express that because the fertilized ovum is considered a juridical person, it cannot be destroyed person, but that if the embryo is being preserved using cryopreservation, or if the embryo fails to develop further during a window if 36 h, it no longer retains the juridical person status because it is

---

[878] *Id.*

[879] *Id.*

[880] *Id.*

[881] *See* LSA-R.S. 9 § 121–133 (1986).

[882] *id.* at § 121.

[883] *See generally id.* at §123.

[884] *id.* at §126.

now deemed unviable, and can then be destroyed.[885] When research is banned on embryos created through *IVF*, the typical situation could involve researchers seeking to clone embryos, but Louisiana also has passed HB 370 which bans all state and federal monies from being used for any human cloning purposes.[886]

4.7.3.2 Florida

Another state which has codified bans on embryonic experimentation is Florida, whose Statute states that:

> [n]o person shall use any live fetus or live, premature infant for any type of scientific, research, laboratory, or other kind of experimentation either prior to or subsequent to any termination of pregnancy procedure except as necessary to protect or preserve the life and health of such fetus or premature infant.[887]

Florida's voters have attempted to add amendments on both sides of the issue.[888] The conservatives in the state sought an amendment that would ban embryonic stem cell research known as the "Florida Anti-Stem Cell Research Initiative," which banned "any revenue of the state from experimentation that involved the destruction of a live human embryo."[889] On the more progressive side, was the Palm Beach County group Floridians for Stem Cell Research and Cures, who sought to add an amendment to the 2008 ballot that called for $20 million dollars a year, for a period of 10 years to be used for embryonic stem cell research, known as the Florida funding of embryonic stem cell research (2008).[890] In the end, neither of these measures appeared on the ballot, but as one can see by the current state of Florida's laws, a more conservative approach was taken within the state.[891]

Even further, Governor Ron DeSantis signed what is known as the "Heartbeat Protection Act" (or "SB 300") in Florida on April 13, 2023.[892] This Law's purpose is to effectively remove any possibilities of harm to a fetus once a heartbeat is able to be detected, thus further expanding the rights of an unborn child.[893] Prior to Governor DeSantis's signing of this Law, Florida's legislature passed it with the protections that take effect of 6-weeks into gestation as that is when a heartbeat is typically first able to be detected in a fetus.[894] The law states makes it that anyone who "any person who willfully performs or actively participates in a termination of pregnancy" would

[885] *id.* at §129.

[886] *See* LA HB 370 (2008).

[887] Fla. Stat. 390.0111-(6) (2023).

[888] *See* Hafenbrack (2005), p. 1A.

[889] Kennedy and Garcia (2005).

[890] *See* Bioethics Today Sysadmin (2007).

[891] *Id.*

[892] Richard (2023).

[893] *Id.*

[894] *Id.*

face felony charges unless one of the exceptions is met.[895] These exceptions include when times when carrying the baby to full term could subject the mother to a risk of losing her life, as well as situations prior to exceeding the 15 weeks of pregnancy, of rape, incest, or human trafficking.[896] While this does not directly create stem cell laws, it does further address the State's attitude towards the rights of the unborn, at least once the fetus begins to show signs of a heartbeat.[897]

4.7.3.3 Indiana

The state of Indiana has implemented some of the strongest protections for fetuses, including aborted fetuses, and prevents virtually all scientific examinations on them.[898] The law of the state also criminalizes[899] the transportation of an aborted fetus across state lines. Statute IC 35-46-5-1.5 directly implicates embryonic stem cells, although the term "fetal stem cells" is used interchangeably, but protects against what the state considers the unlawful transfer of fetal tissues coming from any source, and establishes criminal penalties for those who violate this section.[900] Indiana also has a provision that bans cloning, and only allows for exceptions for those undergoing legitimate *IVF* procedures, but this ban includes those who transfer embryos for scientific purposes and those in violation of this law, can expect to be charged with a felony.[901] The ban on the cloning of embryos directly prohibits state funds from contributing to this area of science, yet leaves open the ability for private funds to create clones.[902] Statute IC 16-34-1-2 also stipulates that public funds cannot be used for anyone seeking an abortion, with a narrow exception for those who require an abortion to protect their own health.[903] This Statute is so strict, that there are even bans on abortions for those who learn of health problems that the fetus will face if it is carried to the full term, such as in the case of fetuses with down syndrome, as well as banning abortions for other reasons, as set forth in IC 16-34-2-1.1 (K) which specifically states: "That Indiana does not allow a fetus to be aborted solely because of the fetus's race, color, national origin, ancestry, sex, or diagnosis or potential diagnosis of the fetus having Down syndrome or any other disability."[904] The laws of the state also require a more explicit version of informed, voluntary consent for women seeking abortions, which mandates that the doctor preforming

[895] *Id.*; Yilek (2023).

[896] *Id.*

[897] The Heartbeat Protection Act (SB 300) (2023).

[898] *See* Ind. Code § 16-34-2-6.

[899] *See id.* at (b).

[900] *See* Ind. Code § 35-46-5-1.5.

[901] Ind. Code § 35-46-5-2 (c).

[902] Acosta and Golub (2018).

[903] Ind. Code § 16-34-1-2.

[904] Ind. Code § 16-34-2-1.1(K).

the abortion procedure must expressly present, in writing or orally, explain that: "human physical life begins when a human ovum is fertilized by a human sperm;" the likely age of the fetus, including a picture of the fetus, its dimensions, and relevant information on the potential survival of the fetus.[905] In the grand scheme of this topic, one can see that this State's political background has played a major part in their laws, in fact, Indiana has long been considered a republican stronghold.[906]

4.7.3.4 Oklahoma

Oklahoma also has very restrictive laws that have an interesting caveat that is more reminiscent of President George W. Bush's Executive Order and was codified with the passing of HB 3126 Advancement in Stem Cell Cures and Therapies Act.[907] In 2008 when this law was passed, it contained the express statement that "This bill was enacted to encourage stem cell research in the State of Oklahoma because stem cells are recognized to have great potential for treatment of life-threatening and debilitating diseases."[908] Yet, the law only allows for experimentation on embryos and embryonic cell lines that were created "prior to August 1, 2001" and must be in accordance with the 2007 federal law, meaning that there are very few hESC lines that could theoretically be used for research purposes.[909] If the experiments are constrained to the letter of the law, then public funds can be allocated and granted to such experiments, but if not, then public funds are banned from being granted.[910] The statute also bans the cloning of embryos for scientific studies and has an informed consent provision similar to many of the other states.

4.7.3.5 North Dakota

Another state that has very strict laws pertaining to fetal and embryonic research while also providing for criminal punishment for violations of said laws is North Dakota.[911] The controlling law, N.D. Cent. Code § 14-02.2-02 specifically prohibits any and all experimentation on deceased fetuses, aside from the occurrence of an induced abortion.[912] The law also prohibits the use of:

> [A] fetus or fetal organs or tissue resulting from an induced abortion in animal or human research, experimentation, or study, or for animal or human transplantation except for

905 *See* Ind. Code § 16-34-2-1.1.

906 Electoral Ventures (2020).

907 *See* Okla. Admin. Code § 310:551 H. B. No. 3126.

908 *Id.*

909 *Id.* at (2).

910 *Id.* at (C).

911 *See* N.D. Cent. Code Ann. § 14-02.2-02 (1989); N.D. Cent. Code Ann. § 12.1-32-01 (2013).

912 N.D. Cent. Code Ann. § 14-02.2-01.

> diagnostic or remedial procedures, the purpose of which is to determine the life or health of the fetus or to preserve the life or health of the fetus or mother, or pathological study.[913]

The implication created from the previous quoted text essentially makes it impossible for any form of embryonic stem cell research to be conducted within the state, and this is further bolstered by the fact that cloning as also been outlawed, thus removing the potential sources of embryonic stem cells from research.[914] The law also bans the transfer of fetal tissue between parties, and this includes anything derived from human cloning.[915] Seemingly, the only exception to this rule is for the actual purpose of *IVF*, and a violation of either of the two cited statutes can result in a penalty of up to 5 years in jail because they are considered a "Class C" felony.[916]

When dealing with states with such strict laws, it is imperative to look to the policy behind the laws, and for North Dakota, one can look to the voting history and see that their citizens' beliefs are more aligned with that of the American conservative, and the state has actually been called one of the "most conservative."[917] North Dakota has also been one of the most vigilant in prosecuting stem cell clinics that fail to live up to their citizens' expectations, even when dealing with strictly adult stem cells which has led to several rules that apply to all stem cell clinics.[918] The story unfolded when a stem cell clinic, which had yet to be operating for more than 2 weeks, received customer complaints as to any lack of improvement in their conditions from injected "amniotic" stem cells.[919] Essentially, amniotic stem cells are a form of adult stem cell that is taken from the amniotic sac, and does not require any harm to an embryo, so without complaints, the clinic was well within the constraints of the law, yet, this situation led to the clinic being charged over $20,000 in refunds, and they were forced to cease providing their so-called "treatments."[920] The State's Attorney General also required non-clinic owners, such as two nurses pay fines for their part in working at the clinic.[921]

4.7.3.6 South Dakota

South Dakota is another state that has been a conservative stronghold for many years which clearly has influenced their laws regarding embryonic stem cells.[922] The State's statute does not speak directly to stem cells, but does specifically state:

[913] D. Cent. Code Ann. § 12.1-39-02 (2017).

[914] *Id.*

[915] *Id.*

[916] N.D. Cent. Code Ann. § 14-02.2-05; N.D. Cent. Code Ann. § 12.1-39-03.

[917] Newport (2017).

[918] Knoepfler (2018).

[919] *Id.*

[920] *Id.;* Loukogeorgakis and De Coppi (2017), pp. 1663–1673.

[921] Emerson (2018).

[922] Hawkins (2020).

"No person may knowingly conduct nontherapeutic research that destroys a human embryo. A violation of this section is a Class 1 misdemeanor."[923] In South Dakota, a Class 1 misdemeanor is punishable by a 1-year prison sentence, and/or a $2000 fine.[924] South Dakota's laws can be seen as extremely conservative, especially when one looks to the adopted definition of "human embryo," found within their statutes.[925] The definition reads: "For purposes of §§ 34-14-16 to 34-14-20, inclusive, the term, human embryo, means a living organism of the species Homo sapiens at the earliest stages of development (including the single-celled stage) that is not located in a woman's body."[926] Clearly, this State has adopted the more conservative, religious view as to when life begins, and thus is due protections that are supposed to be enjoyed by all human beings within the state. Death and destruction of the embryo is not the State's only concern, there is also a law that prohibits "Research subjecting human embryo to substantial risk."[927] This ban actually is an attempt to prevent anyone from "knowingly" transferring a human embryo nontherapeutic research, and a violation is also considered a Class 1 misdemeanor.[928] This view can even be seen in state university policies regarding the exploration of stem cells.[929] In the University of South Dakota's policy regarding stem cells compiled by the Office of Research and Sponsored Programs, actual language taken directly from the Statutes were used to emphasize the ban and criminality of engaging in such banned conduct.[930] There have been attempts by some to change this law, such as those who proposed South Dakota Embryonic Stem Cell Research Measure which was a proposal to ease restrictions on embryonic stem cell research, and was poised to be added to the 2010 ballots, but seems to have failed for the reason of lacking enough supporting signatures.[931]

4.7.3.7 Georgia

While Georgia has not actually passed a specific law targeting embryonic stem cell research, Georgia's Senate did pass a bill that in fact did attempt to do so in a way.[932] While this Bill never made it into law, it did in fact seek to ban *IVF* for any reason other than for implantation into a woman for pregnancy purposes.[933] This would not

[923] S.D. Codified Laws § 34-14-16 (2000).

[924] S.D. Codified Laws § 22-6-2 (2005).

[925] S.D. Codified Laws § 34-14-20 (2000).

[926] *Id.*

[927] S.D. Codified Laws § 34-14-17 (2000).

[928] *Id.*

[929] *See* University of South Dakota (2011).

[930] *Id.*

[931] Ballotpedia (2010).

[932] *See* Legiscan (2010).

[933] *Id.*

be the end of the situation for Georgia in relation to the unborn, as in November of 2022, Georgia's Supreme Court reinstated Georgia's 2019 ban on abortions after 6 weeks of pregnancy.[934] The 6-week period is given as typically, a fetal heartbeat is detectable at 6 weeks, which is why these bills are often termed "heartbeat" bills, and this means that at this time, the unborn obtains the rights of a fully born person.[935] While this law does may not directly impact embryonic research on its face, it in fact does, as it effect it dramatically, as it directly impacts biomedical research including research on some embryonic stem cells.[936] When asked for an article about how this law will impact biomedical research, Suzanna Sherry, who is considered a "constitutional law scholar at Vanderbilt University," stated that "They're not thinking through the other possible consequences of declaring that an embryo is a person from the moment of fertilization...", which was elaborated on by research attorney Mark Barnes for the same article when he said: "If states grant personhood to embryos without specific exceptions for derivation of stem cells from embryos, then there can be no stem cells on which to do research or develop therapeutic products."[937] This statement can be said for any and every state that enacts "personhood" laws, as those laws treat the unborn as full citizens with the obligation to prevent harm from coming to them.[938]

4.7.3.8 Texas

While the state of Texas currently has no specific laws prohibiting embryonic stem cell research, the State did in fact pass what is known as the Texas Heartbeat Act which bans abortions after the fetus develops a detectable heartbeat.[939] What this law effectively acts to ban abortions for women once they are past the 6-week gestation period (as this is typically when a fetal heartbeat is first detectable), unless there is a medical emergency that necessitates the abortion to prevent harm to the mother.[940] The actual law explains that the State of Texas is to be deemed as having "...compelling interests from the outset of a woman's pregnancy in protecting the health of the woman and the life of the unborn child..."[941] This issue has come up again recently in Texas, when a woman, Kate Cox ("Ms. Cox"), who was pregnant with a baby that was found to have Trisomy 18 or Edwards syndrome which is known to cause severe fetal development in many parts of the body, and often life threatening, usually surviving only a short time post-birth, sought to obtain an

---

[934] The Associated Press (2022).

[935] *Id.*

[936] *Id.*

[937] Baumann (2022).

[938] *Id.*

[939] Simmons-Duffin and Feibel (2022).

[940] *See* Texas Health & Safety Code §§ 171.201–171.212; Haining et al. (2022), pp. 535–541.

[941] Texas Health & Safety Code (§§ 171.201–171.212).

exception to Texas's abortion law.[942] As discussed above, the current abortion law in Texas only allows abortions up to week six of the pregnancy, with the only exceptions being when the pregnant mother's life is at risk due to the pregnancy or to "prevent substantial impairment of major bodily function," and in this case, the Ms. Cox was 19 weeks pregnant when she found out that her unborn child had this chromosomal abnormality, therefore precluding her from being able to obtain an abortion under Texas law without having an exception granted.[943] As a result the Center for Reproductive Rights filed a suit in an effort to obtain an abortion for Ms. Cox in the state of Texas, and Ms. Cox's doctor offered her "good faith belief and medical recommendation" that an abortion should be granted as the current pregnancy coupled with the fact that Ms. Cox was required to have two cesarean operations, and that she was suffering from severe pain due to the pregnancy with signs of trouble as these are signs that can point to Ms. Cox's life potentially being threatened by carrying the unborn child to full term, as well as risks that could preclude her from ever becoming pregnant again.[944] This was all supported by information provided by Ms. Cox's doctor Dr. Karsan who stated that she believes that Ms. Cox met the necessary requirements to obtain the necessary exception to have an abortion after the first 6 weeks already passed.[945] Upon the initial filing, the lower court issued a temporary restraining order ("TRO") which would work to block Texas's abortion law as it applied to Ms. Cox in this situation.[946] Immediately after, Texas Attorney General Ken Paxton ("Mr. Paxton") issued a statement stating that the TRO "will not insulate hospitals, doctors, or anyone else, from civil and criminal liability for violating Texas' abortion laws."[947] This statement was followed up with threats to the hospital where Dr. Karsan works by stating that it is up to hospitals to credential their physicians and put them in their roles, and if Dr. Karsan's employer allowed her to perform an abortion, the hospital may be found liable for regulatory and civil violations.[948] Next, Ken Paxton pointed out that Dr. Karsan had failed to truly prove that Ms. Cox qualified as having an exception because a second doctor was not sought to back up Dr. Karsan's claims as is needed under Texas law.[949] Finally, Mr. Paxton explained that under Tex. Health & Safety Code §170A.002(b) there is a requirement that abortions that fall in a time frame after the first 6 weeks of pregnancy only if the pregnancy "places the female at risk of death or poses a serious risk of substantial impairment of a major bodily function unless the abortion is performed or induced."[950] Mr. Paxton went on to challenge the

---

[942] *See* O'Driscoll (2023) (referencing *Cox v. Texas*, No. 23-994).

[943] *Id.*

[944] *Id.*

[945] *Id.*

[946] El-Bawab (2023).

[947] Ruiz and Linton (2023).

[948] *See* Paxton (2023).

[949] *Id.*

[950] *Id.*

TRO that was issued as it failed to point out what the life-threatening condition is that is being aggravated by Ms. Cox's pregnancy, and likewise that it never mentions what the so-called "serious risk of a major bodily function" that Ms. Cox would suffer absent having the abortion.[951] Finally Mr. Paxton closed by stating that "A TRO is no substitute for medical judgment."[952] Only a few days later Texas's Supreme Court blocked the TRO and Ms. Cox was ruled to not qualify for her claimed exception.[953] This is likely not the last time that this issue will be heard as *Zurawski v. Texas* is pending, which was brought by five women who claim that Texas's abortion law put their live at risk by denying them proper medical care.[954]

With this information it is easy to imagine that restrictions on embryonic stem cell research maybe impacted, but it is of note that once this law passed, legal experts, as well as doctors and researchers explained that the law should not impact IVF treatments, but it has raised questions as to what to do with the surplus embryos.[955] For now stem cell research of all kinds are still not technically impacted, and it would seem that there is currently little legislative interest in creating new stem cell laws as one recently failed to get enough votes in 2019.[956] This proposed Bill, sought to prohibit the state of Texas from contracting with "vendors that perform elective abortions, destructive embryonic stem cell research, or human cloning or that conduct research on human fetal tissue."[957]

4.7.3.9 Alabama

While Alabama does not have any laws specifically relating to the study of embryos, Alabama Code § 6-5-391 which was originally passed in 1975 states:

> When the death of a minor child is caused by the wrongful act, omission, or negligence of any person, persons, or corporation, or the servants or agents of either, the father, or the mother as specified in Section 6-5-390, or, if the father and mother are both dead or if they decline to commence the action, or fail to do so, within six months from the death of the minor, the personal representative of the minor may commence an action.[958]

While this law fails to mention what would happen if a fetus is killed while in the womb, the court has extended this cause of action to be applicable if the fetus is dies due to any of the above quoted reasons, whether viable or not and has specifically addressed this in the Alabama Supreme Court's ruling released on February

[951] *Id.*

[952] *Id.*

[953] Klibanoff (2023).

[954] *See* Center for Reproductive Rights (2023).

[955] Mendez (2022).

[956] *Id.*

[957] *See*: TX HB 4407 (2019).

[958] Ala. Code § 6-5-391 (1975).

16, 2024.[959] This holding came about after three couples brought wrongful death law suits against a hospital that was storing the couples' frozen embryos after they were accidentally destroyed by a patient who had somehow accessed the area where the frozen embryos were being stored.[960] In the majority ruling, Justice Jay Mitchell wrote that "Unborn children are "children" under the Act, without exception based on developmental stage, physical location, or any other ancillary characteristics."[961] Effectively, this holding makes it illegal to discard unused frozen embryos that were created for *IVF* treatments as they are to be treated as children who have the right to life attached to them, which caused many *IVF* clinics to stop or at best pause their services.[962] It must be noted that at the time of this law's passing, *IVF* was not even a real or viable option as it was not until 1978 that the first *IVF* baby was born.[963] This issue caused much panic in Alabama for those seeking *IVF* services and for those providing them, so Alabama's House and Senate passed a bill which was then signed by Alabama's governor Kay Ivey that protects *IVF* treatment and procedure providers and patients from criminal and civil liability in relation to destroyed or damaged embryos created for *IVF* purposes.[964] While this is promising in relation to *IVF* treatments, it would seem that anything outside the scope of such a treatment, such as research using discarded embryos would not be protected and the researchers could face both civil and criminal liability.[965] Due to the law being so new, there is very little information as to how it will be applied and who will be deemed immune from such liability, but it is something that will more than likely be challenged in the near future.

References

Abas R, Masrudin SS, Harun AM, Omar NS (2022) Gastrulation and body axes formation: a molecular concept and its clinical correlates. Malays J Med Sci 29:6–14

Acosta ND, Golub SH (2018) The new federalism: state policies regarding embryonic stem cell research. J Law Med Ethics 44:419–436

Agren D (2017) Mexico drug cartel's grip on politicians and police revealed in Texas court files. Available via The Guardian https://amp.theguardian.com/world/2017/nov/10/mexico-drug-cartels-grip-on-politicians-and-police-revealed-in-texas-court-files. Accessed 26 Mar 2024

Agyeman NK, Momodu A (2019) Universal human rights 'versus' cultural relativism: the mediating role of constitutional rights. Afr J Law Stud 12:23–46

---

[959] Chandler (2024).

[960] Callahan (2024).

[961] *James LePage, Emily LePage, William Tripp Fonde, and Caroline Fonde v The Center for Reproductive Medicine, P.C. and Mobile Infirmary Association d/b/a Mobile Infirmary Medical Center*, SC-2022-0515 (2024), p. 11.

[962] Mohammad et al. (2024).

[963] Eskew and Jungheim (2017).

[964] Mascarenhas and Rosales (2024).

[965] *Id.*

Alahmad G, Aljohani S, Fahmi Najjar M (2020) Ethical challenges regarding the use of stem cells: interviews with researchers from Saudi Arabia. BMC Med Ethics 21:1–7

Aleksandrova-Yankulovska S (2007) The Bulgarian Health Care Reform and Health Act 2004. Med Law 26:1–14

Al-Hayani FA (2008) Muslim perspectives on stem cell research and cloning. Zygon 43:783–795

Al-Tabba' A, Dajani R, Al-Hussaini M (2020) Stem cell statute in Jordan: leading the way. Front Genet 11:657

Alvarez Manninen B (2023) A critical analysis of Dobbs v. Jackson Women's Health Organization and the consequences of fetal personhood. Camb Q Healthc Ethics 32:357, 362. Available via Cambridge University Press & Assessment https://www.cambridge.org/core/journals/cambridge-quarterly-of-healthcare-ethics/article/critical-analysis-of-dobbs-v-jackson-womens-health-organization-and-the-consequences-of-fetal-personhood/A8803B46F4A0C9157C639289F463A83C. Accessed 26 Mar 2024

Amaris M, Sivaraman F (2019) Ethical guiding principles of "do no harm" and the "intention to save lives" in relation to human embryonic stem cell research: finding common ground between religious views and principles of medical ethics. Asian Bioeth Rev 11:409–435

American Bar Association (2021) Executive orders. General FAQs on executive orders. Available via the American Bar Association https://www.americanbar.org/groups/public_education/resources/teacher_portal/educational_resources/executive_orders. Accessed 26 Mar 2024

American Civil Liberties Union (2024) Treaty ratification. Available via American Civil Liberties Union https://www.aclu.org/issues/human-rights/treaty-ratification. Accessed 6 Jan 2024

Amy J (2022.) EXPLAINER: what's the role of personhood in abortion debate? Available via the Associated Press https://apnews.com/article/abortion-us-supreme-court-health-government-and-politics-constitutions-93c27f3132ecc78e913120fe4d6c0977. Accessed 26 Mar 2024

Andorno R (2005) The Oviedo Convention: a European legal framework at the intersection of human rights and health law. J Int Biotech Law 2:133–143

Appleby JB, Bredenoord AL (2018) Should the 14-day rule for embryo research become the 28-day rule? EMBO Mol Med 10:1–4. Available via the National Library of Medicine [hereinafter NLM] https://www.ncbi.nlm.nih.gov/pmc/articles/PMC6127884. Accessed 26 Mar 2024

Ashford P, Chapman J, Hildebrandt M, Noël L, Pruett T, Slaper-Cortenbach I, Wilson D (2021) Establishing consensus on ethics, traceability and biovigilance for medical products of human origin. Bull World Health Organ 99:907–909

Australian Stem Cell Centre (2010) Ethics & law of stem cell research. Available via the National Stem Cell Foundation of Australia http://www.stemcellfoundation.net.au/docs/fact-sheets/fact-sheet-6%2D%2D-law-and-ethics-of-stem-cell-research.pdf?sfvrsn=13. Accessed 16 Jan 2024

Bahadur G (2001) The Human Rights Act (1998) and its impact on reproductive issues. Hum Reprod 16:785–789, 787

Baker RB (2010) Customary international law in the 21st century: old challenges and new debates. Eur J Int Law 21:173–204

Ballotpedia (2010) South Dakota embryonic stem cell research measure. Available via Ballotpedia https://ballotpedia.org/South_Dakota_Embryonic_Stem_Cell_Research_Measure. Accessed 26 Mar 2024

Barrios J (2018) Joe and Jill Biden among several honored at world stem cell summit gala. Available via the World Stem Cell Summit https://www.worldstemcellsummit.com/2018/01/26/joe-jill-biden-among-several-honored-world-stem-cell-summit-gala. Accessed 22 Mar 2024

Baumann J (2022) State 'personhood' laws threaten embryonic stem cell research. Available via Bloomberg Law https://news.bloomberglaw.com/health-law-and-business/state-personhood-laws-threaten-embryonic-stem-cell-research. Accessed 26 Mar 2024

Bell J, Bordag N, Borry P, Bovenberg J, Dagher G et al (2018) Governing stem cells: regenerative medicine in Europe - the vision and recommendations from the EUCelLEX project, pp 28–32. Available via Elsevier https://ssrn.com/abstract=3347724. Accessed 26 Mar 2024

Bellayr IH, Li Y (2009) Stem cells: it's good to have choices. J Am Col Certif Wound Spec 1:92–94

Biden J (2004) Letter to President George W. Bush. Available via Vote Smart https://justfacts.votesmart.org/public-statement/40867/letter-to-president-george-w-bush. Accessed 26 Mar 2024

Biden J (2006) Letter to majority leader asks for Senate to take up bill that passed the House nearly a year ago. Available via Vote Smart https://justfacts.votesmart.org/public-statement/170433/letter-to-majority-leader-asks-for-senate-to-take-up-bill-that-passed-the-house-nearly-a-year-ago. Accessed 26 Mar 2024

Binder G (1999) Cultural relativism and cultural imperialism in human rights law. Buff Hum Rights Law Rev 5:211–221, 214

Bioethics Today Sysadmin (2007) Florida may require and ban stem cell funding. Available via BioEthics Today https://bioethicstoday.org/blog/florida-may-require-and-ban-stem-cell-funding. Accessed 26 Mar 2024

Blackwell T (2012) Canadian stem cell trial impeded by federal regulations, doctors say. Available via the National Post http://news.nationalpost.com/news/canada/canadian-stem-cell-trials-impeded-by-federal-regulations-doctors-say. Accessed 26 Mar 2024

Board on Health Sciences Policy; Board on Life Sciences; Division on Earth and Life Studies; Institute of Medicine; National Academy of Sciences (2014) Stem cell therapies: opportunities for ensuring the quality and safety of clinical offerings: Summary of a joint workshop. Chapter 4. Washington (DC): National Academies Press (US); 2014 Jun 18. PMID: 25057697. Available via NLM https://www.ncbi.nlm.nih.gov/books/NBK223198/. Accessed 26 Mar 2024

Board on Health Sciences Policy Institute of Medicine (2010) Final report of the National Academies' Human Embryonic Stem Cell Research Advisory Committee and 2010 amendments to the National Academies' Guidelines for Human Embryonic Stem Cell Research. Washington (DC): National Academies Press (US); 2010. Appendix C, National Academies' Guidelines for Human Embryonic Stem Cell Research Amended as of May 2010. Available via NLM https://www.ncbi.nlm.nih.gov/books/NBK210064/. Accessed 12 Jan 2024

Boggio A, Romano CPR, Almqvist J (2019) Human germline genome modification and the right to science: a comparative study of national laws and policies introduction. Available via Elsevier https://papers.ssrn.com/sol3/papers.cfm?abstract_id=3443797. Accessed 26 Mar 2024

Bradley CA, Goldsmith JL (1997) Customary international law as federal common law: a critique of the modern position. Harv Law Rev 110:815–876, 860–861

British Broadcasting Corporation (2013) Sunnis and Shia in the Middle East. Available via the British Broadcasting Corporation. https://www.bbc.com/news/world-middle-east-25434060. Accessed 26 Mar 2024

Bross PF, Fan C, George B, Shannon K, Joshi BH, Puri RK (2015) Regulation of biologic oncology products in the FDA′s center for biologics evaluation and research. Urol Oncol 33:133–136

Busardò FP, Gulino M, Napoletano S, Zaami S, Frati P (2014) The evolution of legislation in the field of medically assisted reproduction and embryo stem cell research in European Union members. BioMed Res Int 2014:1–15. Available via Hindawi http://downloads.hindawi.com/journals/bmri/2014/307160.pdf. Accessed 26 Mar 2024

Busby H, Hervey T, Mohr A (2008) Ethical EU law? The influence of the European Group on Ethics in Science and New Technologies. Eur Law Rev 33:804–842

Bush D, Walsh D (2006) Bush vetoes embryonic stem cell bill. Available via Cable News Network http://www.cnn.com/2006/POLITICS/07/19/stemcells.veto. Accessed 4 Jan 2024

Butler D (2011) France mulls embryo research reform. Nature 469:277

California's Institute of Regenerative Medicine (CIRM) (2004) California's Institute of Regenerative Medicine (CIRM) Prop 71. Available via California Institute for Regenerative Medicine https://www.cirm.ca.gov/board-and-meetings/prop-71. Accessed 26 Mar 2024

Callahan M (2024) Alabama court ruling that frozen embryos are children and its impact on the future of IVF treatment. Available via Boston University https://www.bu.edu/articles/2024/alabama-court-ruling-impact-on-future-ivf-treatment/. Accessed 3 Mar 2024

Canadian Institute of Health and Research (2010) Stem cell research policy, and oversight committees. Available via Canadian Institutes of Health Research http://www.cihr-irsc.gc.ca/e/15255.html. Accessed 26 Mar 2024

Canadian Institutes of Health Research (2023) Stem cell research. Available via Canadian Institutes of Health Research https://cihr-irsc.gc.ca/e/15255.html. Accessed 4 Jan 2024

Canadian Institutes of Health Research Natural Sciences and Engineering Research Council of Canada, and Social Sciences and Humanities Research Council of Canada (2022) Tri-council policy statement: ethical conduct for research involving humans, December 2022. Available via Canadian Institutes of Health Research https://ethics.gc.ca/eng/policy-politique_tcps2-eptc2_2022.html. Accessed 26 Mar 2024

Cappelletti M, Cohen W (1979) Comparative constitutional law. Bobbs- Merrill Co., Indianapolis, and Kommers, Const. Juris

Cavaliere G (2017) A 14-day limit for bioethics: the debate over human embryo research. BMC Med Ethics 18:1–12. Available via BioMed Central https://bmcmedethics.biomedcentral.com/articles/10.1186/s12910-017-0198-5. Accessed 26 Mar 2024

Center for Reproductive Rights (2012) Inter-American Court of Human Rights declares Costa Rica's ban on in-vitro fertilization a human rights violation. Available via Center for Reproductive Rights http://www.reproductiverights.org/press-room/inter-american-court-of-human-rights-declares-costa-rica%E2%80%99s-ban-on-in-vitro-fertilization-. Accessed 26 Mar 2024

Center for Reproductive Rights (2023) Texas abortion ban emergency exceptions case, Zurawski v. State of Texas. Available via the Center for Reproductive Rights https://reproductiverights.org/case/zurawski-v-texas-abortion-emergency-exceptions/zurawski-v-texas. Accessed 26 Mar 2024

Chan PCW (2007) The protection of refugees and internally displaced persons: non-refoulement under customary international law? Int J Human Rights 10:231–239, 235

Chan S (2018) How and why to replace the 14-day rule. Curr Stem Cell Rep 4:228–234

Chandler K (2024) Warnings of the impact of fertility treatments in Alabama rush in after frozen embryo ruling. Available via AP News https://apnews.com/article/alabama-supreme-court-from-embryos-161390f0758b04a7638e2ddea20df7ca. Accessed 3 Mar 2024

Charalambouse G, Genakritis M (2013) Bioethics and European Union legal records regarding stem cells. Health Sci J 7:155–168

Check E (2006) Election fever inflames the US stem-cell debate. Nature 441:391

Chemerinsky E (2016) The rational basis test is constitutional (and desirable). Georgetown J Law Pub Policy 14:401

Chen Z, Wang H-G, Wen Z-J, Wang Y (2007) Life sciences and biotechnology in China. Philos Trans R Soc B 362:947–957, 951. Available via NLM https://www.ncbi.nlm.nih.gov/pmc/articles/PMC2435562/pdf/rstb20072025.pdf. Accessed 26 Mar 2024

Chen AC-H, Lee KF, Yeung WSB, Lee YL (2020) Human embryonic stem cells as an in-vitro model for studying developmental origins of type 2 diabetes. World J Stem Cells 8:761–775

Chouchane L, Mamtani R, Al-Thani MH, Al-Thani A-AM, Ameduri M, Sheikh JI (2011) Medical education and research environment in Qatar: a new epoch for translational research in the Middle East. J Transl Med 9:1–8

Claiming Human Rights (2010) African Court on Human and Peoples' Rights. Available via Claiming Human Rights www.claiminghumanrights.org/au_court.html. Accessed 26 Mar 2024

Clay AS (2013) South Korea's Bioethics and Biosafety Act (2005). Available via Arizona State University Center for Biology and Society https://embryo.asu.edu/pages/south-koreas-bioethics-and-biosafety-act-2005. Accessed 26 Mar 2024

Codey RJ (2003) Correcting misconceptions about New Jersey's stem cell legislation. Available via the New Jersey Office of Legislative Services https://www.njsendems.org/correcting-misconceptions-about-new-jerseys-stem-cell-legislation. Accessed 26 Mar 2024

Cohen MH (2015) Physician stem cell therapies legally ambiguous, while regenerative clinics prosper. Available via LinkedIn https://www.linkedin.com/pulse/physician-stem-cell-therapies-legally-ambiguous-while-cohen. Accessed 16 Jan 2024

Cohen C, Feigenbaum J, Adashi EY (2015) Sherley v. Sebelius and the future of stem cell research. JAMA 308:2087 (2012). Harvard Public Law Working Paper No. 14-38. Available via Journal of the American Medical Associate https://jamanetwork.com/journals/jama/article-abstract/1386781. Accessed 12 Jan 2024

Commission of the European Communities (1999) The European Commission. Commission of the European Communities, Brussels. Available via Reference for Business https://www.referenceforbusiness.com/encyclopedia/Clo-Con/Commission-of-the-European-Communities.html#ixzz6V6cI456N. Accessed 26 Mar 2024

Commission of The European Communities (2003) Commission staff working paper, report on human embryonic stem cell research. Brussels, 3.4.2003, SEC (2003) 441. Available via the European Commission https://ec.europa.eu/research/press/2003/pdf/sec2003-441report_en.pdf. Accessed 26 Mar 2024

Congressional Research Services (2022) The bioeconomy: A primer (CRS Report No. R46881). Available via Congressional Research Services https://crsreports.congress.gov/product/pdf/R/R46881. Accessed 26 Mar 2024

Connecticut State Department of Public Health (2024) Connecticut stem cell research grants-in-aid program. https://portal.ct.gov/DPH/Stem-Cell/Stem-Cell-Research-Program%2D%2D-Grants. Accessed 9 Jan 2024

Coombs A (2008) Stem cells drafted for war on wounds, Nat Rep Stem Cells. Available via Springer Nature https://www.nature.com/articles/stemcells.2008.148. Accessed 26 Mar 2024

Cop B, Eymirliolu D (2005) The right of self-determination in international law towards the 40th anniversary of the adoption of ICCPR And ICESCR. Perceptions J Int Affairs 10:115–146

Copelon R, Zampas C, Brusie E, Devore J (2005) Human rights begin at birth: international law and the claim of fetal rights. Reprod Health Matters 13:120–129

Corrigan O, Liddell K, McMillan J, Stewart A, Wallace S (2006) Ethical legal and social issues in stem cell research and therapy, pp 1–58 (Cambridge Genetics Knowledge Park, 2nd edn, 2006)

Council of Europe (2003) Print Committee on Social, Health and Family Affairs of the Parliamentary Assembly of the Council of Europe (CoE) RESOLUTION 1352 (2003) – Human stem cell research (October 2003; Rapporteur: Mr Wodarg; Doc. 9902. Opinion by the Committee on Social Affairs; Rapporteur: Mr H?ie; Doc. 9942)

Council of Europe (1950) European Convention for the Protection of Human Rights and Fundamental Freedoms, as amended by Protocols Nos. 11 and 14. 4 November 1950, ETS 5. Available via Ref World https://www.refworld.org/docid/3ae6b3b04.html. Accessed 5 Jan 2024

Council of Europe (1988) Protocol to the Convention for the Protection of Human Rights and Fundamental Freedoms (European Convention on Human Rights)

Council of Europe (2001) Additional Protocol to the Convention for the Protection of Human Rights and Dignity of the Human Being with regard to the Application of Biology and Medicine, on the Prohibition of Cloning Human Beings, ETS No. 168, 01/03/2001

Council of Europe (2024) Oviedo Convention and its Protocols reference texts. Available via Council of Europe https://www.coe.int/en/web/bioethics/oviedo-convention. Accessed 4 Jan 2024

Council of Europe Parliamentary Assembly (2003a) Human stem cell research, assembly debate on 2 Oct. 2003 (33rd Sitting) Resolution 1352. Available via Council of Europe http://www.assembly.coe.int/nw/xml/XRef/Xref-XML2HTML-en.asp?fileid=17158&lang=en. Accessed 26 Mar 2024

Council of Europe Parliamentary Assembly (2003b) Res. 1352. Available via Available via Council of Europe http://www.assembly.coe.int/nw/xml/XRef/Xref-XML2HTML-en.asp?fileid=17158&lang=en. Accessed 6 Jan 2024

Crook JM, Peura TT, Kravets L, Bosman AG, Buzzard JJ, Horne R, Hentze H, Dunn NR, Zweigerdt R, Chua F, Upshal A, Colman A (2007) The generation of six clinical-grade human embryonic stem cell lines. Cell Stem Cell 1:490–494

Cyranoski D (2009) Japan relaxes human stem-cell rules. Available via Springer Nature https://www.nature.com/news/2009/090821/full/4601068a.html. Accessed 26 Mar 2024

Cyranoski D (2012) China's stem cell rules go unheeded. Nature 484:149–150

Cyranoski D (2015) China announces stem-cell rules. Available via Springer Nature https://www.nature.com/news/china-announces-stem-cell-rules-1.18252. Accessed 26 Mar 2024

Cyranoski D (2019) The potent effects of Japan's stem-cell policies. Available via Springer Nature https://www.nature.com/articles/d41586-019-02847-3. Accessed 26 Mar 2024

Daniels FB (2010) D.C. Court of Appeals upholds FDA jurisdiction in regenerative sciences. Available via Lexology https://www.lexology.com/library/detail.aspx?g=e693da9c-e966-4dbb-854e-f0a12bdc9cb0. Accessed 26 Mar 2024

de Arzuaga FC (2013) Stem cell research and therapies in Argentina: the legal and regulatory approach. World Stem Cell Rep 22:40–43

de Jesus Medina Arellano M (2012) The rise of stem cell therapies in Mexico: inadequate regulation or unsuccessful oversight? Revista Redbioetica 3:63–78. Available via Redbioetica https://redbioetica.com.ar/wp-content/uploads/2018/11/RevistaBioetica6b-63a78.pdf. Accessed 26 Mar 2024

de Sá Mde F, de Oliveiro Naves BT (2008) Embryonic stem cell research in Brazil and the (un)constitutionality of Article 5 of the Biosafety Act (part I). Rev Derecho Genoma Hum 28: 177–191

del Bosque M (2017) Human rights lawyer on how government is complicit in Mexico's drug war. Available via the Texas Observer https //www.texasobserver.org/human-rights-lawyer-on-how-government-is-complicit-in-mexicos-drug-war. Accessed 26 Mar 2024

den Exter A (2016) Embryonic stem cell patents at European top court. Eur J Hum Genet 24:311

Denoon A (2011) Brüstle v. Greenpeace: implications for stem cell research. Regen Med 6:85–87

Doherty K (2012) Regulation of stem cell research in France. Available via EuroStemCell http://www.eurostemcell.org/regulations/regulation-stem-cell-research-france. Accessed 26 Mar 2024

Dr. Hany Sleem's Presentation to the 3rd Regional Conference (2012) American University of Beirut Faculty of Medicine, September 14–15

Drabiak-Syed K (2013) New president, new human embryonic stem cell research policy: comparative international perspectives and embryonic stem cell research laws in France. Biotechnol Law Rep 32:349–356

Duffy DT (2002) Background and legal issues related to stem cell research. Cong Research Serv, 12 June 2002, pp 1–6. Available via University of North Texas https://digital.library.unt.edu/ark:/67531/metacrs3268/. Accessed 26 Mar 2024

Dunn K (2005) The politics of stem cells. Available via Public Broadcast Service http://www.pbs.org/wgbh/nova/body/stem-cells-politics.html. Accessed 26 Mar 2024

Duong A, Steinmaus C, McHale CM, Vaughan CP, Zhang L (2011) Reproductive and developmental toxicity of formaldehyde: a systematic review. Mutat Res 3:118–138

Economic and Social Council (2000) General Comment No. 14, paragraph 3 (2000) on Article 12 of the International Covenant on Economic, Social and Political Rights, on the Right to Life (Aug. 11, 2000). Available via Ref World https://www.refworld.org/docid/4538838d0.html. Accessed 26 Mar 2024

El-Awady N (2008) Gulf states embrace stem cell technologies at home and abroad. Available via Springer Nature https://www.nature.com/articles/stemcells.2008.21. Accessed 26 Mar 2024

El-Bawab N (2023) Texas Supreme Court temporarily blocks ruling that allowed woman emergency abortion. Available via ABC News https://abcnews.go.com/US/hearing-underway-texas-woman-court-emergency-abortion/story?id=105455849. Accessed 26 Mar 2024

Electoral Ventures (2020) Indiana. Available via Electoral Ventures https://www.270towin.com/states/Indiana. Accessed 26 Mar 2024

Ellis BA (2017) "Time enough" for scrutiny: the second amendment, mental health, and the case for intermediate scrutiny. William Mary Bill Rights J 25:1325, 1333

Elon M (1969) The sources and nature of Jewish law and its application in the state of Israel, Part IV. Israel Law Rev 4:80–140

Emerson B (2018) ND stem cell clinic to halt injections, repay patients. Available via the Grand Forks Herald https://www.grandforksherald.com/news/science-and-nature/4440046-nd-stem-cell-clinic-halt-injections-repay-patientswyo. Accessed 26 Mar 2024

Equality and Human Rights Commission (2019) International Covenant on Economic, Social and Cultural Rights (ICESCR). Available via Equality and Human Rights Commission https://www.equalityhumanrights.com/our-work/our-human-rights-work/monitoring-and-promoting-un-treaties/international-covenant-economic. Accessed 26 Mar 2024

Eskew A, Jungheim ES (2017) A history of developments to improve in vitro fertilization. Mo Med 114:156–159. Available via NCBI https://www.ncbi.nlm.nih.gov/pmc/articles/PMC6140213. Accessed 16 Mar 2024

European Commission (2024) Policies, information, and services. Available via European Commission https://ec.europa.eu/info/aid-development-cooperation-fundamental-rights/your-rights-eu/eu-charter-fundamental-rights/application-charter_en. Accessed 2 Jan 2024

European Group on Ethics in Science and New Technologies (EGE) (2017) Hat is the EGE? Available via the European Commission https://ec.europa.eu/info/research-and-innovation/strategy/support-policy-making/scientific-support-eu-policies/ege_en. Accessed 26 Mar 2024

European Parliament (2004) Directive 2004/23/EC of the European Parliament and of the Council of 31 March 2004 on setting standards of quality and safety for the donation, procurement, testing, processing, preservation, storage and distribution of human tissues and cells

European Parliament and Council (1998) Directive 98/79/EC. Available via the European Union https://eur-lex.europa.eu/eli/dir/1998/79/oj. Accessed 26 Mar 2024

European Union, Parliament and Council (1998) Decision of 22 Dec. 1998 in the 5th Framework Programme of the European Community for Research, Technological Development and Demonstration, O. J. L. 026, 01/02/1999 P. 0001 – 0033 (1999)

European Union: Council of the European Union (1997) Treaty of Amsterdam Amending the Treaty on European Union, The Treaties Establishing the European Communities and Related Acts, 10 November 1997. Available via Ref World https://www.refworld.org/docid/51c009ec4.html. Accessed 5 Jan 2024

EuroStemCell (2008) Regulation of stem cell research in europe. Available via EuroStemCell http://www.eurostemcell.org/stemcell-regulations. Accessed 26 Mar 2024

Evenson D (2005) The right to health care and the law. Available via Medical Education Cooperation with Cuba https://www.medicc.org/publications/medicc_review/0905/mr-features1.html. Accessed 26 Mar 2024

Executive Order 13505 (2009) 74 FR 74, 46, 10667-10668, (Mar. 11, 2009) (Updated Dec. 2022). Available via NIH http://stemcells.nih.gov/policy/pages/2009guidelines.aspx. Accessed 26 Mar 2024

Executive Order 14081 (2023) 88 FR 25711 "Advancing Biotechnology and Biomanufacturing Innovation for a Sustainable, Safe, and Secure American Bioeconomy-Request for Information; National Biotechnology and Biomanufacturing Initiative-Measuring the Bioeconomy"

Exec. Order No. 13435, 3 C.F.R. 222 (2007) Expanding approved stem cell lines in ethically responsible ways. 34951-34953. Available via U.S. Government Publishing Office http://edocket.access.gpo.gov/2007/pdf/07-3112.pdf. Accessed 4 Jan 2024

Fadel HE (2012) Developments in stem cell research and therapeutic cloning: Islamic ethical positions. A review, Bioethics 26:128–135, 134

Farajkhoda T (2017) An overview on ethical considerations in stem cell research in Iran and ethical recommendations: a review. Int J Reprod Biomed 15:67–74

Farha L (2016) The right to life & the right to housing. Presentation to the Human Rights Committee. July 11, 2016. Available via the Office of the High Commissioner for Human Rights https://www.ohchr.org/sites/default/files/Documents/Issues/Housing/TB/HousingPresentation.pdf. Accessed 26 Mar 2024

Federation of State Medical Boards (2018) About FSMB. Available via the Federation of State Medical Boards [hereinafter FSMB] https://www.fsmb.org/about-fsmb. Accessed 26 Mar 2024

Federation of State Medical Boards' Workgroup to Study Regenerative and Stem Cell Therapy Practices (2019) Regenerative and stem cell therapy practices. Report and recommendations of the Workgroup to Study Regenerative and Stem Cell Therapy Practices. Adopted as policy by the Federation of State Medical Boards. Available via FSMB https://www.fsmb.org/siteassets/advocacy/policies/fsmb-stem-cell-workgroup-report.pdf. Accessed 26 Mar 2024

Felix M, Sobel L, Salganicoff A (2023) Legal challenges to state abortion bans since the Dobbs decision. Available via KFF/Kaiser Family Foundation https://www.kff.org/womens-health-policy/issue-brief/legal-challenges-to-state-abortion-bans-since-the-dobbs-decision. Accessed 26 Mar 2024

Fink DW Jr (2009) FDA regulation of stem cell based products. Science 324:1662–1663. Available via Research Gate https://www.researchgate.net/profile/Donald_Fink/publication/26320774_FDA_Regulation_of_Stem_Cell-Based_Products/links/0c960522f4efdb752b000000/FDA-Regulation-of-Stem-Cell-Based-Products.pdf. Accessed 26 Mar 2024

Fleck F (2004) Swiss voters back stem-cell research. Available via the New York Times https://www.nytimes.com/2004/11/28/international/europe/swiss-voters-back-stemcell-research.html. Accessed 26 Mar 2024

Flynn J, Matthews K (2010) Stem cell research in the greater Middle East: the importance of establishing policy and ethics, interoperability to foster international collaborations. Stem Cell Rev Reps 6:143–150. Available via the Rice Fondren Library https://repository.rice.edu/server/api/core/bitstreams/58c28e5e-1543-49ea-b095-7970769d98cd/content. Accessed 26 Mar 2024

Fon V, Parisi F (2008) The hidden bias of the Vienna Convention on the International Law of Treaties. Rev Law Econ 4:383–406, 404

Ganeva V, Small S (2008) Bulgaria's stem cell research laws. Available via EuroStemCell https://www.eurostemcell.org/regulation-stem-cell-research-bulgaria. Accessed 26 Mar 2024

Gaobotse G (2018) Stem cell research in Africa: legislation and challenges. J Regen Med 7:1. Available via SciTechnol https://www.scitechnol.com/peer-review/stem-cell-research-in-africa-legislation-and-challenges-1E3H.php?article_id=7753. Accessed 26 Mar 2024

Gardner C (2010) That American stem cell ruling. Available via Head of Legal http://www.headoflegal.com/2010/08/25/that-american-stem-cell-ruling. Accessed 20 Jan 2024

Gardner R, Watson T (2005) A patchwork of laws. Available via Scientific American https://www.scientificamerican.com/article/a-patchwork-of-laws. Accessed 26 Mar 2024

Garfinkel M (2004) Stem cells policies and players. Available via Genome News Network www.genomenewsnetwork.org/resources/policiesandplayers. Accessed 26 Mar 2024

Gennarini S (2013) Inter-American Court draws fire from doctors and lawyers. Available via the Center for Family and Human Rights https://c-fam.org/friday_fax/inter-american-court-draws-fire-from-doctors-and-lawyers-2. Accessed 26 Mar 2024

Gittleman R (1982) The African Charter on Human and Peoples' Rights: a legal analysis. Va J Int Law 22:667–714

Goldstein LSB (2017) In the trenches: lessons for scientists from California's Proposition 71 campaign. Mol Biol Cell 22:3943–3944

Goody M (2006) Stem cell research: how the states stand. Available via National Public Radio https://www.npr.org/templates/story/story.php?storyId=5563162. Accessed 26 Mar 2024

Gorby JD (1976) Introduction to the translation of the abortion decision of the Federal Constitutional Court of the Federal Republic of Germany, West German abortion decision: a contrast to Roe v. Wade: introduction. J Marshall J Pract Proc 9:557, 570

Hafenbrack J (2005) Stem cell showdown heats up. South Florida Sun-Sentinel 6, 2005, p 1A

Haining CM, Keogh LA, Savulescu J (2022) The unethical Texas heartbeat law. Prenat Diagn 5: 535–541

Hallermalm K (2002) Leading the way: stem cell research in Sweden. Available via American Association for the Advancement of Science https://www.sciencemag.org/careers/2002/08/leading-way-stem-cell-research-sweden. Accessed 26 Mar 2024

Hamilton J (2021) Biden scraps human fetal tissue research limits while New York stops stem cell research funding. Available via Progress Educational Trust https://www.progress.org.uk/biden-scraps-human-fetal-tissue-research-limits-while-new-york-stops-stem-cell-research-funding. Accessed 26 Mar 2024

Hammond-Browning N (2009) Legal and moral aspects of human embryonic stem cell research. Available via Cardiff University https://orca.cardiff.ac.uk/id/eprint/54951/. Accessed 4 Jan 2024

Hannum H (1996) The status of the Universal Declaration of Human Rights in national and international law. Ga J Int Comp Law 25:289

Hannum H (1998) The UDHR in national and international law. Health Hum Rights 3:144–158

Hanschmidt F, Kaiser J, Stepan H, Kersting A (2020) The change in attitudes towards abortion in former West and East Germany after reunification: a latent class analysis and implications for abortion access. Geburtshilfe Frauenheilkd 80:84–94

Hawkins M (2020) The top 7 conservative states in the U.S. Available via ThoughtCo https://www.thoughtco.com/top-conservative-states-in-the-us-3303802. Accessed 26 Mar 2024

Healio (2013) Cuba on 'cutting edge' of many areas in health care. Available via Healio https://www.healio.com/news/hematology-oncology/20130125/10_3928_1081_597x_20130101_00_999133. Accessed 6 Jan 2024

Henkin L (1984) International law as law in the United States. Mich Law Rev 82:1555–1569, 1561

Higgins R (1978) Conceptual thinking about the individual in international law. N Y Law Sch Law Rev 24(11):16

Hildreth C (2018) Japan to supply human embryonic stem cells (hESC) for clinical research. Available via Bioinformant https://bioinformant.com/human-embryonic-stem-cells-japan. Accessed 26 Mar 2024

Hiltzik M (2017.) California moves against unlicensed stem cell treatments — but is it doing enough? Available via the Los Angeles Times www.latimes.com/business/hiltzik/la-fi-hiltzik-stemcell-regulation-20171013-story.html. Accessed 26 Mar 2024

Holm S (2004) Stem cell transplantation and ethics: a European overview. Fetal Diagn Ther 19: 113–118

Holmes E (2018) Biden faults Palin on stem cells. Available via the Wall Street Journal https://www.wsj.com/articles/SB122100778731917579. Accessed 26 Mar 2024

Hopkins Tanne J (2006) Bush vetoes bill to expand stem cell research. BMJ 333:216

House of Lords (2002) Select Committee on Stem Cell Research report. 2001-02. Chapter IV, The status of the early embryo. Available via Parliament UK https://publications.parliament.uk/pa/ld200102/ldselect/ldstem/83/8301.htm. Accessed 26 Mar 2024

Hovatta O, Stojkovic M, Nogueira M, Varela-Nieto I (2010) European scientific, ethical and legal issues on human stem cell research and regenerative medicine. Stem Cells 28:1005–1007

Hudson KL, Collins FS (2017) The 21st Century Cures Act — a view from the NIH. N Engl J Med 376:111–113

Hug K (2006) Therapeutic perspectives of human embryonic stem cell research versus the moral status of a human embryo-does one have to be compromised for the other? Medicina (Kaunas) 42:107–114

Human Rights Act (1998) Corporate insolvency and governance act 2020, s 12(3)(b)

Human Rights Committee (2017) Observations of the United States of America On the Human Rights Committee's Draft General Comment No. 36 On Article 6 - right to life. Available via U.S. Department of State https://www.state.gov/wp-content/uploads/2019/05/U.S.-observations-on-Draft-General-Comment-No.-36-on-Article-6-Right-to-Life-.pdf. Accessed 26 Mar 2024

Human Rights Committee, General Comment No. 36 (2017) on Article 6 of the International Covenant on Civil and Political Rights, on the Right to Life. (Oct. 06, 2017)

Human Rights Committee, General Comment No. 36, paragraph 2 (2019) on Article 6 of the International Covenant on Civil and Political Rights, on the Right to Life. (Sept. 03, 2019). Available via the Office of the High Commissioner for Human Rights https://www.ohchr.org/en/calls-for-input/general-comment-no-36-article-6-right-life. Accessed 26 Mar 2024

Human Rights Watch (2022) Q&A: access to abortion is a human right. Available via Human Rights Watch https://www.hrw.org/news/2022/06/24/qa-access-abortion-human-right. Accessed 26 Mar 2024

Human Tissue Authority (2024) Regenerative medicine and the regulation of advanced therapies medicinal products. Available via the Human Tissue Authority https://www.hta.gov.uk/guidance-professionals/regulated-sectors/human-application/regenerative-medicine-and-regulation. Accessed 3 Jan 2024

Hunt P (2016) Interpreting the international right to health in a human rights-based approach to health. Health Hum Rights 18:109–130, 114. Available via NLM https://www.ncbi.nlm.nih.gov/pmc/articles/PMC5394996/#r34. Accessed 26 Mar 2024

International Bioethics Committee (2001) United Nations Educational, Scientific and Cultural Organization, "The Use of Embryonic Stem Cells" in Therapeutic Research, BIO-7/00/GT-1/2 (Rev. 3), Paris, 6 April 2001. (While the meeting was held over November 7-9, 2000, the final document was not considered accepted until January 29 – February 2, 2001)

Isasi RM, Knoppers BM (2006) Mind the gap: policy approaches to embryonic stem cell and research in 50 countries. Eur J Health Law 13:9–25

Jackson D (2007) Bush cites 'moral line' in second veto of stem cell bill. Available via USA Today http://usatoday30.usatoday.com/news/washington/2007-06-20-bush-stem-cells_N.htm. Accessed 26 Mar 2024

Johnson H (2017.) What is the EU Charter of Fundamental Rights? Available via Impress https://rightsinfo.org/eu-charter-fundamental-rights. Accessed 26 Mar 2024

Johnston J (2002) New rules for embryo research: Bill C-56 and the Canadian Institutes of Health Research Stem Cell Research Guidelines. J Obstet Gynaecol Can 24:722–726

Jones JM (2016) Red states outnumber blue for first time in Gallup tracking. Available via Gallup https://news.gallup.com/poll/188969/red-states-outnumber-blue-first-time-gallup-tracking.aspx. Accessed 26 Mar 2024

Jorgenson L (2016) NIH policy on informed consent for human fetal tissue research. NOT-OD-16-033. Available via NIH https://grants.nih.gov/grants/guide/notice-files/not-od-16-033.html

Jung KW (2010) Regulation of human stem cell research in South Korea. Stem Cell Rev 6:340–344

Kaiser J, Wadman M (2019) Trump administration releases details on fetal tissue restrictions. Available via ScienceMag https://www.sciencemag.org/news/2019/07/trump-administration-releases-details-fetal-tissue-restrictions. Accessed 26 Mar 2024

Kallman TJ (1995) The German Basic Law and the Federal Constitutional Court: an impact analysis. Available via https://apps.dtic.mil/sti/tr/pdf/ADA294609.pdf. Accessed 4 Jan 2024

Kaplan K (2012) Stem cells created without destroying embryos. Available via the Los Angeles Times https://www.latimes.com/health/la-sci-stemcells11jan11-story.html. Accessed 5 Jan 2024

Kassah JE (2011) Stem cell legislation in Africa. Available via Stem Cell Research Around the World https://embrybros.wordpress.com/2011/03/16/stem-cell-legislation-in-south-africa. Accessed 26 Mar 2024

Katsnelson A (2010) US court suspends research on human embryonic stem cells. Available via Nature Springer Nature https://www.nature.com/news/2010/100824/full/news.2010.428.html. Accessed 26 Mar 2024

Kearl M (2010a) 45 CFR 46 protection of human subjects under United States law (1974). Available via Embryo Project Encyclopedia https://embryo.asu.edu/pages/45-cfr-46-protection-human-subjects-under-united-states-law-1974. Accessed 12 Jan 2024

Kearl M (2010b) Dickey-Wicker Amendment (1996). Available via the Embryo Project Encyclopedia https://embryo.asu.edu/pages/dickey-wicker-amendment-1996. Accessed 22 Jan 2024

Kennedy J, Garcia J (2005) Controversial stem-cell issue making its way to Tallahassee. Available via the Orlando Sentinel https://www.orlandosentinel.com/news/os-xpm-2005-08-07-capview07-story.html. Accessed 26 Mar 2024

Khokhar A (2010) George W. Bush Executive Order 13455, June 2007. Available via the Embryo Project Encyclopedia https://embryo.asu.edu/pages/george-w-bush-executive-order-13455-june-2007. Accessed 6 Jan 2024

Kiecker C, Bates T, Bell E (2016) Molecular specification of germ layers in vertebrate embryos. Cell Mol Life Sci 73:923–947, 923–925

Kiely E (2019) Is right to try saving a 'lot of people'? Available via FactCheck.org https://www.factcheck.org/2019/09/is-right-to-try-saving-a-lot-of-people. Accessed 26 Mar 2024

Kington RS (2009) NIH stem cell information. NIH Guidelines for Human Stem Cell Research. Available via the National Institute of Health [hereinafter NIH] https://stemcells.nih.gov/research-policy/guidelines-for-human-stem-cell-research. Accessed 26 Mar 2024

Klibanoff E (2023) Texas Abortion Case: State Supreme Court halts ruling that would allow woman to terminate pregnancy. Available Via ABC News https://abc7chicago.com/texas-supreme-court-halts-abortion-woman-case-kate-cox/14158751/. Accessed 26 Mar 2024

Knoepfler P (2018) Top 10 takeaways from unique North Dakota action on stem cell clinic. Available via The Niche https://ipscell.com/2018/05/top-10-takeaways-from-unique-north-dakota-action-on-stem-cell-clinic/. Accessed 26 Mar 2024

Knoll AP (1970) International executive agreements: their constitutionality, scope and effect. Case West Res J Int Law 2:94

Knowles L (2009) Religion and stem cell research. Available via Research Gate https://www.researchgate.net/publication/239556476_Religion_and_Stem_Cell_Research. Accessed 15 Jan 2024

Knowles LP, Kaebnick GE (2007) Reprogenetics: law, policy, and ethical issues (JHU Press). https://doi.org/10.1353/book.3420. Accessed 26 Mar 2024

Koeferl Puorger UPS, Buergin M, Wunder D, Crazzolara S, Birkhaeuser MH (2006) Surplus embryos in Switzerland in 2003: legislation and availability of human embryos for research. Reprod Biomed Online 13:772–777

Kostenzer J, de Bont A, van Exel J (2021) Women's viewpoints on egg freezing in Austria: an online q-methodology study. BMC Med Ethics 22:1–12. Available via NLM https://www.ncbi.nlm.nih.gov/pmc/articles/PMC7789674. Accessed 26 Mar 2024

Kramer S (2021) Key findings about the religious composition of India. Available via Pew Research https://www.pewresearch.org/short-reads/2021/09/21/key-findings-about-the-religious-compo sition-of-india. Accessed 26 Mar 2024

Kratochvil J (2011) The inflation of the margin of appreciation by the European Court of Human Rights. Netherlands Q Hum Rights 29:324–357, 327–329

Kunz JL (2017) The nature of customary international law. Am J Int Law 47:662

Kurien P (2015) Hinduism in the modern world. In: Hinduism in the United States. Routledge

Lafontaine D (2014) Army invests in 3-D bioprinting to treat injured soldiers. Available via the Joint Base Langley-Eustis https://www.jble.af.mil/News/Article-Display/Article/843965/army-invests-in-3-d-bioprinting-to-treat-injured-soldiers. Accessed 3 Jan 2024

Lahiry S, Sinha R (2017) Indian national guidelines for stem cell research. Perspect Clin Res 10: 148–154, 149

Lahiry S, Sinha R (2019) Indian national guidelines for stem cell research (2017): summary. Indian J Med Paediatr Oncol 40:153–154

Larijani B, Zahedi G (2007) Ethical and religious aspects of gamete and embryo donation and legislation in Iran. J Religion Health 46:399–408

Lee T-L, Lysaght T, Lipworth W, Hendl T, Kerridge I, Munsiec M, Stewartd C (2017) Regulating the stem cell industry: needs and responsibilities. Bulletin of the World Health Organisation. Available via Yong Loo Lin School of Medicine http://nusmedicine.nus.edu.sg/newsletter/issue24/ethically-speaking/regulating-the-stem-cell-industry-needs-and-responsibilities. Accessed 26 Mar 2024

Legiscan (2010) The Ethical Treatment of Human Embryos Act, GA SB169 (2009-2010) (Regular Session). Available via Legiscan https://legiscan.com/GA/text/SB169/2009. Accessed 26 Mar 2024

Lenoir N (2000) Europe confronts the embryonic stem cell research challenge. Science 287:1425–1427

Levine A (2015) State funding boosts stem cell research in California, other states. Available via Georgia Tech https://news.gatech.edu/2015/02/11/state-funding-boosts-stem-cell-research-california-other-states. Accessed 26 Mar 2024

Liao L, Li L, Zhao RC (2007) Stem cell research in China. Philos Trans R Soc Lond B Biol Sci 362: 1107–1112

Loukogeorgakis SP, De Coppi P (2017) Concise review: amniotic fluid stem cells: the known, the unknown, and potential regenerative medicine applications. Stem Cells 35:1663–1673

Mahalatchimy A (2017) Regulation of stem cell research in the United Kingdom. Available via EuroStemCell https://www.eurostemcell.org/regulation-stem-cell-research-united-kingdom. Accessed 26 Mar 2024

Mahmoud A (2022) Islamic bioethics: national regulations and guidelines of human stem cell research in the Muslim world. Available via Chapman University Digital Commons. https://digitalcommons.chapman.edu/cgi/viewcontent.cgi?article=1015&context=international_studies_theses. Accessed 26 Mar 2024

Malta M, Wells S, LeGrand S, Seixas M, Baptista A, Furtado Passos da Silva CM, Kalume C, Whetten K (2019) Abortion in Brazil: the case for women's rights, lives, and choices. Lancet 4: E552. Available via The Lancet https://www.thelancet.com/journals/lanpub/article/PIIS2468-2667(19)30204-X/fulltext. Accessed 26 Mar 2024

Mandavilli A (2017) Biden administration ends limits on use of fetal tissue for research. Available via the New York Times https://www.nytimes.com/2021/04/17/health/fetal-tissue-abortion-biden.html. Accessed 26 Mar 2024

Marchant J (2011) Qatar sets sights on stem cells. Available via Springer Nature https://www.nature.com/news/2011/110708/full/news.2011.405.html. Accessed 26 Mar 2024

Margulies A (2006) Israel leads world in stem cell research papers per capita. Available via Israel21C https://www.israel21c.org/israel-leads-world-in-stem-cell-research-papers-per-capita. Accessed 26 Mar 2024

Mascarenhas L, Rosales I (2024) Alabama governor signs law protecting IVF providers from legal liability after embryo court ruling. Available via ABC7 Chicago https://abc7chicago.com/alabama-ivf-ruling-law-kay-ivey/14498091/. Accessed 12 Mar 2024

McClusky T (2020) Biden's long history of voting against Catholic values. Available via Real Clear Religion https://www.realclearreligion.org/articles/2020/09/25/bidens_long_history_of_voting_against_catholic_values_578579.html. Accessed 26 Mar 2024

Medical Council (2016) Guide to professional conduct and ethics for registered medical practitioners, 8th edn. Available via Comhairle na nDochtuiri Leighs Medical Council http://www.medicalcouncil.ie/News-and-Publications/Reports/Guide-to-Professional-Conduct-and-Ethics-8th-Edition-2016-.pdf. Accessed 26 Mar 2024

Mendez M (2022) IVF treatment can continue under Texas' current abortion law, experts say. Available via the Texas Tribune https://www.texastribune.org/2022/07/13/texas-ivf-treatments. Accessed 26 Mar 2024

Meyer JR (2000) Human embryonic stem cells and respect for life. J Med Ethics 26:166–170

Min CY (2014) A home for world's best scientists. Available via the Straits Times https://www.straitstimes.com/singapore/a-home-for-worlds-best-scientists. Accessed 26 Mar 2024

Mittal S (2013) Stem cell research: the India perspective. Perspect Clin Res 4:105–107

Mohammad L, Handel S, Shapiro A (2024) New Alabama law protects IVF, but still identifies embryo as a child. Available via NPR https://www.npr.org/2024/03/07/1236721035/new-alabama-law-protects-ivf-but-still-identifies-embryo-as-a-child. Accessed 12 Mar 2024

Molloy K, Smith D (2018) Ethics of embryonic stem cell research in Ireland. R Coll Surg Ireland 1: 73–76

Mulligan SP (2023) International law and agreements: their effect upon U.S. law. Available via UNT Digital Library https://digital.library.unt.edu/ark:/67531/metadc1311927. Accessed 26 Mar 2024

Murnaghan I (2019) Stem cell research around the world. Available via Explore Stem Cells (UK) www.explorestemcells.co.uk/stemcellresearcharoundworld.html. Accessed 26 Mar 2024
Murugan V (2009) Embryonic stem cell research: a decade of debate from Bush to Obama. Tale J Biol Med 82:101–103
Muslim World League (2023) Islamic Fiqh Council. Available via the Muslim World League https://themwl.org/en/node/34237. Accessed 26 Mar 2024
Mutua M (1993) The African human rights system in a comparative perspective. Rev Afr Commun Hum Peoples' Rights 3:5, 7
Mutua M (2000) The African human rights system, a critical evaluation. Prepared for United Nations Development Programme. Human Development Report 2000. Available via Digital Commons @ University at Buffalo School of Law https://digitalcommons.law.buffalo.edu/cgi/viewcontent.cgi?article=1015&context=other_scholarship. Accessed 26 Mar 2024
National Bioethics Advisory Commission [NBAC] (1999) Ethical issues in human stem cell research: executive summary. Available via Georgetown Library https://repository.library.georgetown.edu/handle/10822/546393. Accessed 26 Mar 2024
National Institute of Allergy and Infectious Diseases (2021) HHS removes an administrative requirement for human fetal tissue research proposals. Available via National Institute of Allergy and Infectious Diseases https://www.niaid.nih.gov/grants-contracts/hft-research-policy-change. Accessed 8 Jan 2024
National Institute of Health (1994) Report of the Human Embryo Research Panel. Available via Georgetown Library https://repository.library.georgetown.edu/bitstream/handle/10822/559352/human_embryo_vol_1.pdf. Accessed 4 Jan 2024
National Institute of Health (2009) NIH grants policy statement. 4.1.13 Human Stem Cell Research (Updated Dec. 2022). Available via the NIH https://grants.nih.gov/grants/policy/nihgps/HTML5/section_4/4.1.13_human_stem_cell_research.htm. Accessed 26 Mar 2024
National Institute of Health (2015) About the NIH. Available via the NIH https://www.nih.gov/about-nih/what-we-do/nih-almanac/about-nih. Accessed 26 Mar 2024
National Institute of Health (2016) Implementation of funding plan for the NIH innovation projects under the 21st Century Cures Act. Available via the NIH https://www.nih.gov/sites/default/files/research-training/initiatives/nih-cures-innovation-plan.pdf. Accessed 26 Mar 2024
National Institute of Health (2019) Changes to NIH requirements regarding proposed human fetal tissue research. Available via the NIH https://grants.nih.gov/grants/guide/notice-files/NOT-OD-19-128.html. Accessed 26 Mar 2024
National Institute of Health (2021a) Update on changes to NIH requirements regarding proposed human fetal tissue research. NOT-OD-21-111. Available via NIH https://grants.nih.gov/grants/guide/notice-files/NOT-OD-21-111.html?utm_source=dlvr.it&utm_medium=twitter. Accessed 26 Mar 2024
National Institute of Health (2021b) NIH stem cell information. NIH guidelines for human stem cell research FAQs. Available via NIH https://stemcells.nih.gov/research-policy/stem-cells-faqs. Accessed 26 Mar 2024
National Stem Cell Foundation of Australia (2019) Science and ethics committee. Available via the National Stem Cell Foundation of Australia https://www.stemcellfoundation.net.au/science_and_ethics_committee. Accessed 26 Mar 2024
Nelson RM (2010) Overview of FDA perspective on the ethics of stem cell therapy. Presented on April 4, 2010. Available via the U.S. Food and Drug Administration https://www.fda.gov/media/85776/download. Accessed 26 Mar 2024
New Jersey Legislature (2018) New Jersey Office of Legislative Services. Available via New Jersey Office of Legislative Services https://www.njleg.state.nj.us/members/bio.asp?Leg=82. Accessed 26 Mar 2024
New York State (2020a.) What is the empire state stem cell board? Available via New York state stem cell science https://stemcell.ny.gov/faqs/what-empire-state-stem-cell-board. Accessed 26 Mar 2024

New York State (2020b) Our mission. Available via New York state stem cell science https://stemcell.ny.gov/bylaws. Accessed 26 Mar 2024

New York State (2020c) About us. Available via New York state stem cell science https://stemcell.ny.gov/about-us-0. Accessed 26 Mar 2024

Newport F (2017) Wyoming, North Dakota and Mississippi most conservative, Gallup. Available via Gallup https://news.gallup.com/poll/203204/wyoming-north-dakota-mississippi-conservative.aspx. Accessed 26 Mar 2024

Nordic Council of Ministers (2007) NordForsk, stem cell research in the Nordic countries, science, ethics, public debate & law. Nordforsk policy briefs 2007-2, 55. Available via NordPub http://norden.diva-portal.org/smash/record.jsf?pid=diva2%3A707832&dswid=-5700. Accessed 26 Mar 2024

Nugraha IY (2016) Embryonic stem cells in an international human rights framework. New J Euro Crim Law 34:227–249, 234, 238

O'Brien C (1999) Ethical issues in human stem cell research: executive summary. By the US National Bioethics Advisory Commission. Available via Embryo Project Encyclopedia. https://embryo.asu.edu/pages/ethical-issues-human-stem-cell-researchexecutive-summary-1999-us-national-bioethics-advisory. Accessed 26 Mar 2024

O'Brien C (2004) California Proposition 71 (2004). Available via the Embryo Project Encyclopedia https://embryo.asu.edu/pages/california-proposition-71-2004. Accessed 26 Mar 2024

O'Brien C (2014a) The Singapore Bioethics Advisory Committee. Available via Available via Yong Loo Lin School of Medicine http://nusmedicine.nus.edu.sg/newsletter/issue24/ethically-speaking/regulating-the-stem-cell-industry-needs-and-responsibilities. Accessed 26 Mar 2024

O'Brien C (2014b) Ethical issues in human stem cell research: executive summary (1999), by the US National Bioethics Advisory Commission. Available via Embryo Project Encyclopedia https://embryo.asu.edu/pages/ethical-issues-human-stem-cell-research-executive-summary-1999-us-national-bioethics-advisory. Accessed 9 Jan 2024

O'Driscoll S (2023) Outrage as Texas Supreme Court halts Kate Cox's abortion. Available via News Week https://www.newsweek.com/outrage-texas-supreme-court-kate-cox-abortion-1850981. Accessed 26 Mar 2024

Oakley J (2002) Democracy, embryonic stem cell research, and the Roman Catholic Church. J Med Ethics 28:228–231

Office of the United Nations High Commissioner for Human Rights (2008a) The right to health. Fact sheet No. 31, GE.08-41061. Available via the Office of the High Commissioner for Human Rights https://www.ohchr.org/sites/default/files/Documents/Publications/Factsheet31.pdf. Accessed 23 Jan 2024

Office of the United Nations High Commissioner for Human Rights (2008b) Frequently asked questions on economic, social and cultural rights. Fact sheet no. 33, 1-46, 13. Available via the Office of the High Commissioner for Human Rights https://www.ohchr.org/sites/default/files/documents/publications/factsheet33en.pdf

Office of the United Nations High Commissioner for Human Rights (2018) Universal declaration of human rights at 70: 30 articles on 30 articles - article 3. Available via the Office of the High Commissioner for Human Rights https://www.ohchr.org/en/press-releases/2018/11/universal-declaration-human-rights-70-30-articles-30-articles-article-3. Accessed 22 May. 2024

Office of the United Nations High Commissioner for Human Rights (2019) Born free and equal: sexual orientation, gender identity and sex characteristics in international human rights law, 2nd edn, p 9. Available via the Office of the High Commissioner for Human Rights https://www.ohchr.org/sites/default/files/Documents/Publications/Born_Free_and_Equal_WEB.pdf. Accessed 26 Mar 2024

Oh SS, Galanter J, Thakur N, Pino-Yanes M, Barcelo NE, White MJ, de Bruin DM, Greenblatt RM, Bibbins-Domingo K, Wu AHB, Borrell LN, Gunter C, Powe NR, Burchard EG (2015) Diversity in clinical and biomedical research: a promise yet to be fulfilled. PLoS Med 12:1–9. Available via NLM https://www.ncbi.nlm.nih.gov/pmc/articles/PMC4679830/. Accessed 26 Mar 2024

Ollstein AM (2021) Biden administration reverses Trump restrictions on fetal tissue research. Available via Politico https://www.politico.com/news/2021/04/16/biden-reverses-fetal-tissue-research-restrictions-482503. Accessed 26 Mar 2024
Organization of African Unity (OAU) (1982) African Charter on Human and Peoples' Rights ("Banjul Charter"), 27 June 1981, CAB/LEG/67/3 rev. 5, 21 I.L.M. 58. Available via the African Union https://au.int/en/treaties/african-charter-human-and-peoples-rights. Accessed 26 Mar 2024
Organization of American States (1969) American convention on human rights. Treaty Series, No. 36
Österman T (2021) Cultural relativism and understanding difference. Lang Commun 80:124–135
Oxner R, Borha N (2021) U.S. justice department sues Texas over new abortion law that Attorney General Merrick Garland calls unconstitutional. Available via the Texas Tribune https://www.texastribune.org/2021/09/09/texas-abortion-ban-federal-challenge. Accessed 26 Mar 2024
Pacific Bridge Medical (2017) Embryonic stem cells in Japan. Available via Pacific Bridge Medical https://www.pacificbridgemedical.com/news-brief/embryonic-stem-cells-japan. Accessed 26 Mar 2024
Palca J (2007) States take lead in funding stem-cell research. Available via National Public Radio http://www.npr.org/templates/story/story.php?storyId=9244363. Accessed 26 Mar 2024
Palma V, Pitossi FJ, Rehen SK, Touriño C, Velasco I (2015) Stem cell research in Latin America: update, challenges and opportunities in a priority research area. Regen Med 10:785–797
Pattinson SD (2005) Some problems challenging the UK's Human Fertilisation and Embryology Authority. Med Law 24:391–401
Paul A (2012) Controversial conceptions: the unborn and the American Convention on Human Rights. Loyola Univ Chic Int Law Rev 9:208–247, 210
Paust JJ (1990) Customary international law: its nature, sources and status as law of the United States. Mich J Int Law 12:59
Paust JJ (1999) Customary international law and human rights treaties are law of the United States. Mich J Int Law 20:301–336, 310
Paxton K (Attorney General of Texas) (2023) Re: Cox v. St. of Tex., Cause No. D-1-GN-23-008611, pending in the 200th Judicial District Court, Travis County, Texas. Received by The Methodist Hospital, The Women's Hospital of Texas, Texas Childrens Hospital, 7 December 2023
Pepper MS, Scholefield J, Slabbert M, Ballo R (2015) Why South Africa needs better laws for stem cell research and therapy. Available via The Conversation https://theconversation.com/why-south-africa-needs-better-laws-for-stem-cell-research-and-therapy-45498. Accessed 26 Mar 2024
Pera MF, de Wert G, Dondorp W, Lovell-Badge R, Mummery CL, Munsie M, Tam PP (2015) What if stem cells turn into embryos in a dish? Nat Methods 12:917–919
Petersen N (2005) The legal status of the human embryo in vitro: general human rights instruments. ZaöRV 65:447–466, 457
Pew Research Center (2008) Stem cell research around the world. Available via Pew Research Center http://www.pewforum.org/2008/07/17/stem-cell-research-around-the-world. Accessed 26 Mar 2024
Pineo R (2019) Cuban public healthcare: a model of success for developing nations. J Dev Soc 35:16–61
Plomer A (2011) EU ban on stem cell patents is a threat both to science and the rule of law. Available via The Guardian https://www.theguardian.com/science/blog/2011/dec/12/eu-ban-stem-cell-patents. Accessed 26 Mar 2024
Pollack A (2009) F.D.A. approves a stem cell trial. Available via the New York Times https://www.nytimes.com/2009/01/23/business/23stem.html. Accessed 26 Mar 2024
Prainsack B (2006) 'Negotiating life': the regulation of human cloning and embryonic stem cell research in Israel. Soc Stud Sci 36:173–205

Progress Educational Trust (2017) Cell nuclear replacement. Available via Progress Educational Trust https://www.bionews.org.uk/page_2356. Accessed 4 Jan 2024
Puppinck G (2013) Abortion and the European Convention on Human Rights. Irish J Leg Stud 3: 142–193
Ralston M (2008) Stem cell research from around the world. Pew forum on religion & public life. Available via Pew Forum https://www.pewforum.org/2008/07/17/stem-cell-research-around-the-world. Accessed 26 Mar 2024
Raoux J (2008) McCain bucks party platform, affirms support for embryonic stem cell research. Available via Wired https://www.wired.com/2008/09/as-president-jo. Accessed 26 Mar 2024
Rauch T (2013) Defense health program research & development overview: focus on clinical & rehabilitative medicine & genitourinary/lower abdomen reconstruction. Office of the Assistant Sec. of Defense for Health Affairs. Ref 13-S-1488, pp 1–16. Available via the Recovering Warrior Taskforce https://rwtf.defense.gov/Portals/22/Documents/Meetings/m14/031gurd.pdf. Accessed 6 Jan 2024
Raya A, Izpisúa Belmonte JC (2009) Stem cell research in Spain: if only they were windmills. Cell Stem Cell 4:483–486
Regeringens Proposition (1990) 1990/91:52
Regeringens proposition 2003/04:148 (2004) Stamcellforskning. 27 toukokuuta
Resmovits J (2011) Two countries, two approaches to regulating embryonic stem-cell use. Available via the Jewish Daily Forward https://forward.com/news/134819/two-countries-two-approaches-to-regulating-embryon. Accessed 26 Mar 2024
Resnik DB (2007) Embryonic stem cell patents and human dignity. Health Care Anal 15:211–222
Restatement (Third) of the Foreign Relations Law of the United States § 102 (Am Law Inst 1987)
Richard L (2023) Florida governor Ron DeSantis signs six-week 'heartbeat bill' into law to limit abortion. Available via Fox News https://www.foxnews.com/politics/florida-governor-ron-desantis-signs-six-week-heartbeat-bill-into-law-limit-abortion. Accessed 26 Mar 2024
Roberts EFS (2007) Extra embryos: the ethics of cryopreservation in Ecuador and elsewhere. J Am Ethno Soc 34:181–189
Rosemann A, Sleeboom-Faulkner M (2016) New regulation for clinical stem cell research in China: expected impact and challenges for implementation. Regen Med 11:5–9
Roumi H (2012) Qatar turns spotlight on religious options for stem cell science. Available via Gulf News https://gulfnews.com/world/gulf/qatar/qatar-turns-spotlight-on-religious-options-for-stem-cell-science-1.964929. Accessed 26 Mar 2024
Ruiz J, Linton C (2023) Texas Supreme Court temporarily halts ruling allowing woman to have emergency abortion. Available via CBS News https://www.cbsnews.com/news/texas-abortion-ban-supreme-court-hold-kate-cox/. Accessed 26 Mar 2024
Russo E (2005) Follow the money—the politics of embryonic stem cell research. PLoS Biol 3: 1167–1171
Ryan KA, Sanders AN, Wang DD, Levine AD (2009) Tracking the rise of stem cell tourism. Regen Med 5:27–33
Saniei M (2013) Human embryonic stem cell science and policy: the case of Iran. Soc Sci Med 98: 345–350, 349
Santoro N, Caplan A, Strauss J, Winn VD (2021) A letter to President Biden and Secretary Designate of HHS Xavier Becerra: remove barriers to federal funding of human embryo and fetal tissue research. Reprod Sci 28:933–935
Sautier B (2012) Oliver Brüstle vs. Greenpeace: how to read the moral compass? LLM, King's College London; PhD candidate in IP law. EDSJ Grenoble. Available via Blogs at King's College London https://blogs.kcl.ac.uk/kslreuropeanlawblog/?p=227. Accessed 26 Mar 2024
Savage D (2013) Supreme Court rejects challenge to Obama stem cell policy. Available via the Los Angeles Times http://articles.latimes.com/2013/jan/07/nation/la-na-nn-supreme-court-obama-stem-cells-20130107. Accessed 26 Mar 2024
Schwebel SM (1979) The effect of resolutions of the U.N. General Assembly on customary international law. Am Soc Int Law 73:301–309, 303

Scoville RM (2016) Finding customary international law. Iowa Law Rev 101:1893

Sedlak-Hevener A (2019) Inside the Mexican cartel's unspoken child organ harvesting ring. Available via Ranker https://www.ranker.com/list/mexico-child-organ-harvesting/amandasedlakhevener. Accessed 26 Mar 2024

Semeniuk I (2016) Funding surges in Canada as Trump win throws U.S. stem cell research in doubt. Available via The Globe and Mail https://www.theglobeandmail.com/news/national/canadian-stem-cell-funding-surge-underscores-us-uncertainty/article33089117. Accessed 26 Mar 2024

Serour GI (2013) Ethical issues in human reproduction: Islamic perspectives. Gynecol Endocrinol 29:949–952

Sevrick K (2021) Biden administration scraps human fetal tissue research restrictions. Available via Science https://www.science.org/content/article/biden-administration-scraps-human-fetal-tissue-research-restrictions. Accessed 26 Mar 2024

Shalev C (2002) Human cloning and human rights: a commentary. Health Human Rights 6:137–151. Available via Harvard School of Public Health https://www.hsph.harvard.edu/hhrjournal/wp-content/uploads/sites/2469/2013/07/8-Shalev.pdf. Accessed 26 Mar 2024

Shanks P (2019) Cash problems at California's stem cell program. Available via the Center for Genetics and Society https://www.geneticsandsociety.org/biopolitical-times/cash-problems-californias-stem-cell-program. Accessed 26 Mar 2024

Sharma A (2009) Stem cell research and policy in India: current scenario and future perspective. J Stem Cells 4:133–140

Simmons-Duffin S, Feibel C (2022) The Texas abortion ban hinges on 'fetal heartbeat.' Doctors call that misleading. Available via National Public Radio https://www.npr.org/sections/health-shots/2021/09/02/1033727679/fetal-heartbeat-isnt-a-medical-term-but-its-still-used-in-laws-on-abortion. Accessed 26 Mar 2024

Sinclair AH, Schofield PR (2007) Human embryonic stem cell research: an Australian perspective. Cell 128:221–223

Sithole S (2011) Stem cell research – the regulatory framework in South Africa. S Afr J Bioethics Law 4:55

Small S (2011) Regulation of stem cell research in Germany. Available via EuroStemCell http://www.eurostemcell.org/regulations/regulation-stem-cell-research-germany. Accessed 26 Mar 2024

Small S (2012) Austria's stem cell laws. Available via EuroStemCell https://www.eurostemcell.org/regulation-stem-cell-research-austria. Accessed 26 Mar 2024

Small S (2020) Regulation of stem cell research in Sweden. Available via EuroStemCell https://www.eurostemcell.org/regulation-stem-cell-research-sweden. Accessed 26 Mar 2024

Society for the Protection of Unborn Children (2017) Response to draft General Comment 36 on Article 6 of the ICCPR. Available via the Office of the High Commissioner for Human Rights https://www.ohchr.org/sites/default/files/Documents/HRBodies/CCPR/GCArticle6/ProtectionUnbornChildren.doc. Accessed 26 Mar 2024

Spiegel AM (2013) The stem cell wars: a dispatch from the front. Trans Am Clin Climatol Assoc 124:94–110

State of Israel (1999) Prohibition of genetic intervention law (human cloning and genetic change in reproductive cells) (amendment no. 3), 5776-2016

Staunton C (2013) Stem cell research requires legislation and guidelines. Available via the Irish Times https://www.irishtimes.com/opinion/stem-cell-research-requires-legislation-and-guidelines-1.1397844. Accessed 26 Mar 2024

Stearns M (2017) Obergefell, fisher, and the inversion of tiers. J Con Law 19:1043, 1046. Available via University of Maryland Digital Commons @ UM Carey Law https://digitalcommons.law.umaryland.edu/fac_pubs/1589. Accessed 26 Mar 2024

Stein R (2017) FDA cracks down on stem-cell clinics selling unapproved treatments. Available via Wyoming Public Media https://www.wyomingpublicmedia.org/post/fda-cracks-down-stem-cell-clinics-selling-unapproved-treatments#stream/0. Accessed 26 Mar 2024

Study Country (2024) Religious beliefs and spirituality in Qatar. Available via Study Country https://www.studycountry.com/guide/QA-religion.htm. Accessed 1 Jan 2024

Svendsen CN, Ebert AD (2008) Encyclopedia of stem cell research, vol 2. Sage, pp 454–462

The Associated Press (2022) Georgia's highest court reinstates ban on abortions after 6 weeks. Available via National Public Radio https://www.npr.org/2022/11/23/1139039767/georgia-supreme-court-reinstates-abortion-ban. Accessed 26 Mar 2024

The Consortium Bereshit for Cell Therapy (2024) Mission statement relating to the potential public benefits afforded by embryonic stem cells. Available via the Israel Innovation Authority https://innovationisrael.org.il/en/reportchapter/when-1-1-1-equals-10. Accessed 26 Mar 2024

The European Group on Ethics in Science and New Technologies to the European Commission (2000) Opinion No. 15. Available via Publications Office of the European Union https://op.europa.eu/en/publication-detail/-/publication/45da0341-687d-4486-8a99-6b625de26d30/language-en/format-PDF/source-77404507. Accessed 26 Mar 2024

The Inter-American Commission On Human Rights (1981) Annual report of the Inter-American Commission On Human Rights 1980-1981, resolution 23/81, case 2141 (UNITED STATES), Mar. 6, 1981, para. 30. Available via Global Health & Human Rights Database https://www.globalhealthrights.org/wp-content/uploads/2013/01/IAComHR-1981-White-and-Potter-Baby-Boy-v.-United-States.pdf. Accessed 26 Mar 2024

The Irish Council of Bioethics (2004) Guidance on operational procedures for research ethics committees. Irish Council for Bioethics, 2004; and report of the Commission on Assisted Human Reproduction, March, 2014. Available via An Roinn Slainte Department of Health https://health.gov.ie/wp-content/uploads/2014/03/Report-of-The-Commission-on-Assisted-Human-Reproduction.pdf. Accessed 26 Mar 2024

The White House (2022a) Executive order on advancing biotechnology and biomanufacturing innovation for a sustainable, safe, and secure American bioeconomy. Briefing Room, 12 September 2022. Available via the White House https://www.whitehouse.gov/briefing-room/presidential-actions/2022/09/12/executive-order-on-advancing-biotechnology-and-biomanufacturing-innovation-for-a-sustainable-safe-and-secure-american-bioeconomy. Accessed 26 Mar 2024

The White House (2022b) The President and First Lady's cancer moonshot: ending cancer as we know it. Available via the White House https://www.whitehouse.gov/cancermoonshot. Accessed 26 Mar 2024

The White House (2022c) The cancer cabinet. Available via the White House https://www.whitehouse.gov/cancermoonshot/cancer_cabinet. Accessed 26 Mar 2024

The White House (2022d) President Biden to nominate Dr. Arati Prabhakar to lead Office of Science and Technology policy. Available via the White House https://www.whitehouse.gov/briefing-room/statements-releases/2022/06/21/president-biden-to-nominate-dr-arati-prabhakar-to-lead-office-of-science-and-technology-policy. Accessed 26 Mar 2024

Thompson Reuters Practical Law (2024) Margin of appreciation. Available via Thomson Reuters Practical Law https://uk.practicallaw.thomsonreuters.com/2-500-7514?transitionType=Default&contextData=(sc.Default)&firstPage=true. Accessed 5 Jan 2024

Totenberg N (2021) Mississippi is trying to get the Supreme Court to reverse Roe. Available via National Public Radio https://www.npr.org/2021/07/23/1019746478/on-abortion-mississippi-swings-for-the-fences-asks-the-supreme-court-to-reverse-. Accessed 26 Mar 2024

Tragakiss T (2009) Connecticut's tobacco windfall: a billion dollars up in smoke. Available via the Yankee Institute for Public Policy https://www.ct.gov/sustinet/lib/sustinet/taskforces/tobaccotaskforce/07012010report/appendix_4tobaccostudy.pdf. Accessed 16 Jan 2024

U.S. Department of Defense (2008) Regenerative medicine. Available via U.S. Department of Defense https://www.defense.gov/Multimedia/Photos/igphoto/2001986941. Accessed 7 Jan 2024

U.S. Food & Drug Administration (2017) Statement from FDA Commissioner Scott Gottlieb, M.D. on the FDA's new policy steps and enforcement efforts to ensure proper oversight of stem cell therapies and regenerative medicine Available via U.S. Food & Drug Administration

https://www.fda.gov/news-events/press-announcements/statement-fda-commissioner-scott-gottlieb-md-fdas-new-policy-steps-and-enforcement-efforts-ensure. Accessed 26 Mar 2024
U.S. Food & Drug Administration (2018) FDA-NIH Joint Leadership Council Charter. Available via the U.S. Food & Drug Administration https://www.fda.gov/science-research/advancing-regulatory-science/fda-nih-joint-leadership-council-charter. Accessed 26 Mar 2024
U.S. Food & Drug Administration (2019) FDA consumer warning regarding stem cell treatments. Available via U.S. Food & Drug Administration https://www.fda.gov/consumers/consumer-updates/fda-warns-about-stem-cell-therapies. Accessed 26 Mar 2024
U.S. Food & Drug Administration (2020a) Right to Try Act. Available via the U.S. Food and Drug Administration https://www.fda.gov/patients/learn-about-expanded-access-and-other-treatment-options/right-try. Accessed 26 Mar 2024
U.S. Food & Drug Administration (2020b) Right to try. FDA fact sheet. Available via the U.S. Food and Drug Administration https://www.fda.gov/media/133864/download. Accessed 26 Mar 2024
U.S. Government Accountability Office (2011) Operation warp speed: accelerated COVID-19 vaccine development status and efforts to address manufacturing challenges. GAO-21-319. Available via the U.S. Government Accountability Office https://www.gao.gov/products/gao-21-319. Accessed 26 Mar 2024
U.S. Legal (2024) American Convention on Human Rights Law and Legal Definition. Available via US Legal https://definitions.uslegal.com/a/american-convention-on-human-rights. Accessed 4 Jan 2024
U.S. Senate Committee on Commerce, Science, and Transportation Nominations (2022) Questions for the record – minority. Hearing on July 20, 2022, p 11. Available via U.S. Senate Committee on Commerce, Science, and Transportation https://www.commerce.senate.gov/services/files/11B7E115-F3A5-44EE-ADB4-56C368FD5521. Accessed 19 Jan 2024
United Nations (2023) Universal declaration of human rights: history of the declaration. Available via the United Nations https://www.un.org/en/about-us/udhr/history-of-the-declaration. Accessed 26 Mar 2024
United Nations Declaration on Human Cloning, United Nations General Assembly, (March 8, 2005), UN Doc. GA 10333
United Nations Educational, Scientific and Cultural Organisation (UNESCO) (1997) Universal Declaration on the Human Genome and Human Rights. Available via Ref World https://www.refworld.org/docid/404226144.html. Accessed 26 Mar 2024
United Nations Educational, Scientific and Cultural Organisation (UNESCO) (2000) Universal Declaration on the Human Genome and Human Rights. Available via the Office of the High Commissioner for Human Rights https://www.ohchr.org/en/instruments-mechanisms/instruments/universal-declaration-human-genome-and-human-rights. Accessed 26 Mar 2024
United Nations Educational, Scientific and Cultural Organisation (UNESCO) (2005) Universal Declaration on Bioethics and Human Rights. 19 Oct 2005. Available via United Nations Educational, Scientific and Cultural Organisation [hereinafter UNESCO] https://unesdoc.unesco.org/ark:/48223/pf0000146180. Accessed 26 Mar 2024
United Nations Educational, Scientific and Cultural Organization (2011) Ethics and law in biomedicine and genetics: an overview of national regulations. Chapter 4, "Stem Cell Transfer in Egypt," 19 (UNESCO, Cairo Office, 2011). Available via the Bioethics Network https://bioethicsnetwork.files.wordpress.com/2011/05/ethics-and-law.pdf. Accessed 5 Jan 2024
United Nations General Assembly (1989) Second Optional Protocol to the International Covenant on Civil and Political Rights, aiming at the abolition of the death penalty. 15 December 1989. A/RES/44/128. Available via the Office of the High Commissioner for Human Rights https://www.ohchr.org/en/instruments-mechanisms/instruments/second-optional-protocol-international-covenant-civil-and. Accessed 26 Mar 2024
United Nations General Assembly (1966) International Covenant on Civil and Political Rights. 16 December 1966. United Nations, Treaty Series, vol 999, p 171

United Nations. "Vienna Convention on the Law of Treaties." Treaty Series, vol 1155, May 1969, p 331

United States Mission Geneva (2017) U.S. explanation of position on the realization of economic, social and cultural rights. Available via U.S. Mission to International Organizations in Geneva https://geneva.usmission.gov/2017/03/23/u-s-explanation-of-position-on-the-realization-of-economic-social-and-cultural-rights. Accessed 26 Mar 2024

Universal Declaration of Human Rights art. 21, G.A. Res. 217A, U.N. GAOR, 3d Sess., 1st Plen. Mtg., U.N. Doc A/810 (Dec. 12, 1948a)

Universal Declaration of Human Rights, G.A. Res. 217 (III) A, U.N. Doc. A/RES/217(III) (Dec. 10, 1948b) Art. 1

University of South Dakota (2011) Policy no. 6.011. Use of human stem cells in research. Office of Research and Sponsored Programs. Available via the University of South Dakota https://www.usd.edu/-/media/Project/USD/DotEdu/Policies/Research-and-Intellectual-Property/6011-Use-of-Stem-Cells-in-Research.pdf?rev=a9db69a9a4aa4a6bae8feca643958898&hash=E34860D8FF5E651C82C8DD9EDD56A108. Accessed 26 Mar 2024

Vatanoğlu-Lutz EE (2012) Research on embryos in Turkey with ethical and legal aspects. J Turk Ger Gynecol Assoc 13:191–195

Verginer L, Riccaboni M (2021) Stem cell legislation and its impact on the geographic preferences of stem cell researchers. Eurasian Bus Rev 11:163–189

Vickery K (2010) Generating potential for regenerative medicine in Cuba: Porfirio hernández MD DrSc. MEDICC Rev 12:10–14

Villa Y (2020) CIRM Board approves third clinical trial for COVID-19. Available via the Stem Cellar: The Official Blog of CIRM, California's Stem Cell Agency https://blog.cirm.ca.gov/2020/06/26/cirm-board-approves-third-clinical-trial-for-covid-19. Accessed 26 Mar 2024

Viney S (2012) Stem cell research goes national, ethical issues linger. Available via Egypt Independent https://egyptindependent.com/stem-cell-research-goes-national-ethical-issues-linger-highlighted-news. Accessed 26 Mar 2024

Vote Smart (2021a) Joe Biden, Jr.'s political summary on issue: stem cell research. Available via Vote Smart https://justfacts.votesmart.org/candidate/53279/joe-biden-jr?categoryId=77&type=V,S,R,E,F,P. Accessed 26 Mar 2024

Vote Smart (2021b) S 1601 - Human Cloning Research bill - National Key Vote. Available via Vote Smart https://justfacts.votesmart.org/bill/2954/8615/53279/joe-biden-voted-nay-cloture-vote-s-1601-human-cloning-research-bill#8615. Accessed 26 Mar 2024

Wadman M (2013) High court ensures continued US funding of human embryonic-stem-cell research. Available via Springer Nature https://www.nature.com/news/high-court-ensures-us-funding-of-human-embryonic-stem-cell-research-1.12171. Accessed 6 Jan 2024

Wadman M (2020) New U.S. ethics board rejects most human fetal tissue research proposals. Available via the American Association for the Advancement of Science https://www.science.org/content/article/new-us-ethics-board-rejects-most-human-fetal-tissue-research-proposals. Accessed 26 Mar 2024

Wadman M, Kaiser J (2020) Antiabortion ethicists and scientists dominate Trump's fetal tissue review board. Available via American Association for the Advancement of Science https://www.sciencemag.org/news/2020/07/antiabortion-ethicists-and-scientists-dominate-trump-s-fetal-tissue-review-board. Accessed 26 Mar 2024

Wadvalla B-A (2012) Stem cell research and Islamic regulations. Available via Springer Nature http://www.natureasia.com/en/nmiddleeast/article/10.1038/nmiddleeast.2012.125. Accessed 26 Mar 2024

Walker LM, McClelland-Descalzo DLV, zur Nieden NI (2018) The validated embryonic stem cell test with murine embryonic stem cells. Teratogenicity Testing 1797:97–124

Waskey A (2008) Saudi Arabia. In: Encyclopedia of stem cell research. SAGE Publications, Inc., pp 488–489. https://doi.org/10.4135/9781412963954

Weissman I (2012) Stem cell therapies could change medicine if they get the chance. Cell Stem Cell 10:663–665

Whelan D (2007) Frozen embryos in the high court, Irish law updates. Available via Irish Law Blog https://irishlawblog.blogspot.com/2007/12/frozen-embryos-in-high-court.html. Accessed 26 Mar 2024

White BD, Gelinas LC (2016) Balancing the surgeon's responsibility to individuals and society. In: Stain SC et al (eds) The SAGES manual ethics of surgical innovation. Springer International Publishing, Cham, pp 191–211

Williams MLK, Solnica-Krezel L (2018) Regulation of gastrulation movements by emergent cell and tissue interactions. Curr Opin Cell Biol 48:33–39

Winkler A (2006) Fatal in theory and strict in fact: an empirical analysis of strict scrutiny in the federal courts. Vand Law Rev 59:793, 815

Winter SF (1997) The cornerstones for a prohibition of cloning human beings laid down in the European Convention on Human Rights and Biomedicine. Eur J Health Law 4:189–193, 192

Women's Link Worldwide (2024) Court rulings. Available via Women's Link Worldwide https://www.womenslinkworldwide.org/en/gender-justice-observatory/court-rulings-database/baby-boy-v-united-states. Accessed 9 Jan 2024

World Health Organization (2023) Constitution. Available via World Health Organization https://www.who.int/about/governance/constitution. Accessed 26 Mar 2024

Xinhua (2019) China introduces product standard for human embryonic stem cells. Available via Xinhua http://www.xinhuanet.com/english/2019-02/26/c_137852641.htm. Accessed 26 Mar 2024

Yilek C (2023) DeSantis says he does not support criminalizing women who get abortions. Available via CBS News https://www.cbsnews.com/news/ron-desantis-abortion-florida-law-women-penalities. Accessed 26 Mar 2024

Zahra SA, Muzavir SR, Ashraf S, Ahmad A (2015) Stem cell research in Pakistan; past, present and future. Int J Stem Cells 8:1–8

Zamfir I (2018) The Universal Declaration of Human Rights and its relevance for the European Union. European Parliamentary Research Service, PE 628.295 (Nov. 2018). Available via the European Parliament https://www.europarl.europa.eu/thinktank/en/document/EPRS_ATA(2018)628295. Accessed 26 Mar 2024

Zeldovich L (2016) What Iran may be able to teach us about stem cells. Available via Mental Floss http://www.mentalfloss.com/article/76349/what-iran-may-be-able-teach-us-about-stem-cells. Accessed 9 Jan 2024

Zhai J, Xiao Z, Wang Y, Wang H (2022) Human embryonic development: from peri-implantation to gastrulation. Trends Cell Biol 32:18–29

Chapter 5
Predictions

In considering the above laws, policies and rules relating to embryonic stem cell research in relation to those that are arguing on either side of it, it is difficult to make a single solid prediction. There are so many potentialities pertaining to this new area of scientific research, and yet there is so little being done currently due to the current ethical considerations. This situation has a lot of conditioning factors that affect the laws that are created. Previously, with President Obama in the position of power there was still hope that the restrictions set by Bush that were not overturned could have been changed. This is made more difficult though by the new majority in the House of Representatives which is made up of a majority of political conservatives who have tended to lean toward a more restrictive approach regarding funding and loosening stem cell laws. If, however, more information is brought to light, such as the fact that embryonic stem cells do not have to be retrieved through the destruction of embryos, maybe this stance will be softened.[1] More results from international studies, as well as the few American studies will lead the American people to seek changes in the law to allow for federal funding so that we too can enjoy the same potential benefits that these cells can bring about.[2]

So much unpredictability comes from the fact that it would seem that depending on the political party of the current president, the laws change. One need only look to President George W. Bush's law being overturned by President Barak Obama's executive order, followed by current President Donald Trump's new restrictions on the study of human fetal tissue. To further illustrate this issue, one can need only look to the benefit that was intended for the American public with President Obama's 2009 Executive Order was that if a case is brought to prevent further funding of embryotic stem cells, the case would likely be lost. Under these new laws, the Courts would be forced to interpret the law as President Obama envisioned. However, due to the fact that this law was created through the use of an executive order, any further

[1] Soler (2009), pp. 22–28.
[2] *Id.* p. 28.

© The Author(s), under exclusive license to Springer Nature Switzerland AG 2024
J. Weiser, *Embryonic Stem Cells and the Law*,
https://doi.org/10.1007/978-3-031-59750-3_5

Presidents would be able to reverse, or even further restrict embryonic stem cell funding through at the very least, an executive order as well. This also brings another issue into focus which is that of the major questions doctrine. The major questions doctrine essentially presumes that Congress would not delegate certain issues that would be deemed to be "major questions" in political and economic arenas.[3] The major questions doctrine was cited to in recent case where the Environmental Protection Agency (the "EPA") under President Biden sought to regulate existing power plants using generation shifting approach to reduce the old dirtier methods into newer, cleaner ones.[4] The Court held that due to the severe economic impact that this would have, that this was an issue that was best for Congress to resolve as it deals with a major question.[5] The majority opinion specifically stated that

> …in certain extraordinary cases, both separation of powers principles and a practical understanding of legislative intent make us 'reluctant to read into ambiguous statutory text' the delegation claimed to be lurking there… To convince us otherwise, something more than a merely plausible textual basis for the agency action is necessary. The agency instead must point to "clear congressional authorization" for the power it claims.[6]

Essentially, this holding acts to disallow such broad delegations of power to the Executive Branch.

One issue that will probably never receive an international, or even a national agreement comes in the form of when life truly begins. Based on cultural, and religious beliefs, as well as scientific information, this question will always remain. Even if there is a certain time period that is determined through scientific research and experiments, the time will likely not conform to some, most, or all of the religious views. With this question always in place, there will always be challenges to embryonic stem cell research. Once more and more people realize that the destruction of embryos is not required to acquire stem cells, this view may change, because the life of the embryo is not threatened, or ended with what some believe to be complete disregard to human life.

In a 2022 publication from Sweden, an interview study was conducted by the authors who sought the views of people regarding embryonic stem cell research in regards to a treatment for Parkinson's disease.[7] The authors understandably discovered a vast set of beliefs among the participants, Parkinson's disease patients themselves, as to "[t]he factual beliefs and moral concerns related to the human embryo;" with the same participants' expressing similarly diverse views on whether the use of embryonic stem cells should be used in medical treatments.[8] Yet, the study did find that:

---

[3] *See* Deacon and Litman (2023).

[4] *See generally West Virginia v. EPA*, June 30, 2022, 597 U.S. __ (2022).

[5] *Id.*

[6] *Id.* p. 19.

[7] Drevin et al. (2022).

[8] *Id.* at 5.

> overall, the participants were positive towards the use of hESC for treatment of PD. They found the treatment interesting and exciting and thought that it should be used if it could be beneficial to others. Using surplus embryos for medical product development was seen as something positive, "natural" or heart-warming, as it increases their utility, compared to discarding them.[9]

The participants were even more interested if it was they themself who could benefit through the use of embryonic stem cells for their Parkinson's disease treatment.[10]

> The participants described an interest in having a treatment that did not limit an active everyday life, and that increased their health and gave them a better life with higher quality. An easy everyday life was desirable, without any tubes or cords attached to them as well as not having to take multiple doses of medicines every day.[11]

One of the most interesting findings of this Study affirms the most important issues that embryonic stem cells themselves bring up, such as the inherent rights of the embryo; patient rights; and the need for scientific transparency.[12] One of the main conclusions obtained from this Study specifically states that:

> [t]ransparency is therefore seen as vital. Science, medical advances and healthcare depend on trust from society and ultimately on how development of new treatments affects patients. It is therefore important to take patients' views seriously and consider these when deciding on how human embryos are allowed to be used.[13]

In relation to the overall belief as to how an embryo is to be viewed in relation to when protections should apply, the study found that the embryo is not yet considered a life so human rights have not yet attached.[14] Yet, of great importance is the fact that the Study concludes as to the above statement, that another study is needed to follow "up in quantitative studies with a larger number of patients."[15] It is also important to remember that opinions of this nature have been deemed by the European Court of Human Rights to essentially be decided state parties within their margin of appreciation, depending on the state in question.[16]

As discussed above, another very interesting situation, which is virtually unprecedented has occurred due to the internationally problematic pandemic caused by Covid-19/coronavirus which saw an uptick and various forms of experimentation[17] to both prevent and cure the virus. Due to the novel situation which poses such risks to so many individuals and populations, many novel treatments were considered including the use of fetal stem cells and tissue.[18] Some promising treatments and

---

[9] *Id.* at 6.

[10] *Id.*

[11] *Id.*

[12] *Id.* at 9.

[13] *Id.*

[14] *Id.*

[15] *Id.*

[16] *See generally Vo v. France*: ECHR 2004-VIII.

[17] *See generally* Geier and Geier (2020).

[18] Saldanha-Araujo et al. (2020), pp. 1–4.

vaccines were discovered to be useful that utilized or was made from fetal tissues and cells which led to very similar moral and ethical problems that always come about when discussing embryonic stem cells, and as a result, many people have sought guidance from the Vatican in regards to this situation.[19] On Dec. 21, 2020, the Vatican's Congregation for the Doctrine of the Faith finally answered the question which can change everything, stating that it is in fact "morally acceptable" for Roman Catholics to obtain help to protect against or to cure coronavirus, making the situation for embryonic stem cells potentially less of an issue for this group.[20] A caveat was given in this instance though which states that it is only morally permissible in the event that "ethically irreproachable" vaccines are not available, and when multiple vaccines are created, but the "health authorities do not allow citizens to choose the vaccine with which to be inoculated."[21] Before one gets their hopes up that the Vatican will stand by this policy for all diseases, and treatments, it is important to note that the Dec. 21, 2020 message specifically stated that it "does not and should not in any way imply that there is a moral endorsement of the use of cell lines proceeding from aborted fetuses."[22] While this is by no means a general acceptance or endorsement of medical interventions using fetal cells, this is absolutely a significant deviation from the norm and may eventually lead to more changes in the Vatican's overall position. This can especially be bolstered if a similar groundbreaking treatment, cure or vaccine is created using embryonic stem cells, and where there are no other viable options for the patient.

It would also seem that when viewing the ongoing studies and research using stem cells of all kinds the United States is moving in a direction further from using embryonic stem cells.[23] In one exhaustive study that compiled the information of ongoing stem cell research, it was found that approximately 75% of stem cell research in the United States uses non-embryonic stem cells.[24] While some can read this number to mean that embryonic stem cells are not valued as highly, one must look to the same percentages on a global scale where approximately 74% of all international stem cell studies use embryonic stem cells. Unfortunately, there is no direct explanation given, but it can be inferred that the countries that have restrictive laws like those in the United States, do not attempt to conduct research on embryonic stem cells at the same rate as those that have permissive laws. This makes sense and points to the likely trend at large as there is less funding for studies that can be challenged as impermissible seemingly at random.

---

[19] D'Emilio (2020).

[20] *Id.*

[21] *Id.*

[22] *But see id.*

[23] Deinsberger et al. (2020), pp. 1–5.

[24] *Id.* p. 7.

References

Deacon DT, Litman LM (2023) The new major questions doctrine. Va Law Rev 109:1009. Available via the Virginia Law Review https://virginialawreview.org/articles/the-new-major-questions-doctrine/. Accessed 3 Jan 2024

Deinsberger J, Reisinger D, Weber B (2020) Global trends in clinical trials involving pluripotent stem cells: a systematic multi-database analysis. NPJ Regen Med 15:5, 15. Available via Nature https://www.nature.com/articles/s41536-020-00100-4.pdf. Accessed 26 Mar 2024

D'Emilio F (2020) Vatican: OK to get virus vaccines using abortion cell lines. Available via ABC News https://abcnews.go.com/International/wireStory/vatican-virus-vaccines-abortion-cell-lines-74841735. Accessed 26 Mar 2024

Drevin J, Nyholm D, Widner H, Van Vliet T, Viberg Johansson J, Jiltsova E, Hansson M (2022) Patients' views on using human embryonic stem cells to treat Parkinson's disease: an interview study. BMC Med Ethics 23:1–10. Available via BioMed Central https://bmcmedethics.biomedcentral.com/articles/10.1186/s12910-022-00840-6#citeas. Accessed 26 Mar 2024

Geier MR, Geier DA (2020) Respiratory conditions in Coronavirus disease 2019 (COVID-19): important considerations regarding novel treatment strategies to reduce mortality. Med Hypo 140:1. Available via the National Library of Medicine https://pubmed.ncbi.nlm.nih.gov/32344310/. Accessed 26 Mar 2024

Saldanha-Araujo F, Melgaço Garcez E, Evelin Silva-Carvalho A, Lott Carvalho J (2020) Mesenchymal stem cells: a new piece in the puzzle of COVID-19 treatment. Front Immun 11:1, 4. Available via ReadCube https://www.readcube.com/articles/10.3389/fimmu.2020.01563. Accessed 26 Mar 2024

Soler LA (2009) Federal funding of stem cell research: past, present, and future. World Stem Cell Report:22–28

Tyson C (2017) 'Dirty money'. Available via Inside Higher Ed https://www.insidehighered.com/news/2014/08/28/research-funding-morally-permissible-long-its-legal-and-without-strings-attached. Accessed 26 Mar 2024

Chapter 6
Appraisal

This section has the goal of creating maximum access by all to the process of shaping and sharing all things humans value, of which there are eight unique "values."[1] In no particular order, these values are: (1) power; (2) wealth; (3) affection; (4) respect; (5) skills; (6) well-being; (7) enlightenment; and (8) rectitude.[2] This is also known as the order of human dignity which should be afforded to all.[3]

In light of the above, one must essentially balance the practical, ethical, and moral considerations by looking to the above stated "values" that all humans share, pertaining to this argument and while realizing that the overall goal is to create a situation to maximize self-realization, by forming a situation that allows us all to flourish as opposed to creating something that restricts or harms. This situation requires the analysis of two separate arguments that are the at the source of this issue, the first essentially looks to benefit those in need by broadening the overall knowledge of embryonic stem cells compared with the right to life of an embryo, but the opponents argue against such change to prevent a misuse of science. While power, wealth, affection and skills, are implicated by certain other arguments such as those made by the claimants with monetary desires such as pharmaceutical companies, these arguments hold little weight in relation to the more pressing ones which worry about harming or taking lives, and those that seek to protect lives. This leaves the values of respect, well-being, affection enlightenment, and rectitude as the "values" that should be properly balanced.

The remaining values in regards to the right to the embryo's life can each be viewed multiple different lights depending on which side of the argument one was to fall on. As to the value of "respect," one may look to the dignity of those in need of embryonic stem cell treatments, yet on the other side of the argument, there are those that seek to protect the dignity of the embryo. Even though all people are owed each

[1] See Wallace (2014), pp. 305–306.

[2] Wiessner and Willard (1999), pp. 316–334.

[3] *Id.*

© The Author(s), under exclusive license to Springer Nature Switzerland AG 2024

J. Weiser, *Embryonic Stem Cells and the Law*,
https://doi.org/10.1007/978-3-031-59750-3_6

of the eight values equally, in comparing the two sides in relation to this argument, there seems to be a lesser harm to the embryo if experimentation was conducted, and to ensure this, a law can be created that prohibits the removal of stem cells that lead to the death of the embryo. This is by no means a perfect situation, but as discussed, this is not just possible, but the method used is essentially the same used to test embryos for various diseases when being carried to full term. At the same time, this has the potential to help countless people in various ways, thereby affording them the respect that they are entitled to. It is also important to note that there is a major disagreement as to when a fetus should be afforded this overall treatment. A very similar argument can be made for the area of "well-being" where people are entitled to feel secure in their physical health and if steps can be taken to further this security, without much of a risk of causing a harm to another's well-being than one should strive to do so. "Rectitude," or the value of moral-well-being, is very much at issue for both sides of the argument. There are those who want the expansion of laws relating to embryonic stem cells to allow for new treatments for those in need, which many see as a moral responsibility, and then there are those who believe that the embryo is a living being, which should be afforded all of the same protections as living person. Many proponents of a change in law cite the "Hippocratic Oath," which is sworn to by most graduating doctors in the Western world and in part states that a doctor must help a person to the best of their ability.[4] This oath creates a moral obligation in doctors that they must work to help those in need, and if doctors have the ability to use embryonic stem cells for specific treatments, then they are morally obligated to. The other side of the argument is very different, and looks to the moral obligation in protecting the life and health of the embryo. When looking to the value of enlightenment, it took does not hold as much sway because it is a goal that looks to obtaining knowledge, which is clearly implicated here, and benefits others though implementation of such knowledge, but it seems more indirectly affected than some of the other values.

The second issue of preventing a misuse of science through perverse experimentation, and the harm in potentially "playing God" in creating "designer" babies is almost as important as the above argument, but can more easily be solved. There are certain safeguards in the form of positive law that can be created and put into place that prevent such misuses and punish those who violate said law. This is in fact all that is in use today to protect against the same problematic actions.

In the end, one must determine the value of the proposed legislation in relation to the "degree of contribution to the achievement of a public order of human dignity."[5] This is no easy hurdle to overcome, but as described at length above, the health benefits and the ability to transform lives of those who would otherwise be suffering must take precedent over a hypothetical loss to dignity of an embryo, which typically, would never have been used in the creation of an actual human being.[6]

[4]Tyson (2001).

[5]Reisman (1999), pp. 936–937.

[6]*See generally* Resnik (2007), pp. 211–222.

Why should there be people suffering when there are opportunities, or when there is a potential to help them? One should realize, that there are many potential methods to drastically reduce the potential for issues that can arise which would give concern to certain claimants, thereby reducing their overall fears, such as requiring the creation of certain laws that protect the inherent dignity of a fetus, prevent embryonic stem cell misuse through morally reprehensible, and objectionable experiments, and at the same time, both provide for advances in our collective understanding of embryonic stem cells, and how they can be used to better the lives of those in need. This topic is in no way meant to present the issue as all or nothing, in fact, creating certain prohibitions while allowing for certain experiments would be the most beneficial to all. With this in mind, one must look to the greater good, which in this case, would likely be the fact that embryonic stem cell-based therapies and treatments hold much promise to cure, alleviate, or help those who are suffering, with the proper prohibitions, cause minimal problems. In the most broken-down form, we have an argument pitting two key issues, the right to life against the right to dignity, and based on the above discussed mitigating facts, the right to life must trump the clearly less affected right to dignity. Based on the above stated facts, one must come to the conclusion that embryonic stem cells should be further investigated, and utilized, yet parameters must first be in place that prevent any and all unnecessary harm to embryos, and prohibit scientists and researchers from crossing any morally or ethically created hard lines.

An appraisal of this nature must take a lot of facts and opinions into consideration, and this is by no means an easy task. As discussed above, there are many conflicting claimants with a variety of beliefs, and reasons for such beliefs, and there is also a vast amount of information regarding the potential of embryonic stem cell research, and therapies while still grossly under studied. In light of these facts, and due to the nature of science itself, and the continual rotating of politicians from both sides of the aisle, it is very hard to know what will happen, if anything in terms of the creation of new laws either benefitting embryonic stem cell studies and experimentation, consequences of such experiments, and the potential for new discoveries.[7]

Looking at the laws of the countries discussed above can be very confusing, yet on both sides there is clearly a trend that seems to mimic the current situation in the United States. This is not to say that there are not wholly unique beliefs and arguments that do not occur here but the bottom line is that the countries can be divided into similar groups that we would consider "conservative" and "liberal" in both a political and religious sense.[8] Many of the countries with pro-embryonic stem cell laws seem to be more liberal in their thought process and are willing to possibly push the envelope in science to learn and benefit the future. The anti-embryonic stem cell laws and bans from the anti-embryonic stem cell countries that typically view all stages of life as equals and deserving of protections.[9]

---

[7] *See generally* Mummery et al. (2014), pp. 397–408.

[8] *See generally* Isasi and Knoppers (2006), pp. 9–25.

[9] *Id.*

The exact same can be said of the state laws in America. The states that have passed amendments to fund embryonic stem cell research and lift federal experimental restrictions have very similar views as the drafters behind the internal pro-embryonic stem cell laws.[10] Additionally, the same can be said with the various states that have purposefully not implemented or further restricted funds and access to embryonic stem cells.[11]

The only information that is completely known is that we do not yet know enough about embryonic stem cells. It is also important to understand that the study of something like embryonic stem cells provides a great deal of hope to those in need or to those with family in need.[12] For an accurate appraisal of the future when it comes to embryonic stem cell research, one needs to first attempt to understand other potential successors or alternatives to this area of science.[13] Understanding that embryonic stem cells are progenitor cells that are truly unique will first allow for others to understand that there is a true potential and that further study is not just warranted, but crucial, and one must also learn how other medical advances work with embryonic stem cells to make them safer and more beneficial.[14] This reasoning may seem odd because science and medicine do not usually go hand-in-hand with the law, but once new studies emerge and the general public becomes aware of the potential benefits that embryonic stem cells offer or seem to offer, it would not be unrealistic to expect that the people demand changes to the current law to reflect the potential benefits provided by embryonic stem cells.[15] Yet, one must also remember that time is of the essence to get a successful law passed that allows for a less restrictive approach to embryonic stem cell research in an attempt to prevent any potentially unnecessary suffering.

This particular situation is ripe for a rational decision based on the merits, but such a resolution will virtually never receive unanimous support. In the end as bioethics, and human rights professor George Annas stated in his 2010 article pertaining governmental funding of embryonic stem cell research, "[e]mbryo research was born political."[16] With this being said, it would seem much more beneficial to expand embryonic stem cell laws while seeking the ability to preserve the dignity of the unborn. If the goal was strictly meant to benefit all, and work to restrict any and all contact with embryos that can cause their destruction, then expanding such laws would be a good decision. This may never be feasible to the full extent which the opponents would desire, but it is an important goal nonetheless that should be worked towards. In the meantime, it would likely be harmful to further

---

[10] *Id.* p. 14.

[11] *Id.* p. 16.

[12] Nadig (2009), pp. 131–138.

[13] Ma et al. (2012), pp. 1–7.

[14] Nogrady (2019).

[15] *See generally* Lu et al. (2008), pp. 14–19; *see generally* Healthline (2017); Nature Report Stem Cells (2007); Nature Report Stem Cells, *Who Says it's Right?* (2007); He et al. (2003), pp. 279–287.

[16] Annas (2010), pp. 1687–1689.

restrict, or to not expand embryonic stem cell laws, in that more and more people will be left to needlessly suffer, and even as a country, the United States may fall further behind in this arena.

References

Annas GJ (2010) Resurrection of a stem-cell funding barrier — Dickey–Wicker in court. N Engl J Med 363:1687–1689

He Q, Li J, Bettiol E, Jaconi ME (2003) Embryonic stem cells: new possible therapy for degenerative diseases that affect elderly people. J Gerontol A 58:M279–M287

Healthline (2017) Stem cell research. Available via Healthline https://www.healthline.com/health/stem-cell-research. Accessed 26 Mar 2024

Isasi RM, Knoppers BM (2006) Mind the gap: policy approaches to embryonic stem cell and cloning research in 50 countries. Eur J Health Law 13:9–25

Lu H, Wang YC, Cong XQ (2008) Human embryonic stem cells and liver diseases: from basic research to future clinical application. Dig Dis 9:14–19

Ma X, Zhang Q, Yang X, Tian J (2012) Development of new technologies for stem cell research. J Biomed Biotech 2012:1–7. Available via the National Library of Medicine https://www.ncbi.nlm.nih.gov/pmc/articles/PMC3518316. Accessed 26 Mar 2024

Mummery C, van de Stolpe A, Roelen BAJ, Clevers H (2014) Chapter 16 - Stem cell perspectives: a vision of the future. In: Stem cells: scientific fact and fiction, 2nd edn. Academic Press, pp 397–408

Nadig RR (2009) Stem cell therapy – hype or hope? A review. J Conserv Dent 12:131–138

Nature Report Stem Cells (2007) What policies govern stem cell research? Nature Reports Stem Cells. Available via Springer Nature https://doi.org/10.1038/stemcells.2007.29. Accessed 26 Mar 2024

Nogrady B (2019) Stem-cell and genetic therapies make a healthy marriage. Available via Springer Nature https://www.nature.com/articles/d41586-019-01442-w. Accessed 26 Mar 2024

Reisman WM (1999) Theory about law: jurisprudence for a free society. Yale Law J 108:935–937

Resnik DB (2007) Embryonic stem cell patents and human dignity. Health Care Anal 15:211–222

Tyson P (2001) The Hippocratic oath today. Available via Public Broadcasting Service https://www.pbs.org/wgbh/nova/article/hippocratic-oath-today. Accessed 26 Mar 2024

Wallace D (2014) Human rights and business: a policy-oriented perspective, pp 305–306. https://doi.org/10.1163/9789004279872. Accessed 26 Mar 2024

Wiessner S, Willard AR (1999) Policy-oriented jurisprudence and human rights abuses in internal conflict: toward a world public order of human dignity. Am J Int L 93:316

Chapter 7
Recommendations

The unfortunate situation is that the two sides of this argument are diametrically opposed to each other on the core issue of whether embryos can be experimented on and with due to their very nature. This problem is clearly the most prominent reason for restricting much of the science in practice and by withholding potential, crucial grant money. While this argument lingers on, there are multiple downsides affecting many of us differently, such as that critical studies cannot be conducted to further our mutual understanding of how embryonic stem cells can benefit those in need; that people seeking medical help put themselves at risk in attempts to obtain embryonic stem cell treatments and that our citizens and country as a whole are missing out on potentially benefiting our national economy and foregoing very attractive business opportunities.[1] In fact, one of the best chances of convincing Congress to act in a positive manner comes from the pharmaceutical companies, and insurance companies who can lobby Congress to further enhance and loosen the current restrictions on stem cell laws. This has actual potential due to the money and power behind these companies. For some time, many large companies have been quietly researching embryonic stem cells, and have put vast sums of money into this research.[2] There are even several lobbying groups that are known to be seeking further funding from the National Institute of Health for studies using embryos.[3] Due to the fact that these two groups are only now actually seeing the potential benefits that can come with more embryonic stem cell research, they may use their political clout and large finances to do so. These companies would also be in the best position to orchestrate an information campaign to illuminate the realities behind embryonic stem cells, such as how they can be obtained without the destruction of an embryo, and the known and potential medical benefits. A campaign of this nature would be very costly, and if it suits these wealthy companies to do so, then they would be in the best position.

[1] *See generally* Mountford (2008), pp. 1–12; Pompe et al. (2005), pp. 297–300.

[2] Regalado (2005).

[3] Taylor-Corbett (2000).

© The Author(s), under exclusive license to Springer Nature Switzerland AG 2024

J. Weiser, *Embryonic Stem Cells and the Law*,
https://doi.org/10.1007/978-3-031-59750-3_7

Insurance companies will likely even push for expanded laws that would allow for the creation of what has been described as designed babies, as there will likely be potential for certain benefits that will could outweigh the cost in that the "designer" people will have a much lower risk of illness, and will likely remain much healthier for their entire life. This fact also helps to lessen the idea that there will be a divide in those with and those without the designer traits because eventually, most people will be able to have the choice due to cost effectiveness.[4]

This is clearly not meant to implicitly advocate for all types of experiments to be conducted. There must be limits to protect everyone. As a country we must still be bound by a so called "standard" that believes in and promotes ethics and morals. Unbridled experimentation will not only not help humanity in any way, it will also lead to us to losing our humanity and the value that we attach to life. Like even the most liberal laws discussed above, all cloning should be banned.

Equal access to the benefits that are potentially provided by and through embryonic stem cell research should be considered paramount.[5] The goal of advancing our understanding of this area of science are for the benefit of all people.[6] This is especially true if federal funding is granted, and the government can stipulate that the people seeking experimental treatments cannot be made to pay for such treatments or associated costs that are foreseeable to occur.

Based on the review of the aforementioned laws from around the world and even within in the United States itself, it would be hard to deny that the government owes a duty to all of its citizens to work to benefit their lives and the lives of their loved ones. Even more importantly though, the government certainly should not place impediments on potential groundbreaking science without a true review of the potential promises and hazards before placing virtually complete ban. Even if one were to argue that there are a number of people with personal, religious and/or ethical concerns pertaining to this research, the government should still open a full investigation on how to work out a possible solution. As explained in depth below, something close to a workable situation may just exist. Unfortunately, there is no true way to ensure that 100% of the population will be satisfied or agree with what is the end result of said compromise. That is why it is extremely important to understand the arguments of the above discussed conflicting claimants and work to ease each and all issues from each group of similar minded claimants.

A federal law based on the Government's commerce and spending powers is needed to accommodate all interests and viewpoints the way suggested here. It would include the following provisions.

---

[4]Conti (2017), pp. 6–9.

[5]*See generally* Nunes et al. (2017), pp. 1–9.

[6]Rull (2014), pp. 919–920.

7.1 Harvesting Cells

One of the most important issues that tends to separate both sides on embryonic stem cell research comes in the form of how these cells are obtained. As discussed above, embryonic stem cells have historically required the destruction of the embryo used to obtain the cells, thus bringing up the major ethical issues. Both sides tend to agree that there is no desire to destroy any embryo, whether the people believe it to be considered a "life" or not. This issue is essentially one of the ultimate ones being that lead people to believe that a life is being taken, so there is no room for compromise on the issue, embryos should not be destroyed.

To find a solution on this important issue, one needs to view the reasons behind the beliefs of those against embryonic stem cell research. So, first, all of those who believe who, as explained above, believe that life begins at conception, and that at all stages of life, human dignity attaches because an embryo is to be considered a life, with all the rights inherent to all people. There are many good counterpoints to this stance, which will be discussed, but it is of the upmost importance to once again explain that this belief is only meant to benefit the life that is being created, and not meant to deny potential medical cures. It is also important to state that this belief should be granted the highest degree of respect because protecting the sanctity of life is tantamount to safeguarding the essence of human rights. Due to this unwavering belief, there are not many viable means of compromise. One can go over the fact that there are many surplus embryos created during *IVF* which will never be used, and will eventually be discarded, or the fact that science has now found a more exact timeframe of when life actually begins, even after conception, but none of this would matter because those against these arguments would argue, as they have, that life must be protected at all costs, and that life is created at the moment of conception. Therefore, the only actual, viable system which will seemingly work is the method of removing embryonic stem cells from an embryo without having to harm, or destroy the embryo in the process. This process not only leaves the embryo unharmed, it only needs to remove a single cell, and it is the same process used to determine embryo health. Many of these beliefs are held by the religious and political conservatives in the United States, as many of them believe that life should be protected at all costs, as all life, even in the form of an embryo should be treated equally, and be protected from harm.[7] Again this view is completely rational, and thoughtful, there is no malice in their beliefs that prevent people from getting help from cells harvested from embryos. It would seem that using a technique that removes a stem cell from an embryo without destroying the embryo would likely work the best. Using this method, the scientists are able to successfully remove an embryonic stem cell, and still fully respect the embryo as a being that is alive. Many conservatives likely undergo very similar procedures used to determine the health of the embryo as mentioned above.

---

[7] *See* Masci (2008).

Yet, this form of what can be deemed "humane" harvesting is still in its infancy, and as a result, not all labs and/or scientists will be able to benefit from this area in the near future, so the solution should also include a provision that ties into the informed consent area where donors will have to have all of their rights explained to them in full, while also being made fully apprised of what will or can happen to their donated embryo(s). Because this will be a more standardized format, created by a committee (discussed in detail below,) all previously donated embryos that were obtained prior to the passage of a standardized law, will only be able to have embryonic stem cells removed through the "humane" method, leaving the embryo intact, and all others donated post-passage of this law will allow for the donors to have an option to allow for harvesting embryonic stem cells in the more traditional manner, or if they would like to opt in to only allowing for the "humane" method.

7.2 Timing

Modern science has informed the public that fetuses do not begin to develop enough to have feelings until they reach the 20-week mark in gestation.[8] This can be deemed as the most conservative view when it comes to hard science, because many scientists and researchers argue that pain fetal pain does not present itself until the third trimester of pregnancy.[9] Yet, even though this can be deemed as supplanting the old belief that pain can be felt at 14-days which was based on older scientific misunderstandings, as well as other factors explained above. If new information is gleaned that changes this belief according to the ISSR, the timing in this law can be changed to reflect the new research, as they are a group of more than 4000 scientists, specializing in this field.[10] This will ensure that there can be very little to no chance of causing any physical harm.

7.3 Cloning Embryos

While it would seem that cloning embryos would solve some issues pertaining to the donation of embryos, there is a strong consensus within most of the pro-embryonic stem cell laws from around the world and within the nation, this may cross too many ethical and moral lines. In fact, this literally can be seen as "playing God" and also quite barbaric in that an embryo is being created for the sole purpose of destroying

[8] Miller (2016) (discussing that the 20-week period has been generally accepted as the earliest a fetus can feel pain, and that is it more likely that they cannot feel pain until the 27th week, which is the beginning of the third trimester).

[9] *Id.*

[10] Felton (2022) and Regalado (2021).

it. We already know that the cloning method most often used to clone embryos is the same that was used to create Dolly the sheep, and understandably there is a deep-rooted fear amongst many groups of people that cloning humans would go too far.[11] Thankfully, with the current state of science, cloning is not necessary, even though it may remain helpful. The costs just do not outweigh the benefits in situations where cloning is permitted, and there are too many more worthwhile options to investigate before taking such drastic steps. Therefore, the law will ban both Somatic cell nuclear transfer ("SCNT") and therapeutic cloning will not be permitted. SCNT seeks to create a direct and genetically identical copy to the donor and in order to obtain the benefits of SCNT one has to first create and then destroy an embryo which crosses all moral and ethical bounds of decency. Those who fail to comply with a law that bans embryo cloning should face severe penalties and lose their licenses and certifications, and can be suspended for a to-be determined time period, or outright banned from receiving federal funds for future research.

7.4 Banning Human Chimeras

This law will act to ban the creation of human-animal hybrids for any and all purposes. This will prevent unethical and oftentimes questionable forms of research and experiments. Creating new forms of life using human cells of any kind, in all forms should be banned. Many view the possibility of creating such unnatural creatures as a benefit to science, but mixing human genetics with that of another form of life crosses many lines that most people would find offensive. It is not that there is an assumption that the chimera will ever grow past a cellular stage, but there is an inherent "unnaturalness" to this area of science, and clearly violates any semblance of human dignity that all humans innately possess.[12]

7.5 Stem Cell Banking and Distribution

Stem cell banks can prove to be very helpful. First, they allow for scientists and researchers to properly store specific hESCs, and can be used to work with the patent office, if it is decided that hESCs are in fact patentable, and other researchers and scientists will be able to purchase approved, and qualified hECs direct from the stem cell banks, thus removing the need for the further harvesting of embryonic stem cells, lessening the potential for any unnecessary interactions with the fetuses.

Prior to transferring any embryonic cells, the NIH will have to certify that the purpose or purposes of the proposed research will meet the laws described by this

[11] Cyranoski (2013), pp. 295–296.

[12] Karpowicz et al. (2015), pp. 107–134.

law. This will not create an undue burden on the NIH as they are already performing such a function, albeit on a potentially smaller scale. This portion of the law will prevent potentially unethical, and even dangerous experiments from being conducted using embryonic stem cells.

7.6 Creation of New hESC Lines

The creations of new embryonic stem cell lines should be permitted, so long as the lines are not derived from embryos that were created for the sole purpose of being used for experimentation and research purposes. Stem cell lines will have to be registered with the NIH who is currently the agency that oversees embryonic stem cell rule promulgation.

7.7 Registry

As mentioned above, the NIH is the agency best suited to ensure that these rules are being followed. Currently, the NIH uses and maintains a registry of embryonic stem cell lines that are allowed to be used in experiments where federal funding can be granted.[13] Adding stem cell lines will increase the amount and type of experiments that will be conducted using embryonic stem cells. Failure to comply with registration requirements could lead to slower progress in this sector, and time is of the essence since people are always in need of medical help, and unregistered stem cell lines cannot be used in clinical trials, or in medical practice.

7.8 Informed Consent

Every treaty, country, state, and non-binding recommendation that allows for embryonic stem cell research, requires informed consent from the donor or donors. This is used to ensure that the donors understand their rights, the nature of the procedure, potential risks and benefits, and the one seeking informed consent must ensure that all parts are understood by the donor.[14] There are also different approaches to providing such consent, such as the subjective method; the reasonable patient method; and the reasonable physician method.[15] The subjective method asks what a specific patient needs to know to make an informed decision, while the

---

[13] *See* National Institute of Health (2009); Exec. Order No. 13,505; 3 C.F.R. 229 (2009).

[14] Shah et al. (2020).

[15] *Id.*

reasonable patient method asks what information a reasonable person would need to know to make an informed decision, and the reasonable physician method asks what information an ordinary physician would believe to be required about the specific procedure.[16] When dealing with minors, there is usually a requirement to seek permission from the minor's parents, unless the minor has been emancipated, married, serving in the military, or can show that they are financially independent from their parents.[17] There are also exceptions to requiring informed consent, such as in times of incapacity, medical emergency, or when the donor has waived consent. For the purposes of creating this law, one can see how important informed consent would be, where a donated embryo can be experimented on, which is exactly why the subjective standard should be used. Using the reasonable patient standard would likely fail to fully protect the donor as it fails to address other, potentially influential information. One can argue that the bare minimum of an explanation is all that is needed for a patient to make an "informed" decision, but when dealing with the donation of an embryo, a higher standard should apply to ensure the donor's decision is based on all that they need to know, and not on what the general public/ordinary person would need to know to make the same decision. The donor should be fully apprised of what their decision means for both themselves and for the embryo, as well as any and all realistic potential risks that can occur for the donor and for the embryo. While in research situations, some information will be far too speculative, but so long as there is a realistic chance of something occurring, it must fully be explained to the donor nonetheless. Once the information is provided, it must be ensured that the donor fully understands what had been explained to them.

7.9 Creation of Embryonic Stem Cell Research Committee, Taskforce, Investigations and Enforcement of Provisions

While many would believe that it would be best for the NIH to handle investigations and enforcement of their rules, currently, the NIH only provides such services on their own campuses and where they have a section in Fort Detrick, so it would likely be a virtual impossibility to grant a much larger amount of responsibility to a particularly small group.[18] This area of science necessarily implicates life and death situations for both the donated embryos as well as prospective patients seeking to undergo new treatments utilizing embryonic stem cells, so it would be wise to establish an independent committee made up of lawyers, scientists, and ethicists who are vastly educated on the subject and can provide their expertise in making determinations as to those who break the rules of the law. The Committee would

[16] *Id.*

[17] *Id.*

[18] National Institute of Health (2024).

be called the "Embryonic Stem Cell Research Committee," (the "Committee") and would then have the ability to order investigations, and create a Taskforce, the "Embryonic Stem Cell Taskforce" (the "Taskforce") who will have the sole purpose of determining if any wrongdoing took place by any researchers, or doctors and report back to the Committee. If any laws were found to be broken, the Committee should have to take a vote which will merely require a majority, to open up a criminal proceeding for those in violation. If the criminal proceeding finds evidence of criminal conduct, or intent, the issue can then be sent to the Federal Bureau of Investigations (the "FBI") to initiate criminal proceedings and charges. The FBI is in the best position to do so as they are a federal police branch that has authority throughout the country, and the FBI "emphasizes close relations and information sharing with other federal, state, local, and international law enforcement and intelligence agencies."[19] Depending on the outcome of the FBI's proceedings, the Committee will be charged with determining the type of punishment that is appropriate for the violation.

The Committee shall also meet yearly to discuss new scientific and medical breakthroughs in the area of embryonic stem cell science which will look to published entries pertaining to this field of science that will help guide the changes and creation of new laws.

The Committee will also need to be directly charged with determining the validity, and efficacy of experiments and researchers seeking federal grant money and approval of their experiments. The Taskforce will also be made up by members who are experts in the same fields as those in the Committee, but will have a more direct approach that will monitor the compliance of the individual research projects with the pertinent laws.

7.10 Equal Access to Groundbreaking Treatments and Experimental Treatments

Unfortunately, throughout much of history, the wealthy have had the ability to obtain experimental and new, groundbreaking treatments that could cure or save them, yet those without the necessary means were left to suffer. Therefore, to prevent this issue from continuing, all those who are seeking federal funding for embryonic stem cell research, that require some form of human testing, will be required to allocate a portion of the federal funding to use to cover treatment costs of all of those involved in trials. This will not be a change to what most clinical trials do, which is to require the person obtaining the trial treatments, or drugs, to report back to the trial doctor, in whatever location they may be, and covering airfare and hotel accommodations for the monthly visits if travel is required.[20] While this is not always the case

[19] Federal Bureau of Investigations (2019).

[20] *See generally* McDowell (2020).

when it comes to medical trials, it is the most logical, fair and equitable solution that benefits all, without granting any greater benefit to those who can pay. Further, all necessary costs that are associated with such experimental treatments will also have to be paid by the receiver of the federal funds. This will ensure that those who are in need of treatment can travel to any location in the United States, and not have to worry about hotel prices, or travel expenses. Because this is extremely beneficial to the pharmaceutical companies that are creating these treatments, they would have to bear the costs. This is the fairest situation as the treatments would be experimental and the information taken from its use would be beneficial to the company itself.

7.10.1 Text of Proposed Law on Embryonic Stem Cell Research

1. Research conducted using and/or involving embryonic stem cells has shown, and likely will continue to show great promise and innovation in medical and other therapeutic applications that can greatly benefit and enhance the lives of many people.
2. Research involving embryonic stem cells has been both privately and publicly, while meeting the stringent rules provided by the National Institute of Health.
3. Currently, a few embryonic stem cell laws have been promulgated by the United States government, but these federal laws allow for states to enact their own legislation. As a result, there is no true federal comprehensive legislative scheme regulating human embryonic stem cells and their uses. This law will effectively codify a new set of rules that will control over all state and local laws in regards to embryonic stem cell research.
4. This law will not overturn any bans of embryonic cloning, and will effectively alter any existing laws by issuing a nationwide ban of such cloning activities.
   (a) Somatic cell nuclear transfer cannot be used under any circumstances involving any human somatic cells or ova; and
   (b) The therapeutic and reproductive cloning of human cells will not be permitted.
5. Under this legislation, only surplus embryos created for the purposes of *in-vitro* fertilization can be used to harvest embryonic stem cells. Accordingly, donors who seek to donate their surplus embryos will be required to give their informed consent.
6. Informed consent shall be provided by the donors of surplus embryos. In order for informed consent to be properly provided, the party seeking the donation must provide all relevant information to the potential donors so that they may make a truly informed decision. The standard to be used is the subjective method, which asks what a specific patient needs to know to make an informed decision.

7. Donors must also specify whether they are consenting to the traditional method of harvesting embryonic stem cells which is the type that leads to the embryo's destruction, or if they are only consenting to what is known as the "humane" method, where the embryo does not have to be destroyed in order to harvest the necessary embryonic stem cells.
8. The party seeking the embryonic stem cell donation is charged with providing the prospective donors with the following information: (a) donors must be informed as to their right to refuse donations; (b) the donors must be notified as to the processes of harvesting embryonic stem cells, and what each method means for the donated embryo; (c) what will happen to the surplus embryos if donation is not chosen; and (d) any risks of potential problems, or dangers that may befall the donor, and the embryo. If the chances of such problems or dangers fall below the chance of 15% or below for occurring, then the party seeking donation does not have to discuss them, unless the potential donor requests this information.
9. Minors and those who are deemed incapacitated for the purposes of providing such consent will be required to seek parental or guardian permission, or if deemed impossible, a judge may rule on the issue on a case-by-case basis.
10. Until the International Society for Stem Cell Research creates a new rule, the 20-week rule will apply to embryos donated for research purposes.
11. Those who seek and receive any portion of federal funding for embryonic stem cell studies, that inevitably will require test subjects, must provide experimental treatments free of charge to all patients, and must be used to compensate patients for their necessary expenses incurred from the patient in receiving such experimental treatment.
12. Those who seek to procure federal funding in the form of a grant, and use embryonic stem cells for research purposes will first have to apply to the National Institute of Health, who will be charged with determining if the research and/or experiments are of the type and caliber that seek to be used for therapeutic purposes, to benefit patients now, and in the future, while not harming, desecrating, or mishandling any embryos or embryonic material. Because this area of science is so new, unique, and promising, grants should be awarded to those who apply, unless there is something inherently inconsistent with the above-written requirements for National Institute of Health grant approval. Examples of what should not be funded would be research into practices that are explicitly banned in this proposed law such as the practice of therapeutic cloning, the creation of chimeras, and the experiments must have an objectively stated scientific value, more than just extending our knowledge for knowledge's sake.
13. This legislation creates two new bodies, an Embryonic Stem Cell Research Committee and a Taskforce that will work under the direction of the National Institute of Health.
14. The Embryonic Stem Cell Research Committee (hereinafter "the Committee") will be charged with drafting, and amending various detailed requirements and procedures pertaining to embryonic stem cell sciences. Information will be

collected by, and then presented by the Taskforce that directly answers to the Committee.

15. The Committee will consist of members that represent various backgrounds which offer unique perspectives to this novel area of both science and law. Lawyers, doctors, scientists, experts, and ethicists will comprise this group and will use their backgrounds to draft laws and amendments based on information collected and presented by the taskforce.
16. The Committee will actively research new embryonic stem cell studies and be charged with making annual presentations to the Committee which the Committee will then therefore be forced to vote on potential changes based on the presentations.
17. Once the Committee has voted to deem the "humane" method of harvesting embryonic stem cells both financially, and scientifically viable then all other methods of harvesting of embryonic stem cells will be outlawed and deemed illegal without prior authorization from the Committee.
18. A federal stem cell registry will be created and maintained by the Committee that will investigate new embryonic stem cell lines, and will certify their provenance, and that they are qualified for scientific use and study in the United States.
19. Embryonic stem cells and stem cell lines successfully registered with the Committee will be deemed to be of the highest quality and can be purchased and sold within the United States for research purposes. The recipient laboratory must obtain a prior authorization by the Committee for their study.
20. Once passed the investigative, research phases, the Food and Drug Administration will use the information collected by the Committee under the Right to Try Act/Trickett Wendler, Frank Mongiello, Jordan McLinn, and Matthew Bellina Right to Try Act, patients will be able to request specific embryonic stem cell-based therapies without government interference. This will work on the basis of the Right To Try Act which does not itself approve or even review "Right to Try" requests, but relies on the doctor and developer to provide their approvals.
21. The Committee will be charged with investigating potential malfeasance and violations of this law through the Taskforce. The Committee can directly order reviews of specific studies, scientists, and laboratories who, based on evidence, are believed to be in violation. If a review is ordered, the Committee will submit a formal recommendation for the Taskforce to open an investigation into the potential offender(s). The Taskforce will report back to the Committee with recommendations, and the Committee will then decide the method of enforcement.

(a) Methods of enforcement: Should any potential issues be found to exist; the Committee will have to decide how to proceed based on the following options:

- Suspension: The Committee can find that a certain specific actor or actors are at fault and that the entire study should not be affected in light of its importance, in this case, they can order the actual offenders suspended from continuing in the particular study.
- Reduction of funding: If the Committee finds a higher level of culpability in a particular study, the Committee will have the ability to reduce grant money that the government was willing to provide.
- Termination of funding: In the most severe cases, the Committee can determine that an entire study or laboratory be barred from receiving any further federal grant funds. Removal of stem cells from national registry: Cells that were illegally obtained or obtained in contravention to the fund granting documents can lead to an entire cell line to be removed from the registry, preventing the creator(s) from profiting from their breach.

References

Conti A (2017) Drawing the line: disability, genetic intervention and bioethics. Laws 6:1–23, 6, 9
Cyranoski D (2013) Human stem cells created by cloning. Nature 497:295–296
Exec. Order No. 13,505, 3 C.F.R. 229 (2009) Removing barriers to responsible scientific research involving human stem cells
Federal Bureau of Investigations (2019) About us. Available via the Federal Bureau of Investigations https://www.fbi.gov/about/faqs/what-are-the-primary-investigative-functions-of-the-fbi. Accessed 26 Mar 2024
Felton A (2022) When can a fetus feel pain in the womb? Available via WebMD https://www.webmd.com/baby/when-can-a-fetus-feel-pain-in-the-womb. Accessed 26 Mar 2024
Karpowicz P, Cohen CB, van der Kooy D (2015) Developing human-nonhuman chimeras in human stem cell research: ethical issues and boundaries. Kennedy Inst Ethics J 2:107–134
Masci D (2008) The case against embryonic stem cell research: an interview with Yuval Levin. Available via the Pew Research Center https://www.pewforum.org/2008/07/17/the-case-against-embryonic-stem-cell-research-an-interview-with-yuval-levin. Accessed 26 Mar 2024
McDowell A (2020) Do clinical trials pay for travel? Available via Antidote Technologies https://www.antidote.me/blog/do-clinical-trials-pay-for-travel. Accessed 26 Mar 2024
Miller SG (2016) Do fetuses feel pain? What the science says. Available via Live Science https://www.livescience.com/54774-fetal-pain-anesthesia.html. Accessed 26 Mar 2024
Mountford JC (2008) Human embryonic stem cells: origins, characteristics and potential for regenerative therapy. Transf Med 18:1–12
National Institute of Health (2009) National Institutes of Health guidelines for human stem cell research. Available via the National Institute of Health https://stemcells.nih.gov/research-policy/guidelines-for-human-stem-cell-research. Accessed 26 Mar 2024
National Institute of Health (2024) Criminal investigations. Available via the National Institute of Health https://www.ors.od.nih.gov/ser/dp/lawenforcement/Pages/Investigations.aspx. Accessed 3 Jan 2024
Nunes R, Nunes SB, Rego G (2017) Health care as a universal right. Z Gesundh Wiss 1:1–9
Pompe S, Bader M, Tannert C (2005) Stem-cell research: the state of the art. EMBO Rep 6:297–300

Regalado A (2005) Big companies quietly pursue research on embryonic stem cells. Available via the Wall Street Journal https://www.wsj.com/articles/SB111325723385303962. Accessed 26 Mar 2024

Regalado A (2021) Scientists plan to drop the 14-day embryo rule, a key limit on stem cell research. Available via Technology Review https://www.technologyreview.com/2021/03/16/1020879/scientists-14-day-limit-stem-cell-human-embryo-research. Accessed 26 Mar 2024

Rull V (2014) The most important application of science. EMBO Rep 9:919–922, 920

Shah P, Thornton I, Turrin D, Hipskind JE (2020) Informed consent [Updated 2020 Jul 23]. In: StatPearls [Internet]. Treasure Island (FL). Available via the National Library of Medicine https://www.ncbi.nlm.nih.gov/books/NBK430827/. Accessed 26 Mar 2024

Taylor-Corbett S (2000) Embryo research: profit vs ethics? Available via the Center for Public Integrity https://publicintegrity.org/health/embryo-research-profit-vs-ethics. Accessed 26 Mar 2024

GPSR Compliance
The European Union's (EU) General Product Safety Regulation (GPSR) is a set of rules that requires consumer products to be safe and our obligations to ensure this.

If you have any concerns about our products, you can contact us on

ProductSafety@springernature.com

In case Publisher is established outside the EU, the EU authorized representative is:

Springer Nature Customer Service Center GmbH
Europaplatz 3
69115 Heidelberg, Germany

www.ingramcontent.com/pod-product-compliance
Ingram Content Group UK Ltd.
Pitfield, Milton Keynes, MK11 3LW, UK
UKHW020246290926
15470UKWH00002B/768
9783031597497